1,1'-BINAPHTHYL-BASED CHIRAL MATERIALS

Our Journey

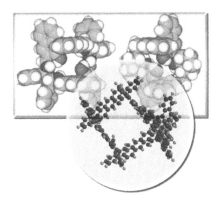

1,1'-BINAPHTHYL-BASED CHIRAL MATERIALS

Our Journey

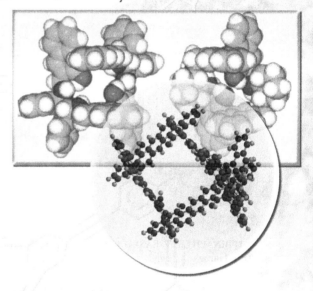

Lin Pu
University of Virginia, USA

Imperial College Press

Published by

Imperial College Press
57 Shelton Street
Covent Garden
London WC2H 9HE

Distributed by

World Scientific Publishing Co. Pte. Ltd.
5 Toh Tuck Link, Singapore 596224
USA office: 27 Warren Street, Suite 401-402, Hackensack, NJ 07601
UK office: 57 Shelton Street, Covent Garden, London WC2H 9HE

British Library Cataloguing-in-Publication Data
A catalogue record for this book is available from the British Library.

1,1'-BINAPHTHYL-BASED CHIRAL MATERIALS
Our Journey

Copyright © 2010 by Imperial College Press

All rights reserved. This book, or parts thereof, may not be reproduced in any form or by any means, electronic or mechanical, including photocopying, recording or any information storage and retrieval system now known or to be invented, without written permission from the Publisher.

For photocopying of material in this volume, please pay a copying fee through the Copyright Clearance Center, Inc., 222 Rosewood Drive, Danvers, MA 01923, USA. In this case permission to photocopy is not required from the publisher.

ISBN-13 978-1-84816-411-6
ISBN-10 1-84816-411-4

Typeset by Stallion Press
Email: enquiries@stallionpress.com

Printed in Singapore.

Preface

1,1'-Binaphthyls represent an important class of chiral organic compounds. In the past several decades, the 1,1'-binaphthyl-based molecules have received very broad research attention. Since early 1970s, Cram and Lehn started the use of the optically active 1,1'-binaphthyls to build the structurally diverse molecular hosts for chiral recognition. In 1980s, Noyori and Takaya initiated the study of the transition metal complexes of the 1,1'-binaphthyl-based phosphorus ligands for the asymmetric hydrogenation of alkenes and carbonyl compounds. These monumental achievements have been awarded with the Nobel Prize in Chemistry. In addition to these studies, tremendous amount of reports by other researchers on the use of the 1,1'-binaphthyls for asymmetric catalysis and molecular recognition have also appeared, and many enantioselective catalysts and molecular receptors have been obtained. Since 1990s, our laboratory has launched a program to use the optically active 1,1'-binaphthyls to build novel chiral materials such as the main chain chiral-conjugated polymers, light harvesting chiral dendrimers, polymeric and monomeric chiral catalysts, and enantioselective fluorescent sensors. This book presents the findings from this research program as well as the fundamental knowledge about 1,1'-binaphthyls.

There are six chapters in this book. In Chapter 1, some basic concepts on the 1,1'-binaphthyl molecules and a general background on the study of 1,1'-binaphthyls are provided. In Chapter 2, the synthesis of the main chain chiral-conjugated polymers is described. In these materials, the optically active 1,1'-binaphthyl units are incorporated into the main chain of conjugated polymers. The conductivity, light emission, electroluminescence and non-linear optical properties of some of these polymers are investigated.

In Chapter 3, a systematic study of the 1,1′-binaphthyl-based rigid and sterically regular polymeric chiral catalysts is discussed. Highly enantioselective polymeric catalysts have been obtained. Chapter 4 presents the application of BINOL and other non-polymeric binaphthyl compounds in asymmetric catalysis. Catalysts with high enantioselectivity have been discovered for diverse organic reactions. Chapter 5 focuses on the development of the 1,1′-binaphthyl-based dendrimers, macrocycles and acyclic molecules for the enantioselective fluorescent recognition of chiral α-hydroxycarboxylic acids, amines, amino alcohols and amino acids. Highly enantioselective as well as sensitive fluorescent recognition of chiral molecules has been achieved. In Chapter 6, some miscellaneous studies on the materials related to 1,1′-binaphthyls are described, including the chiral molecular wires, polybiphenols and the supramolecular chemistry of propargylic alcohols.

This journey of exploration would not have been possible without the support of many people. I thank all of my students, post-doctoral researchers, visiting scholars as well as my collaborators for their intellectual contribution and dedication. I am also grateful to the following funding agencies in the USA for supporting our research: NSF, NIH, ACS-PRF, AFOSR, ONR and NASA.

Lin Pu
Department of Chemistry
University of Virginia
Charlottesville, Virginia 22904
USA

Contents

Preface		v
1.	**Introduction About 1,1′-Binaphthyls**	**1**
	References	10
2.	**Main Chain Chiral-Conjugated Polymers**	**13**
	2.1. Introduction About Chiral-Conjugated Polymers	13
	2.2. Binaphthyl-Based Polyarylenevinylenes	17
	2.3. Binaphthyl-Based Polyarylenes	20
	2.4. Binaphthyl-Based Polyaryleneethynylenes	23
	2.5. Binaphthyl–Thiophene Copolymers	30
	2.6. Copolymers of BINAM and Thiophene-Containing Conjugated Linkers	37
	2.7. Polybinaphthyls Without Conjugated Linkers	39
	2.8. Propeller-Like Polybinaphthyls	51
	2.9. Dipole-Oriented Propeller-Like Polymers	69
	2.10. Binaphthyl-Based Polysalophens	71
	2.11. Helical Ladder Polybinaphthyls	76
	References	85
3.	**Polybinaphthyls in Asymmetric Catalysis**	**87**
	3.1. Introduction about Chiral Polymers in Asymmetric Catalysis	87
	3.2. Synthesis of Major-Groove Poly(BINOL)s	89

vii

3.3.	Application of the Major-Groove Poly(BINOL)s to Catalyze the Mukaiyama Aldol Reaction	95
3.4.	Application of the Major-Groove Poly(BINOL)s to Catalyze the Hetero-Diels–Alder Reaction	99
3.5.	Using the Ti(IV) Complex of the Major-Groove Poly(BINOL) to Catalyze the Diethylzinc Addition to Aldehydes	101
3.6.	Synthesis of the Minor-Groove Poly(BINOL)s	102
3.7.	Application of the Major- and Minor-Groove Poly(BINOL)s to Catalyze the Asymmetric Organozinc Addition to Aldehydes	105
3.8.	Asymmetric Reduction of Prochiral Ketones Catalyzed by the Chiral BINOL Monomer and Polymer Catalysts	122
3.9.	Asymmetric Epoxidation of α,β-Unsaturated Ketones Catalyzed by the Minor- and Major-Groove Poly(BINOL)s	125
3.10.	Asymmetric Diels–Alder Reaction Catalyzed by Poly(BINOL)–B(III) Complexes	133
3.11.	1,3-Dipolar Cycloaddition Catalyzed by the Minor- and Major-Groove Poly(BINOL)–Al(III) Complexes	136
3.12.	Asymmetric Michael Addition Catalyzed by the Poly(BINOL)s	139
3.13.	Synthesis and Study of Poly(BINAP)	141
3.14.	Synthesis and Study of a BINOL–BINAP Copolymer	146
	References	150

4. Asymmetric Catalysis by BINOL and Its Non-polymeric Derivatives   151

4.1.	Introduction	151
4.2.	Asymmetric Alkyne Additions to Aldehydes	151
4.3.	Asymmetric Arylzinc Addition to Aldehydes	198
4.4.	Asymmetric Alkylzinc Addition to Aldehydes	216
4.5.	Asymmetric TMSCN Addition to Aldehydes	225
4.6.	Asymmetric Hetero-Diels–Alder Reaction	233
	References	235

5. **Enantioselective Fluorescent Sensors Based on 1,1′-Binaphthyl-Derived Dendrimers, Small Molecules and Macrocycles**   **239**

    5.1. Introduction 239
    5.2. Using Phenyleneethynylene-BINOL Dendrimers for the Recognition of Amino Alcohols 240
    5.3. Using Phenylene-BINOL Dendrimers for the Recognition of Chiral Amino Alcohols 246
    5.4. Using Functionalized BINOLs for the Recognition of Amino Alcohols and Amines 252
    5.5. Using Acyclic Bisbinaphthyls for the Recognition of α-Hydroxycarboxylic Acids and Amino Acid Derivatives 257
    5.6. Using Diphenylethylenediamine-BINOL Macrocycles for the Recognition of α-Hydroxycarboxylic Acids ... 263
    5.7. Using Cyclohexanediamine-BINOL Macrocycles and Their Derivatives for the Recognition of α-Hydroxycarboxylic Acids and Amino Acid Derivatives 282
    5.8. Application of the Cyclohexanediamine-BINOL Macrocycle Sensor in Catalyst Screening[31] 294
    5.9. Using Monobinaphthyl Compounds for the Recognition of α-Hydroxycarboxylic Acids and Amino Acid Derivatives 299
    5.10. Enantioselective Precipitation and Solid-State Fluorescence Enhancement in the Recognition of α-Hydroxycarboxylic Acids by Using Monobinaphthyl Compounds 303
    References 308

6. **Miscellaneous Studies on Materials Related to 1,1′-Binaphthyls**   **311**

    6.1. Chiral Molecular Wires 311
    6.2. A Biphenol Polymer 316
    6.3. Supramolecular Chemistry of Self-Assembly of Racemic and Optically Active Propargylic Alcohols 318
    References 331

Index   333

Chapter 1

Introduction About 1,1′-Binaphthyls

1,1′-Binaphthyl compounds represent a special class of biaryl molecules. For a simple unsubstituted biphenyl molecule, its rotation barrier around the phenyl–phenyl bond in gas phase was found to be ∼1.4 kcal/mol.[1] 1,1′-Binaphthyl incorporated with fused benzene rings to biphenyl has a greatly increased rotation barrier of 23.5 kcal/mol ($\Delta G^{\neq}$).[2] In addition, 1,1′-binaphthyl no longer possesses the plane of symmetry that biphenyl has at the orthogonal conformation. This allows the isolation of the optically active 1,1′-binaphthyl enantiomers. A chiral 1,1′-binaphthyl molecule contains a chiral axis instead of a chiral center. The racemization half-life of the optically active 1,1′-binaphthyl was 14.5 min at 50°C. 1,1′-Binaphthyls have two possible racemization pathways. One goes through a *syn* interaction state where there are close contacts for the 2,2′-H's and 8,8′-H's, and another goes through an *anti* interaction state where there are close contacts for the 2,8′-H's and 2′,8-H's. When theoretical rigid models are applied, the *anti* pathway gives lower steric hindrance than the *syn* route. Calculations show that 1,1′-binaphthyl favors the *anti* inversion racemization mechanism.[3,4] However, semi-empirical calculations suggest that 2,2′-dibromo-1,1′-binaphthyl should favor a multi-stage *syn* racemization pathway. The aromatic rings in the transition states of this racemization process are significantly distorted.[3]

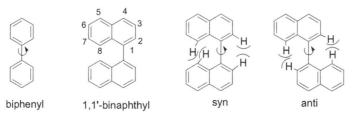

In 1971, Pincock et al. discovered that racemic 1,1′-binaphthyl underwent spontaneous resolution to generate the optically active R or S enantiomer when this compound crystallized from the melt.[5] Out of the 200 crystallization experiments of rac-1,1′-binaphthyl examined, the probability to generate the R or S enriched enantiomer was about the same.

When substituents are introduced to the 2,2′-positions of 1,1′-binaphthyl, the chiral configuration of the 1,1′-binaphthyl compounds becomes very stable. For example, (S)-1,1′-binaphthyl-2,2′-dicarboxylic acid did not racemize at 175°C in N,N-dimethyl formamide,[6] and 2,2′-dimethyl-1,1′-binaphthyl did not racemize after 40 h at 240°C.[7] This is in sharp contrast to substitution at the 8,8′-positions. For example, 1,1′-binaphthyl-8,8′-dicarboxylic acid underwent racemization in a rate similar to that of the unsubstituted 1,1′-binaphthyl.[2] The half-life for the racemization of the optically active 1,1′-binaphthyl-8,8′-dicarboxylic acid was 51.5 min at 50°C ($\Delta G^{\neq} = 24.4$ kcal/mol). This is attributed to the steric repulsion of the 1,8-substituents on the naphthalene rings of 1,1′-binaphthyl-8,8′-dicarboxylic acid, which could raise the ground-state energy and cause the deformation of this molecule.

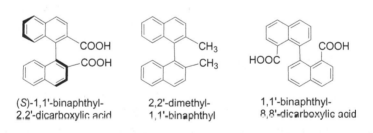

(S)-1,1′-binaphthyl-
2,2′-dicarboxylic acid

2,2′-dimethyl-
1,1′-binaphthyl

1,1′-binaphthyl-
8,8′-dicarboxylic acid

The absolute configurations for chiral binaphthyl compounds were originally proposed by Mislow on the basis of the study of optical properties, stereochemical mechanisms and thermal analysis.[8] This was later confirmed by Yamada and coworkers from the X-ray analysis of (R)-(+)-2,2′-dihydroxy-1,1′-binaphthalenyl-3,3′-dicarboxylic acid dimethyl ester and its chemical correlation with other binaphthyl molecules.[9] In this compound, its dihedral angle between the two naphthalene rings is ∼77°. The absolute configuration of an axially chiral 1,1′-binaphthyl molecule is designated as R or S according to the following steps: (i) start from one of the naphthalene rings arbitrarily and assign the priority of the two carbons directly connected to the carbon 1 as $a > b$; (ii) move to the other naphthalene ring and assign the priority of the two carbons directly connected to the carbon $1'$ as $c > d$; and (iii) determine the R or S configuration on the basis of the arrangement of $a > b > c > d$ according to the Cahn–Ingold–Prelog R–S notation system.

(R)-(+)-2,2'-dihydroxy-1,1'-binaphthalenyl
-3,3'-dicarboxylic acid dimethyl ester

An optically pure 1,1'-binaphthyl molecule can exist in two conformations, *i.e.* cisoid and transoid, as shown in Fig. 1.1. In the (S)-cisoid

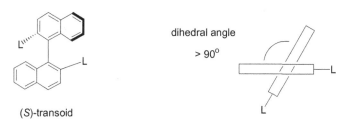

Figure 1.1. Conformations of 2,2'-substituted 1,1'-binaphthyls.

conformation, the dihedral angle between the two naphthalene rings is less than 90°, whereas, in the (S)-transoid conformation, the dihedral angle is greater than 90°. The crystal structure of rac-1,1′-binaphthyl shows that it exists in the cisoid conformation with a dihedral angle of 68°.[10,11] However, its optically active crystals exist in the transoid conformation with a dihedral angle of 103°. The study of various 2,2′-substituted 1,1′-binaphthyl molecules has indicated that when the 2,2′-substituents L are either small or capable of intramolecular hydrogen bonding, e.g. when L = OH, CH_2OH, OCH_2COOH, NH_2, OCH_3, CH_3 or OCH_2Ph, the cisoid conformation is preferred.[12–15] When the 2,2′-substituents are large, e.g. when L = CH_2Br or $CHBr_2$, the transoid conformation is preferred. The CD spectra of the binaphthyl molecules that have the same R or S configuration but with the opposite cisoid or transoid conformation are found to be almost mirror images of each other. That is, the conformation of the binaphthyl molecules strongly influences their CD signals. The dynamics of the atropisomerization of 1,1′-binaphthyl in liquid crystalline solvents was also studied.[16]

Because of the highly stable chiral configuration of the 2,2′-substituted 1,1′-binaphthyls, these molecules have been extensively used to control many asymmetric processes and have demonstrated outstanding chiral discrimination properties.[17–19] A great number of 1,1′-binaphthyl molecules are C_2 symmetric with two identical naphthyl units. The rigid structure and the C_2 symmetry of the chiral binaphthyl molecules are found to be important for their roles in chiral induction. Many binaphthyl-based C_1 symmetric catalysts or reagents have also been prepared and studied.

Since the early 1970s, extensive studies on the use of chiral binaphthyl-based compounds including crown ethers, cyclophanes and cyclic amides as hosts for molecular recognition have been conducted in Cram's laboratory.[20–42] These macrocycles either undergo complexation-induced organization in the presence of guests or have pre-organized cavities based on the rigid structures. They interact with the functional groups of guest molecules through weak forces such as hydrogen bonding, π–π stacking, and van der Waals forces. The chirality of the binaphthyl units in these hosts leads to their enantioselective complexation with chiral guest molecules. Such a chiral recognition has been applied to the resolution of racemic molecules including amino acids, amino esters, amines, sugars and other chiral alkyl or aryl compounds.

(R,R)-**1.1**

For example, the bisbinaphthyl macrocyclic ether (R,R)-**1.1** was synthesized and its specific optical rotation was $[\alpha]_{578} = +152$.[22,23] This compound was found to be an excellent chiral host in the differentiation of the enantiomeric salts of chiral amines, chiral amino acids and amino esters.[22,26,27] The chiral recognition factors, D_A/D_B, (D_A is the distribution coefficient of the enantiomer more complexed; D_B is that of the less complexed.) as high as 52 for $PhCH(NH_3^+)CO_2HClO_4^-$, 48 for p-$HOC_6H_4CH(NH_3^+)CO_2HClO_4^-$, and 31 for $PhCH(CO_2CH_3)\,NH_3^+PF_6^-$ were observed when (R,R)-**1.1** or (S,S)-**1.1** was used to extract these ammonium salts from their water solution into the organic phase (chloroform or chloroform/acetonitrile solution). The D-enantiomers of the amino acid or ester salts bind with the host (R,R)-**1.1** more favorably than the L-enantiomers do. Such a high chiral recognition is due to the complementary complexation between the host and the guest. According to the Corey–Pauling–Koltun (CPK) model, the preferred complex between (R,R)-**1.1** and an R chiral amino ester has a structure as shown by **1.2**. In this host–guest complex, the binding between the crown ether and the amino ester salt involves three O–H–N hydrogen bonds and a π–π attractive interaction of the ester group with a naphthalene ring. This four-point binding is essential for the observed high chiral discrimination.

1.2, a (R,R)-**1.1**-(R)-amino ester complex

(R,R)-1.3

A single crystal X-ray structure of (S,S)-**1.3** complexed with the hexafluorophosphate salt of (R)-phenylglycine methyl ester, the less stable diastereomeric complex, was obtained.[28] Analysis of this structure indicates that the introduction of substituents at the 3,3′-positions of (S,S)-**1.3** would increase the steric interaction between the two binaphthyl units in this unfavorable diastereomeric complex and further destabilize it, leading to the observed much higher chiral recognition capability of (S,S)-**1.1** over (S,S)-**1.3**. The crown ethers made of monobinaphthyl units generally showed lower enantioselectivity than the bisbinaphthyl hosts.[29] A liquid–liquid extraction machine was designed to use (S,S)-**1.1** and (R,R)-**1.1** to catalytically resolve racemic amino ester salts. Up to 90% optical purity of the enantiomers could be continuously removed from this device.[30]

Macrocycle (R,R)-**1.1** was also covalently bound to polymer resins to prepare the polymeric material (R,R)-**1.4** as a chiral stationary phase for enantioselective chromatography.[31] The chromatography columns made of (R,R)-**1.4** gave baseline separation for the enantiomers of a number of amino acid and amino ester salts. The chiral recognition behavior of the immobilized host parallels that of the macrocycle itself.

(R,R)-1.4

Besides the study of the binaphthyl-based crown ethers for chiral recognition, another important milestone in the development of the binaphthyl

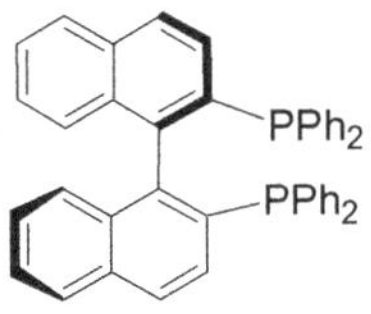

(R)-BINAP

chemistry is the discovery of 2,2′-bis(diphenylphosphino)-1,1′-binaphthyl (BINAP)-based chiral catalysts in asymmetric hydrogenation.[43–46] For example, the rhodium complex of BINAP, (S)-**1.5**, catalyzed the hydrogenation of enanmides to generate the α-amino acid derivatives with excellent enantiomeric excess ($ee = |R - S|/|R + S|$) (Scheme 1.1).[43] The dihydride complex **1.6** was proposed as an intermediate for the asymmetric hydrogenation process. From **1.6**, the hydride and alkene ligands can undergo migratory insertion and then reductive elimination to generate the N-acyl amino acid product. The stereochemistry for the hydride-alkene reaction is governed by the chelated chiral BINAP ligand.

The ruthenium complex of BINAP, (S)-**1.7**, was later found to be a more generally applicable catalyst for the asymmetric hydrogenation of a variety of functional alkenes (Scheme 1.2). Many other binaphthyl-based metal catalysts have been developed for asymmetric catalysis. Scheme 1.2

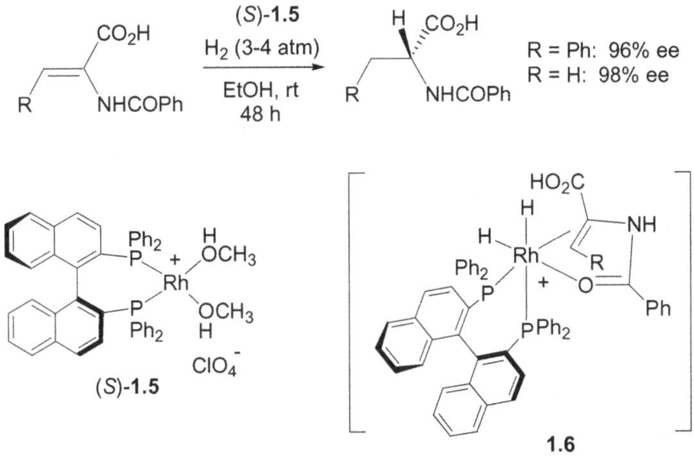

Scheme 1.1. Asymmetric hydrogenation catalyzed by a BINAP complex.

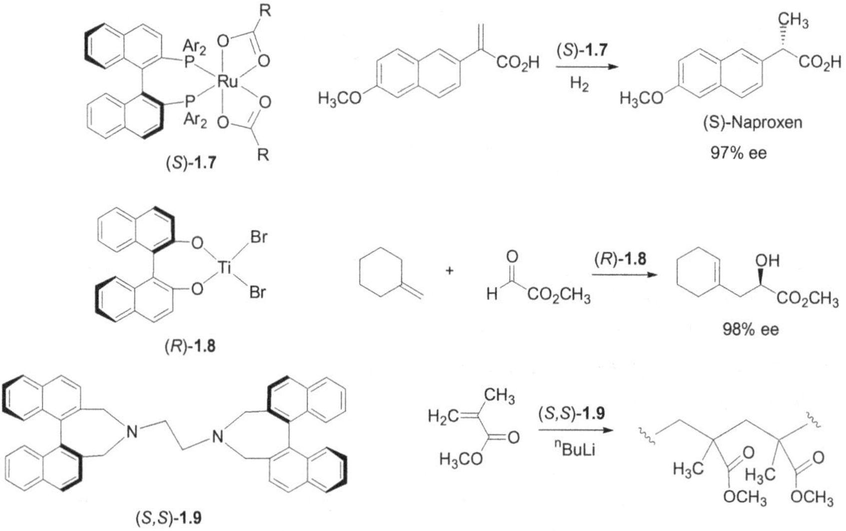

Scheme 1.2. A few asymmetric reactions catalyzed by binaphthyl complexes.

shows three examples where the chiral binaphthyl complexes (S)-**1.7**, (R)-**1.8** and (S,S)-**1.9** are used to carry out asymmetric hydrogenation,[44] asymmetric ene reaction[47] and asymmetric anionic polymerization.[48]

In the study of 1,1′-binaphthy-based chemistry, 1,1′-bi-2-naphthol (BINOL) often serves as the starting material to chiral binaphthyl compounds.[17–19] The specific optical rotation $[\alpha]_D$ of (R)-BINOL is +35.5 (c = 1, THF) and its melting point is 205–211°C. The 2,2′-hydroxyl groups of BINOL can be easily converted to other functional groups. The 3,3′-, 4,4′- and 6,6′-positions of this molecule can also be selectively functionalized to prepare a variety of binaphthyl derivatives. Because of the importance of this molecule, great efforts have been devoted to prepare it in the optically pure forms and many methods have been reported.[49–55] Among these methods, the use of (8S,9R)-(−)-N-benzylcinchonidinium chloride to resolve racemic BINOL into its optically pure (R)- and (S)-enantiomers received significant attention because of its simplicity and efficiency when used in organic laboratory (Scheme 1.3).[56–59] This method was originally developed by Toda et al.[56,57] and was later improved by our laboratory[58] and Cai.[59] Both (R)-BINOL and (S)-BINOL can be obtained in large scale with high optical purity by using this method. The chiral resolving agent is easily recovered after the resolution.[58] This compound is commercially

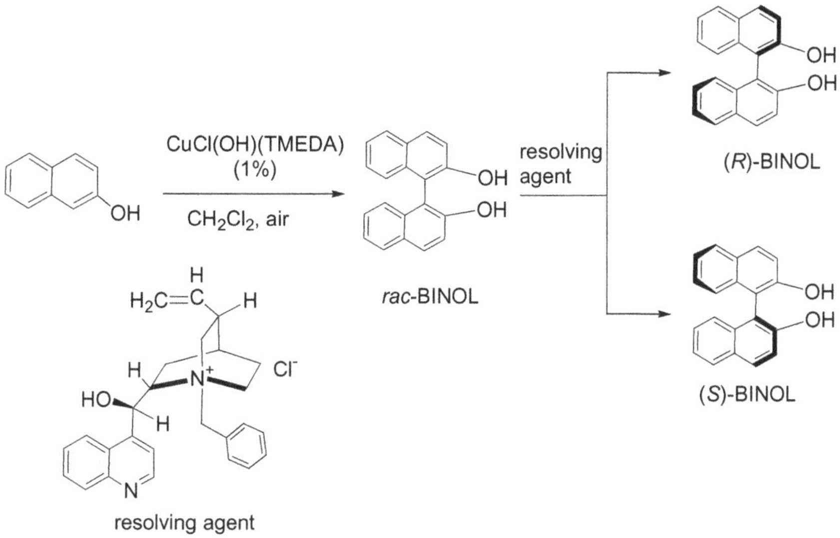

Scheme 1.3. Synthesis of racemic BINOL and its optical resolution.

available and can also be synthesized from the reaction of (−)-cinchonidine with benzyl chloride.[60] Racemic BINOL can be produced in large scale from the oxidative coupling of 2-naphthol in air in the presence of a copper catalyst (Scheme 1.3).[58,61]

The chiral configuration of BINOL is thermally stable. After heated at 100°C for 24 h in a dioxane-water solution, (S)-BINOL showed no sign of racemization.[62] However, acid or base can promote the thermal racemization of BINOL. In 1.2 N HCl solution, there was 72% racemization for (S)-BINOL when it was heated at 100°C for 24 h. In butanol containing 0.67 M KOH, (S)-BINOL also underwent 69% racemization after 23 h at 118°C. At 220°C in diphenyl ether, the racemization half-life of BINOL is 60 min ($\Delta G^{\neq} = 158\,\text{kJ/mol}$).[63] The racemization of BINOL might go through an anti pathway as shown by a density functional calculation.[63,64]

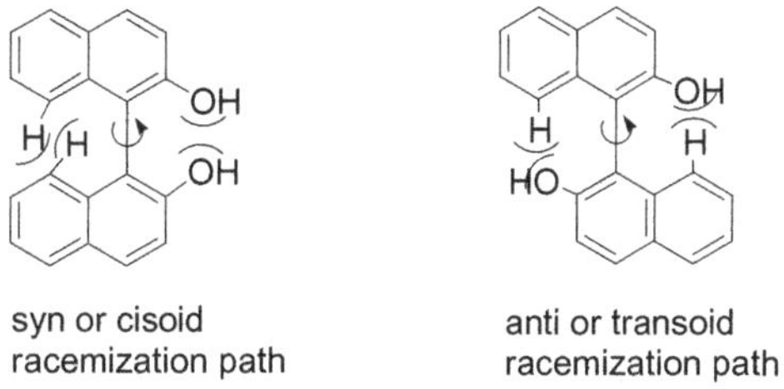

The structures of (+)-(R)- and racemic BINOL was determined by X-ray analysis of the single crystals of these molecules obtained from the slow evaporation of their toluene and ethanol solutions.[65] In the crystals of (R)-BINOL, this molecule has a cisoid conformation with a dihedral angle of 80.8° between the two naphthol rings. In the crystals of racemic BINOL, it has a more orthogonal conformation with a dihedral angle of 91.4°. The C1–C2 bond lengths (1.383–1.367 Å) in both structures are slightly shorter than their C2–C3 bond lengths (1.411–1.407 Å). The C1–C1' bond distances for these two structures are 1.494 and 1.500, respectively. The C1–C2–O bond angles (122.8–123.6°) in both structures are larger than those of C3–C2–O (116.2–114.9°) because of the steric interaction between the OH group and the other naphthol ring at position 1.

While the study of the monomeric chiral binaphthyl compounds continues to be an active research subject in many laboratories, multiple binaphthyl units have also been joined together to extend the chiral differentiation power of the binaphthyl structure and to construct molecular objects and polymers of unique structures and properties. Over a decade ago, our laboratory has stepped on a journey to explore the use of BINOL to construct structurally diverse chiral polymers, dendrimers, macrocycles and functional BINOLs. We have explored the applications of these materials in asymmetric catalysis and chiral sensing and have studied their electrical/optical properties. This is a journey of excitement, dedication and endurance. I am most grateful for the students, post-doctorals and visiting scholars working in my laboratory over these years. Without their intelligence and hardworking, it would be impossible to tell the stories in the following chapters.

References

1. Almenningen A, Bastiansen O, Fernholt L, Cyvin BN, Cyvin SJ, Samdal S, *J Mol Struct* **128**:59–76, 1985.
2. Cooke AS, Harris MM, *J Chem Soc* 2365, 1963.
3. Kranz M, Clark T, Schleyer P von R, *J Org Chem* **58**:3317–3325, 1993.
4. Tsuzuki S, Tanabe K, Kagawa Y, Nakanishi H, *J Mol Struct* **216**:279, 1990.
5. Pincock RE, Perkins RR, Ma AS, Wilson KR, *Science* **174**:1018, 1971.
6. Hall DM, Turner EE, *J Chem Soc* 1242, 1955.
7. Dixon W, Harris MM, Mazengo RZ, *J Cehm Soc B* 775, 1971.
8. Mislow K, *Angew Chem* **70**:683, 1958.
9. Akimoto H, Shioiri T, Iitaka Y, Yamada S, *Tetrahedron Lett* **1**:97, 1968.
10. Mason SF, *Molecular Optical Activity and the Chiral Discriminations*, Cambridge University Press: New York, 1982, pp. 73.

11. (a) Kuroda R, Mason SF, *J Chem Soc Perkin Trans* **2**:167, 1981. (b) Kress RB, Duesler EN, Etter MC, Paul IC, Curtin DY, *J Am Chem Soc* **102**:7709, 1980.
12. Gottarelli G, Hibert M, Samori B, Solladié G, Spada GP, Zimmermann R, *J Am Chem Soc* **105**:7318, 1983.
13. Gottarelli G, Spada GP, Bartsch R, Solladié G, Zimmermann R, *J Org Chem* **51**:589, 1986.
14. Mason SF, Seal RH, Roberts DR, *Tetrahedron* **30**:1671, 1974.
15. Kuroda R, Mason SF, *Tetrahedron* **37**:1995, 1981.
16. Naciri J, Spada GP, Gottarelli G, Weiss RG, *J Am Chem Soc* **109**:4352, 1987.
17. (a) Rosini C, Franzini L, Raffaelli A, Salvadori P, *Synthesis* 503, 1992. (b) Whitesell JK, *Chem Rev* **89**:1581, 1989. (c) Bringmann G, Walter R, Weirich R, *Angew Chem Int Ed Engl* **29**:977, 1990.
18. Pu L, *Chem Rev* **98**:2405–2494, 1998.
19. (a) Chen Y, Yekta S, Yudin AK, *Chem Rev* **103**:3155–3211, 2003. (b) Kočovský P, Vyskočil S, Smrčina M, *Chem Rev* **103**:3213–3245, 2003.
20. Kyba EB, Koga K, Sousa LR, Siegel MG, Cram DJ, *J Am Chem Soc* **95**:2692, 1973.
21. Kyba EP, Gokel GW, de Jong F, Koga K, Sousa LR, Siegel MG, Kaplan L, Sogah GDY and Cram DJ, *J Org Chem* **42**:4173, 1977.
22. Hegeson RC, Timko JM, Moreau P, Peacock SC, Mayer JM, Cram DJ, *J Am Chem Soc* **96**:6762, 1974.
23. Cram DJ, Helgeson RC, Peacock SC, Kaplan LJ, Domeier LA, Moreau P, Koga K, Mayer JM, Chao Y, Siegel MG, Hoffman DH, Sogah GDY, *J Org Chem* **43**:1930, 1978.
24. Helgeson RC, Tarnowski TL, Cram DJ, *J Org Chem* **44**:2538, 1979.
25. Kyba E, Timko JM, de Jong F, Gokel GW, Cram DJ, *J Am Chem Soc* **100**:4555, 1978.
26. Peacock SC, Domeier LA, Gaeta FCA, Helgeson RC, Timko JM, Cram DJ, *J Am Chem Soc* **100**:8190, 1978.
27. Peacock SC, Cram DJ, *J C S Chem Comm* 282, 1976.
28. Goldberg I, *J Am Chem Soc* **99**:6049, 1977.
29. Lingenfelter D, Helgeson RC, Cram DJ, *J Org Chem* **46**:393, 1981.
30. Newcomb M, Toner JL, Helgeson RC, Cram DJ, *J Am Chem Soc* **101**:4941, 1979.
31. Peacock SS, Walba DM, Gaeta FCA, Helgeson RC, Cram DJ, *J Am Chem Soc* **102**:2043, 1980.
32. Sogah GDY, Cram DJ, *J Am Chem Soc* **98**:3038, 1976.
33. Cram DJ, Cram JM, *Science* **183**:803, 1974.
34. Cram DJ, *Science* **240**:760, 1988.
35. Cram DJ, *Nature* **356**:29, 1992.
36. Sogah GDY, Cram DJ, *J Am Chem Soc* **79**:3035, 1979.
37. Sousa LR, Hoffman DH, Kaplan L, Cram DJ, *J Am Chem Soc* **96**:7100, 1974.
38. Sogah GDY, Cram DJ, *J Am Chem Soc* **97**:1259, 1975.

39. Sousa LR, Sogah GDY, Hoffman DH, Cram DJ, *J Am Chem Soc* **100**:4569, 1978.
40. Cram DJ, Sogah GDY, *J Chem Soc Chem Comm* 625, 1981.
41. Cram DJ, Sogah DY, *J Am Chem Soc* **107**:8301, 1985.
42. Helgeson RC, Timko JM, Cram DJ, *J Am Chem Soc* **95**:3023, 1973.
43. Cram DJ, Helgeson RC, Koga K, Kyba EP, Madan, K, Sousa LR, Siegel MG, Moreau P, Gokel GW, Timko JM, Sogah GDY, *J Org Chem* **43**:2758, 1978.
44. Miyashita A, Yasuda A, Takaya H, Toriumi K, Ito T, Souchi T, Noyori R, *J Am Chem Soc* **102**:7932, 1980.
45. Noyori R, Takaya H, *Acc Chem Res* **23**:345, 1990.
46. Takaya H, Ohta T, Noyori R, in *Catalytic Asymmetric Synthesis*, Ojima I (ed.), VCH, New York, 1993, p. 1.
47. Ohta T, Miyake T, Seido N, Kumobayashi H, Takaya H, *J Org Chem* **60**:357, 1995.
48. Mikami K, *Pure Appl Chem* **68**:639–644, 1996.
49. Cram DJ, Sogah DY, *J Am Chem Soc* **107**:8301–8302, 1985.
50. Kazlauskas RJ, *Org Syn* **70**:60, 1991.
51. Chow H-F, Wan C-W, Ng M-K, *J Org Chem* **61**:8712, 1996.
52. Kawashima M, Hirayama A, *Chem Lett* 2299, 1990.
53. Wang M, Liu SZ, Liu J, Hu BF, *J Org Chem* **60**:7364, 1995.
54. Shan ZX, Wang GP, Duan B, Zhao DJ, *Tetrahedron Asymmetry* **7**:2847, 1996.
55. (a) Brunel J-M, Buono G, *J Org Chem* **58**:7313, 1993. (b) Jacques J, Fouguey C, *Org Synth* **67**:1, 1988.
56. (a) Truesdale LK, *Org Synth* **67**:13, 1988. (b) Fabbri D, Delogu G, De Lucchi O, *J Org Chem* **58**:1748, 1993. (c) Smrcina M, Poláková J, Vyskocil S, Kocovsky, P, *J Org Chem* **58**:4534, 1993. (d) Brussee J, Groenendijk JLG, te Koppele JM, Jansen ACA, *Tetrahedron* **41**:3313, 1985. (e) Osa T, Kashiwagi Y, Yanagisawa Y, Bobbitt JM, *J Chem Soc Chem Commun* 2535, 1994.
57. Toda F, Tanaka K, Stein Z, Golderberg I, *J Org Chem* **59**:5748, 1994.
58. Tanaka K, Okada T, Toda F, *Angew Chem Int Ed Engl* **32**:1147, 1993.
59. Hu Q-S, Vitharana DR, Pu L, *Tetrahedron Asymmetry* **6**:2123, 1995.
60. Cai D, Hughes D, Verhoeven TR, Reider DJ, *Tetrahedron Lett* **36**:7991, 1995.
61. Colonna S, Re A, Wynberg H, *J Chem Soc Perkin Trans I* **1**:547, 1981.
62. Noji M, Nakajima M, Koga K, *Tetrahedron Lett* 7983, 1994.
63. Kyba EP, Gokel GW, de Jong F, Koga K, Sousa LR, Siegel MG, Kaplan L, Sogah GDY, Cram DJ, *J Org Chem* **42**:4173–4184, 1977.
64. Meca L, Řeha D, Havlas Z, *J Org Chem* **68**:5677–5680, 2003.
65. Sahnoun R, Koseki S, Fujimura Y, *J Mol Struct* **735–736**:315–324, 2005.
66. Mori K, Masuda Y, Kashino S, *Acta Cryst C* **C49**:1224–1227, 1993.

Chapter 2

Main Chain Chiral-Conjugated Polymers

2.1. Introduction About Chiral-Conjugated Polymers

The study of conjugated polymers such as polyacetylene (PA), poly(p-phenylene) (PPP), poly(p-phenylenevinylene) (PPV) and poly(p-phenyleneethynylene) (PPE) (Fig. 2.1) has become a very exciting field of research since the discovery of the high conductivity of doped PAs in 1977.[1] Intensive research in this area during the past three decades has uncovered many potential applications for conjugated materials. These polymers can be used to prepare organic conductors, electroluminescent devices, sensors, non-linear optical devices, high energy density batteries, solar cells, *etc.*[1–3] Chiral-conjugated polymers are a special class of conjugated materials with great potentials. Electrodes prepared from these chiral polymers can be used to carry out asymmetric electro-reactions to synthesize optically active organic compounds. Scheme 2.1 shows an experiment where an asymmetric electro-oxidation of *t*-butyl phenyl sulfide occurs on a chiral electrode to generate a chiral sulfoxide with high enantioselectivity.[4] The chiral electrode used in this reaction is made by coating a platinum electrode with a conjugated polymer (polypyrrole) and a chiral polymer (poly[L-valine]). Optically active conjugated polymers may be used to replace the dual polymer coatings in this chiral electrode to conduct potentially more efficient asymmetric electrosynthesis. Other applications of chiral-conjugated materials include uses as ferroelectric liquid crystals, non-linear optical materials and polarized light emissions. Chiral sensors may be also made from chiral-conjugated polymers for the detection of chiral molecules, including those of biological interests.

Over the years, various chiral-conjugated polymers have been prepared.[5–9] In 1991, Grubbs reported the synthesis of soluble chiral PAs

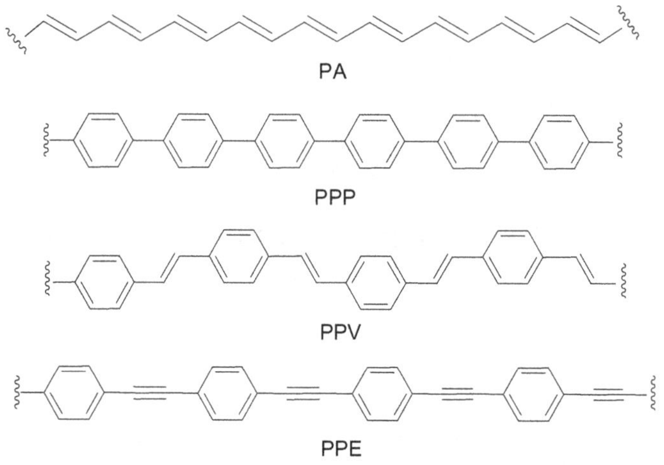

Figure 2.1. Examples of conjugated polymers.

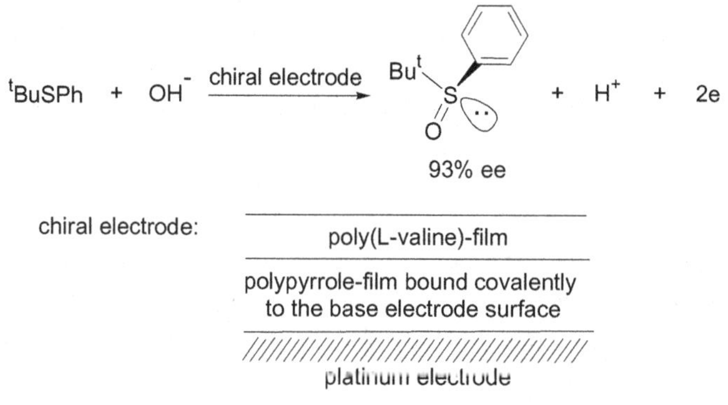

Scheme 2.1. Asymmetric electro-oxidation of a sulfide to a chiral sulfoxide using a chiral electrode.

using a tungsten carbene catalyzed ring-opening-metathesis-polymerization process.[5] In their system, a cyclooctatetraene substituted with an optically active alkyl group was polymerized to generate a soluble chiral-conjugated PA (Scheme 2.2). The originally generated polymer contained mainly *cis* double bonds, which was photochemically isomerized to *trans*. The main chain of this partially substituted PA was shown to be twisted by the chiral substituents to form a helical chain conformation.

Terminal alkynes substituted with chiral substituents were polymerized by using a rhodium catalyst, [RhCl(NBD)]$_2$ (NBD = norbornadiene).[6]

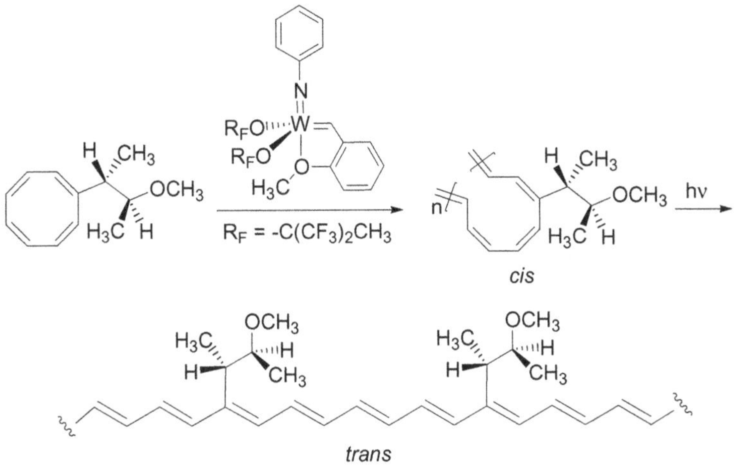

Scheme 2.2. Ring-opening metathesis polymerization of an optically active cyclooctatetraene.

As shown in Scheme 2.3, polymerization of a chiral (carbamoyloxy)phenylacetylene forms a *cis*-substituted PA. Owing to the bulkiness of the substituents, these polymers generate a helical conformation with no extended conjugation in the polymer chain. These materials were shown to have potential as enantioselective permeable membranes to separate racemic amino acids and alcohols in water or in methanol. They can also be used as chiral stationary phase for enantioselective HPLC analysis.

Optically active polythiophenes were synthesized by electropolymerization of thiophene monomers containing chiral substituents. For example, Scheme 2.4 shows the formation of a chiral poly(thiophene) from a chiral alkyl-substituted thiophene.[7] When this chiral poly(thiophene) was oxidized in the presence of chiral anions, there was stereoselective recognition of the anions. Other chiral-conjugated polymers including polypyrroles substituted with chiral amino acid groups[8] and polyazulenes substituted with sugars[9] were also prepared by electropolymerization.

The conjugation of polyisocyanates is known to be completely disrupted by their substitutents on the nitrogen atoms to form a helical conformation. Green found that this class of polymers had very interesting chiral amplification properties.[10] As shown in Scheme 2.5, when a chiral isocyanate ($[\alpha]_D = +0.65$) with its chirality solely derived from isotope substitution was polymerized, the resulting polymer exhibited a dramatically enhanced optical rotation ($[\alpha]_D = -367$).[10b]

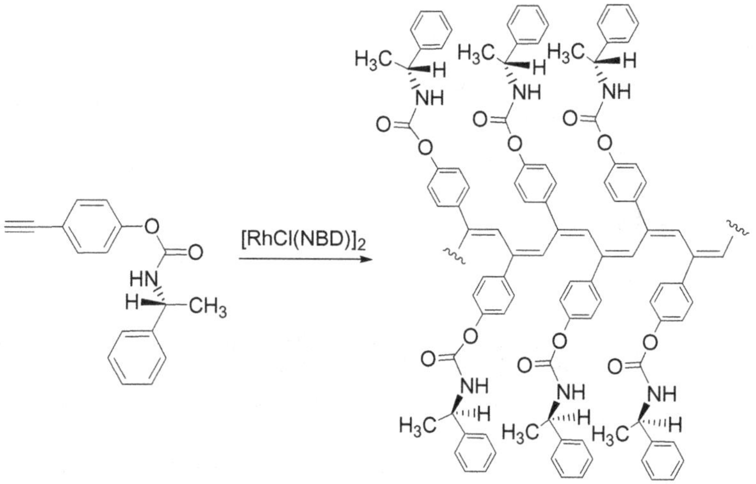

Scheme 2.3. Synthesis of a chiral-substituted PA.

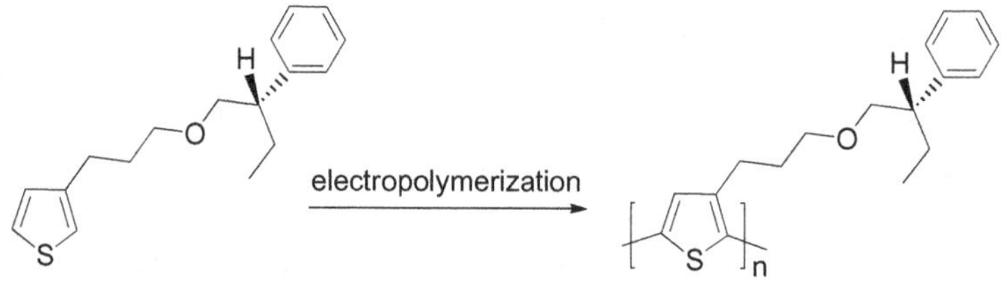

Scheme 2.4. Electropolymerization of thiophene substituted with a chiral substituent.

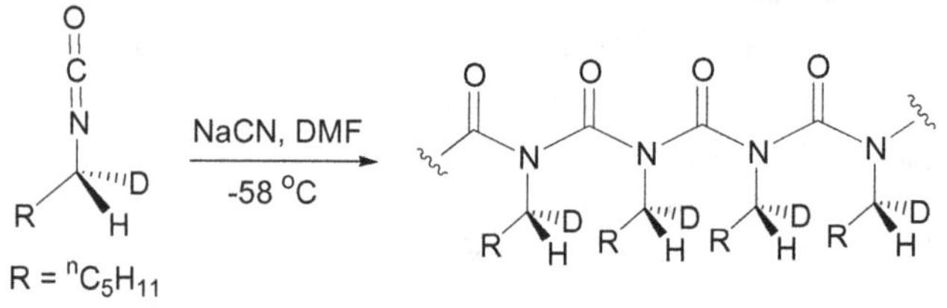

Scheme 2.5. Chiral amplification in a polyisocyanate.

The chiral-conjugated polymers described above share one common feature that the chirality of these materials is derived from their side chain substituents. The main chain of these materials is not inherently chiral. Thus, the chiral chain conformations of these materials are generally not stable and they often undergo significant changes with respect to external factors such as solvent, temperature, pH and others. In our laboratory, we have incorporated the 1,1′-binaphthyl units into the main chain of various conjugated polymers such as arylenevinylenes, arylenes, aryleneethynylenes and thiophenes to build the novel main chain chiral-conjugated polymers. In these polymers, the chiral binaphthyl units allow these materials to have stable main chain chiral configuration. The synthesis and study of these polymers are described in this chapter.

2.2. Binaphthyl-Based Polyarylenevinylenes[11,12]

In order to construct the binaphthyl-based main chain chiral-conjugated polymers, we have prepared a binaphthyl-based 6,6′-diboronic acid monomer (R)-**2.1** (Scheme 2.6). This synthesis utilizes a specific bromination at the 6,6′-positions of (R)-BINOL to generate (R)-6,6′-Br$_2$BINOL. Treatment of this compound with hexyl iodide in the presence of K$_2$CO$_3$ gives (R)-6,6′-dibromo-2,2′-dihexyloxy-1,1′-binaphthyl [(R)-DBDH]. The specific optical rotation $[\alpha]_D$ of (R)-DBDH is 25.9 ($c = 0.52$, THF). This compound is then treated with Mg and B(OMe)$_3$, which upon hydrolysis gives the diboronic acid monomer (R)-**2.1**. The specific optical rotation $[\alpha]_D$ of (R)-**2.1** is 35.6 ($c = 0.22$, DMSO). The chiral configuration of the binaphthyl unit is found to be stable during the course of these reactions.

The phenylenevinylene linker molecule **2.2** is prepared from the reaction of αα′-dichloro-p-xylene with PPh$_3$ followed by treatment with LiOEt and p-bromobenzaldehyde (Scheme 2.7). This compound contains both *trans* and *cis* double bonds with a *trans:cis* ratio of 1:1.2. The Suzuki coupling of **2.2** with (R)-**2.1** in the presence of Pd(PPh$_3$)$_4$ leads to the main chain chiral conjugated polyphenylenevinylene (R)-**2.3** in 95% yield (Scheme 2.8). This polymer is soluble in THF, benzene and chloroform. GPC shows its molecular weight as $M_w = 67,000$ and $M_n = 20,000$ (PDI = 3.4). The specific optical rotation $[\alpha]_D$ of this polymer is

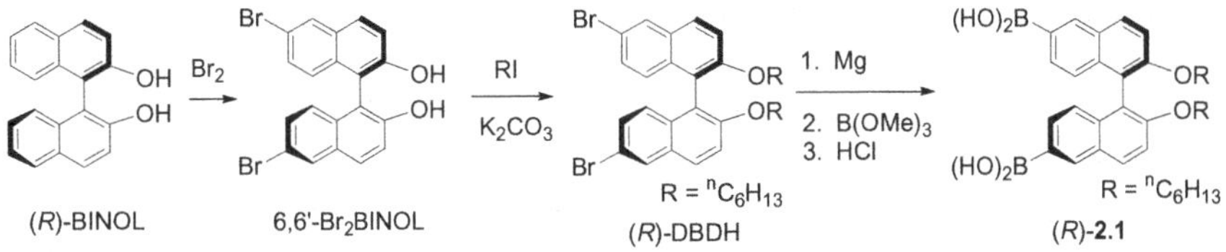

Scheme 2.6. Synthesis of chiral monomer (R)-**2.1**.

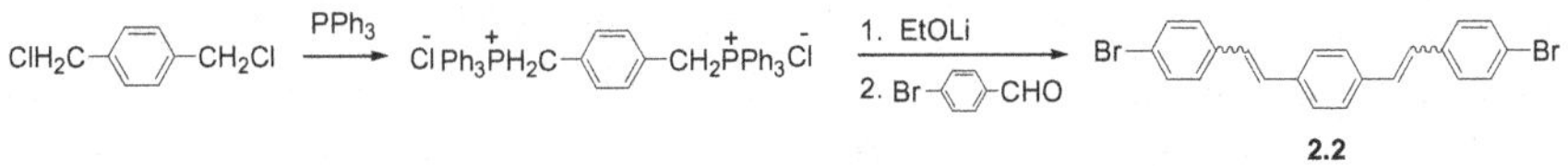

Scheme 2.7. Synthesis of arylene vinylene linker **2.2**.

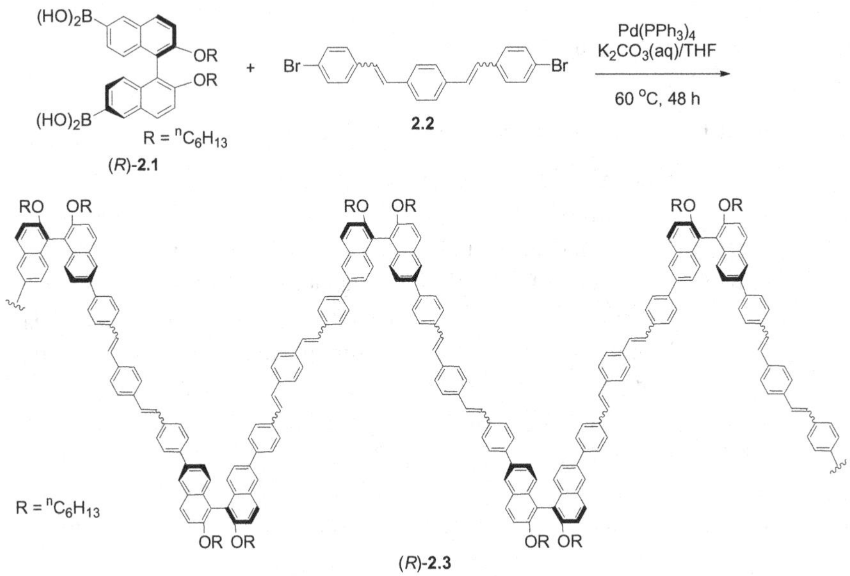

Scheme 2.8. Synthesis of the main chain chiral-conjugated polyarylenevinylene (R)-**2.3**.

−351 ($c = 0.38$, THF). The NMR analysis of the polymer indicates that polymer (R)-**2.3** still contains both *cis* and *trans* double bonds with a *trans:cis* ratio of 1:0.45. That is, some of the *cis* double bonds are converted to *trans* during the polymerization. In the IR spectrum, the *p*-phenylene units in (R)-**2.3** give a characteristic strong C–H out-of-plane bending absorption at 814 cm^{-1}. Polymerization of the racemic monomer *rac*-**2.1** with **2.2** gives polymer *rac*-**2.3**, which most likely contains a random distribution of the (R)- and (S)-binaphthyl units in the polymer chain. GPC shows the molecular weight of polymer *rac*-**2.3** as $M_n = 17{,}000$ and $M_w = 48{,}000$ (PDI = 2.8) relative to polystyrene standards. The ratio of *cis* to *trans* double bonds in *rac*-**2.3** is similar to that in (R)-**2.3**. Both polymers *rac*-**2.3** and (R)-**2.3** are green solids.

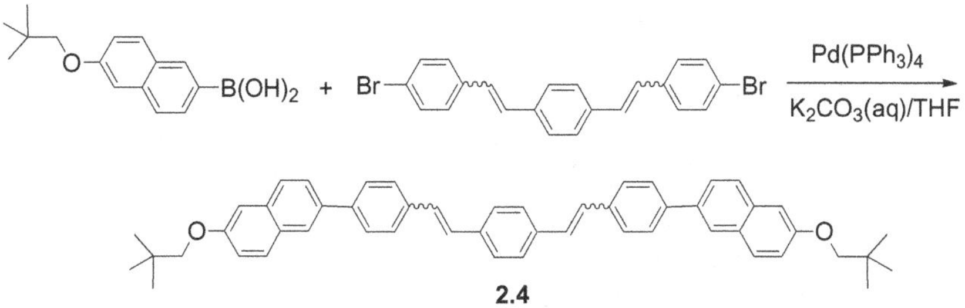

Scheme 2.9. Synthesis of arylene vinylene repeating unit **2.4**.

The arylenevinylene repeating unit of these polymers is prepared as shown in Scheme 2.9. Unlike polymers rac-**2.3** and (R)-**2.3**, which have good solubility in common organic solvents, the arylene vinylene repeating unit has very low solubility. Although the introduction of the bulky neopentyl groups in **2.4** improves its solubility over the one containing linear hexyl groups, the solubility is still not very good. The poor solubility of **2.4** is attributed to its planar conjugated molecular structure, which may resist solvation due to strong intermolecular π–π stacking. The flexible and non-planar binaphthyl units in the polymers increase their solubility. The UV spectrum of **2.4** shows the longest wavelength absorption at $\lambda_{max} = 386$ nm. It is very similar to the absorption of polymer (R)-**2.3** at $\lambda_{max} = 390$ nm. This indicates that the effective conjugation of the polymer is almost the same as that of the repeating unit and there is no extended conjugation across the 1,1'-bond of the binaphthyl units. That is, the conjugation of the polymer is determined by its repeating unit and is independent of the molecular weight of the polymer.

Polymer (R)-**2.3** is a blue light emitting material. When excited at 390 nm, it shows emission at $\lambda = 468$ nm. The fluorescence quantum yield of (R)-**2.3** in toluene solution is about 80% as estimated by comparison with 9-anthracene carboxylic acid. The fluorescence quantum yield of (R)-**2.3** is significantly greater than that of rac-**2.3**, which is found to be 50%. Thus the steric regularity of (R)-**2.3** has enhanced its fluorescence quantum efficiency. The repeating unit **2.4** shows emission at 433, 463 and 494 (sh) nm when excited at 386 nm.

The thin films of the polymers can be obtained by spin-coating their solutions onto glass slides. When these thin films are exposed to the acetonitrile solution of $NOBF_4$, the main chain chiral-conjugated polymers are oxidatively doped. The conductivity of the doped thin films of (R)-**2.3** and rac-**2.3** is in the range of 4×10^{-5} to 7×10^{-5} scm^{-1}. Atomic force microscopy study shows that the thin films of both (R)-**2.3** and rac-**2.3** are amorphous. Polymer (R)-**2.3** is stable up to 340°C under nitrogen as shown by thermogravimetric analysis (TGA). Its glass transition temperature (T_g) is 210°C as determined by differential scanning calorimetry (DSC). The thermal stability of polymer rac-**2.3** is found to be higher than that of polymer (R)-**2.3**. TGA shows that rac-**2.3** is stable up to 390°C under nitrogen. The T_g of rac-**2.3** is at 203°C.

In collaboration with Zhang, the formation and decay dynamics of the photogenerated excitons in (R)-**2.3** in solution are investigated by using femtosecond transient absorption spectroscopy.[13] Photoexcitation initially creates hot excitons that quickly (<200 fs) relax geometrically toward the equilibrium position in the excited state. The exciton subsequently decays following a double exponential with time constants of 6.5 and 420 ps in toluene. In pyridine, the decays become faster (5 and 250 ps). That is, the relaxation process is dependent on the solvent environment. The fast decay is attributed to vibrational relaxation and internal conversion (recombination) of the exciton from the excited to the ground electronic state through tunneling or thermal-activated barrier crossing before thermalization. The slow decay is assigned to conversion of the thermalized exciton to the ground state through both radiative and non-radiative pathways. Anisotropy decay shows a fast component of ∼6 ps in toluene and 10 ps in pyridine and an offset, which persists up to 650 ps. Possible explanations for the fast decay include internal conversion, vibrational relaxation, conformational change and exciton migration. The offset may decay on a longer time scale through local reorientation of the conjugation segments, exciton migration or rotational diffusion of the polymer. It is also found that the excited state of this polymer has a dipole moment rotated with respect to the ground electronic state by about 24°.

2.3. Binaphthyl-Based Polyarylenes[14]

From the Suzuki coupling of (R)-**2.1** with p-dibromobenzene, the main chain chiral-conjugated polyarylene (R)-**2.5** is obtained (Scheme 2.10). This polymer is soluble in THF, chloroform, methylene chloride and benzene. GPC shows its molecular weight as M_w = 41,000 and M_n = 20,000 (PDI = 2.0). The specific optical rotation $[\alpha]_D$ of the polymer is -289.4 ($c = 0.5$, CH$_2$Cl$_2$). When a toluene solution of (R)-**2.5** is heated at reflux under nitrogen for 40 h, there is only 5% decrease in the optical rotation.

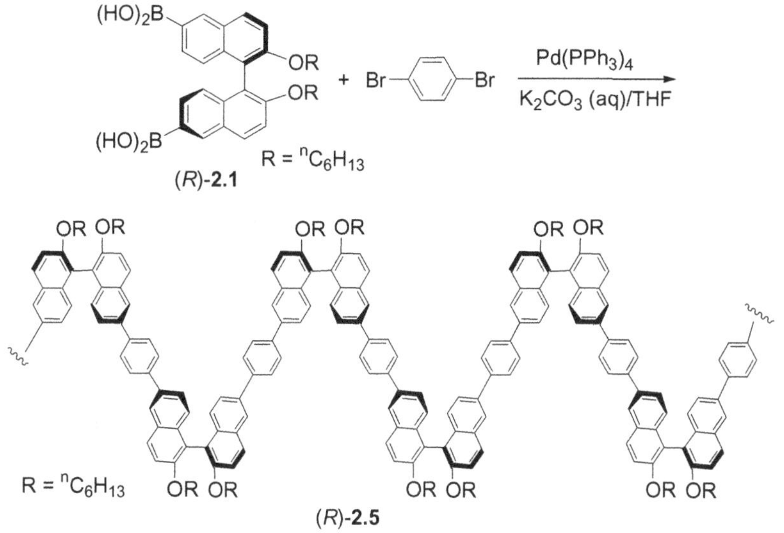

Scheme 2.10. Synthesis of the main chain chiral-conjugated polyarylene (R)-**2.5**.

This demonstrates that the chiral configuration of the polymer is very stable. A well-resolved ^{1}H NMR spectrum of the polymer is obtained, which is consistent with its structure. The methylene chloride solution of this polymer gives UV absorption at $\lambda_{\text{max}} = 328$ nm. The thin film of polymer (R)-**2.5** spin-coated on glass slide shows only a slightly blue shifted UV signal at $\lambda_{\text{max}} = 324$ nm. This indicates that there is no significant π–π interaction for the polymer in the solid state. It can be attributed to the chiral non-planar binaphthyl units, which reduces the intermolecular interaction. When the methylene chloride solution of (R)-**2.5** is excited at 328 nm, it shows emissions at $\lambda_{\text{emi}} = 388$ and 407 nm. TGA of (R)-**2.5** shows that this polymer starts to decompose at 394°C. At 476°C, it loses ∼32% of its mass, which corresponds to the loss of the hexyl groups. DSC shows the T_g of this polymer at 202°C.

Polymer rac-**2.5** is obtained from the coupling of the racemic monomer with p-dibromobenzene. Its molecular weight is $M_w = 37,000$ and $M_n = 22,000$ (PDI = 1.7) as shown by GPC analysis. TGA shows that rac-**2.5** loses its hexyl groups from 386 to 469°C. DSC shows its T_g at 193°C. The UV absorption of rac-**2.5** is observed at $\lambda_{\text{max}} = 328$ nm. While excited at 328 nm, it emits at 387 and 405 nm. These measurements show similar properties between rac-**2.5** and (R)-**2.5**.

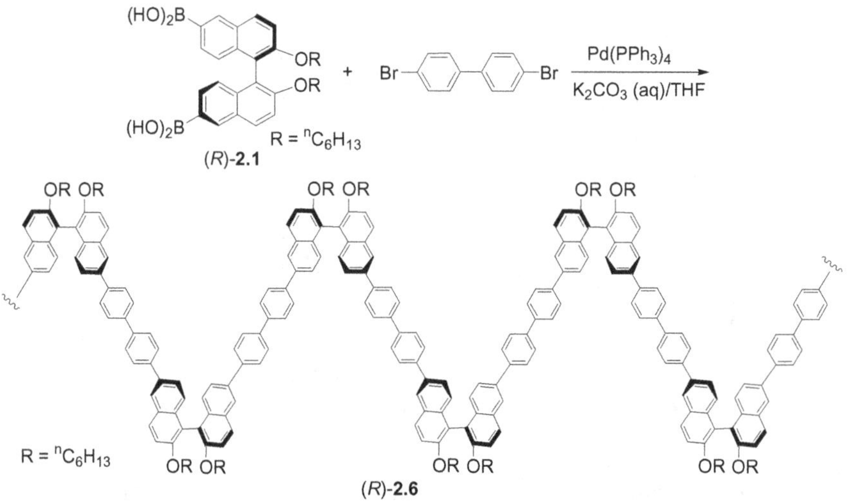

Scheme 2.11. Synthesis of the main chain chiral-conjugated polyarylene (R)-**2.6**.

Polymerization of (R)-**2.1** with 4,4′-dibromobiphenyl gives another main chain chiral-conjugated polyarylene (R)-**2.6** (Scheme 2.11). This polymer is also soluble in common organic solvents. GPC shows its molecular weight as $M_w = 47{,}000$ and $M_n = 14{,}000$ (PDI = 3.4). The specific optical rotation $[\alpha]_D$ of (R)-**2.6** is -275 ($c = 0.22$, CH_2Cl_2). TGA shows that (R)-**2.6** loses its hexyl groups from 396 to 478°C. After the loss of the hexyl groups, this polymer becomes very stable. It shows no significant mass loss from 480 up to 800°C. The T_g of the biphenylene linker polymer (R)-**2.6** is observed at 229°C, which is higher than that of the monophenylene linker polymer (R)-**2.5**. The UV absorption of this polymer in methylene chloride solution is observed at $\lambda_{max} = 334$ nm. The thin film of (R)-**2.6** gives the same UV absorption as the solution, indicating a minimum interchain π–π interaction in the solid state. While excited at 334 nm, the polymer solution emits at 396 and 415 nm. Polymer rac-**2.6** is obtained from rac-**2.1** with $M_w = 92{,}000$ and $M_n = 23{,}000$ (PDI = 4.0). This polymer shows properties similar to those of (R)-**2.6** except the optical rotation.

Both the optically active polymers (R)-**2.5** and (R)-**2.6** display strong Cotton effects in their circular difference (CD) spectra. Table 2.1 summarizes the CD signals of these two polymers. The molar ellipticity is based on the molecular weights of the polymer repeating units.

Table 2.1. The molar ellipticity from the CD spectra of the chiral-conjugated polymers (R)-**2.5** and (R)-**2.6**.

Polymer	λ_{CD} (nm)	$[\theta]_\lambda \times 10^{-5}$ (deg·L/mol·cm)
(R)-**2.5**	228	2.46
	262	2.54
	285	−3.14
	331	−1.04
(R)-**2.6**	232	4.2
	247	−1.23
	273	0.30
	297	−0.98
	340	−0.69
	364	−0.25

The repeating units **2.7** and **2.8** of these polymers are synthesized in a way similar to the preparation of **2.4**. The neopentyl groups in these compounds are necessary to make them soluble. Using other alkyl groups such as methyl and hexyl in place of neopentyl gives insoluble compounds. Compound **2.7** in methylene chloride gives a major UV absorption at $\lambda_{max} = 322$ nm, being only 6 nm shorter than that of polymer (R)-**2.5**. That is, the conjugation of the polymer is mostly determined by the repeating unit and there is little extended conjugation across the binaphthyl units. The repeating unit **2.8** in methylene chloride gives a major UV absorption at $\lambda_{max} = 324$ nm, being only 10 nm shorter than that of polymer (R)-**2.6**. The UV and fluorescence spectroscopic data of the polymers and repeating units are summarized in Table 2.2.

2.7

2.8

2.4. Binaphthyl-Based Polyaryleneethynylenes[15–17]

From the Sonogashira coupling of the alkylated 6,6′-dibromoBINOL (R)-**2.9** $\{[\alpha]_D = +20.5\ (c = 1.0,\ CH_2Cl_2)\}$ with p-diethynylbenzene, the main chain chiral-conjugated polyaryleneethynylene (R)-**2.10a** is obtained (Scheme 2.12). It is found that the long chain octadecyloxy groups in

Table 2.2. The UV absorption and fluorescence spectroscopic data of the polymers and the repeating units.

Polymer or repeating unit	UV absorption (CH$_2$Cl$_2$), λ_{max} (nm)	Fluorescence (CH$_2$Cl$_2$), λ_{emi} (nm)
(R)-**2.5**	234, 282, 328	388, 407
rac-**2.5**	232, 282, 328	387, 405
(R)-**2.6**	238, 293, 334	396, 415
rac-**2.6**	240, 293, 332	396, 415
2.7	234, 272, 322	378, 396, 417
2.8	240, 280, 324	389, 408, 418

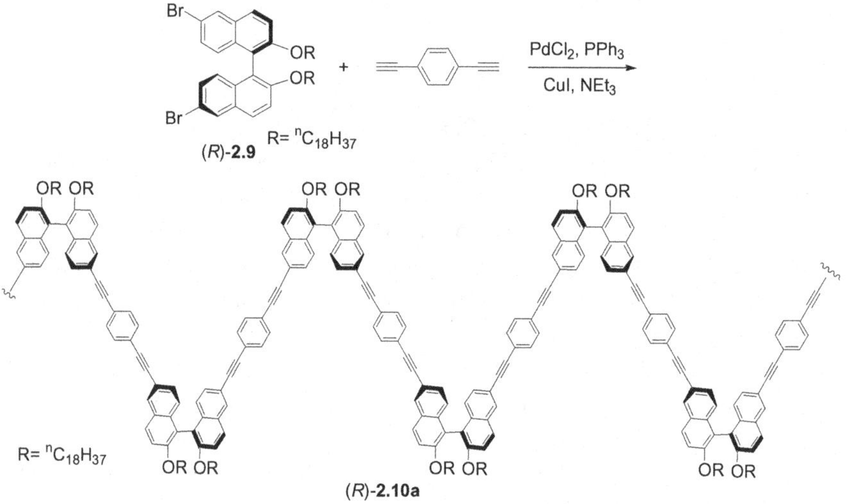

Scheme 2.12. Synthesis of the main chain chiral-conjugated polyaryleneethynylene (R)-**2.10a**.

monomer (R)-**2.9** are necessary to make the resulting polymer soluble in organic solvents. Using the binaphtyl monomers containing shorter chain alkoxy groups such as hexyloxy and decyloxy cannot produce a soluble polymer. Polymer (R)-**2.10a** is soluble in common organic solvents. Polymerization of the racemic monomer rac-**2.9** with p-diethynylbenzene gives rac-**2.10a**.

From the coupling of (R)-**2.9** with 4,4′-diethynyl-biphenyl, polymer (R)-**2.11a** is obtained (Scheme 2.13). Polymer rac-**2.11a** is also obtained by using the racemic monomer rac-**2.9**.

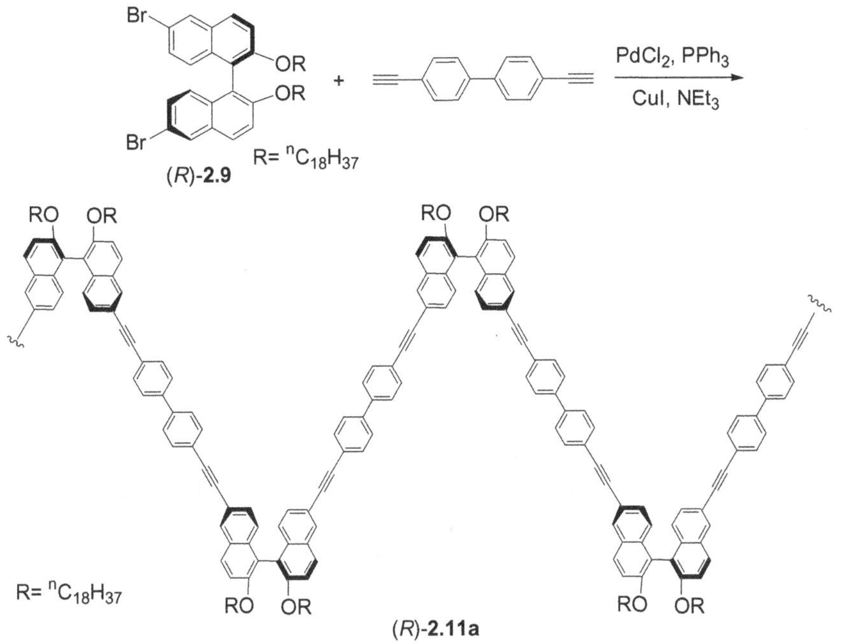

Scheme 2.13. Synthesis of the main chain chiral-conjugated polyaryleneethynylene (R)-**2.11a**.

The 6,6′-diethynyl binaphthyl monomer (R)-**2.12** {$[\alpha]_D = -14.7$ ($c = 1.0$, CH_2Cl_2)} is prepared from the coupling of (R)-**2.9** with 2-methyl-3-butyn-2-ol followed by the base-promoted deprotection as shown in Scheme 2.14. Polymerization of (R)-**2.12** with p-diiodobenzene in the presence of $PdCl_2$ and CuI gives polymer (R)-**2.10b**. Polymer (R)-**2.10b** has the same structure as (R)-**2.10a** except that the molecular weight of (R)-**2.10b** is much higher (Table 2.3). Polymerization of rac-**2.12** with p-diiodobenzene gives polymer rac-**2.10b**. Polymerization of (R)-**2.12** with 4,4′-diiodobiphenyl gives polymer (R)-**2.11b**. Polymer (R)-**2.11b** has the same structure as (R)-**2.11a** but with much higher molecular weight (Table 2.3). Polymer rac-**2.11b** is prepared from the coupling of rac-**2.12** with 4,4′-diiodobiphenyl.

The use of $Pd(PPh_3)_4$ in place of $PdCl_2$ is also tested for the polymerization, which leads to the formation of polymers (R)-**2.10c**, rac-**2.10c**, (R)-**2.11c** and rac-**2.11c**, respectively. In general, using $Pd(PPh_3)_4$ produces higher molecular weight polymer than using $PdCl_2$.

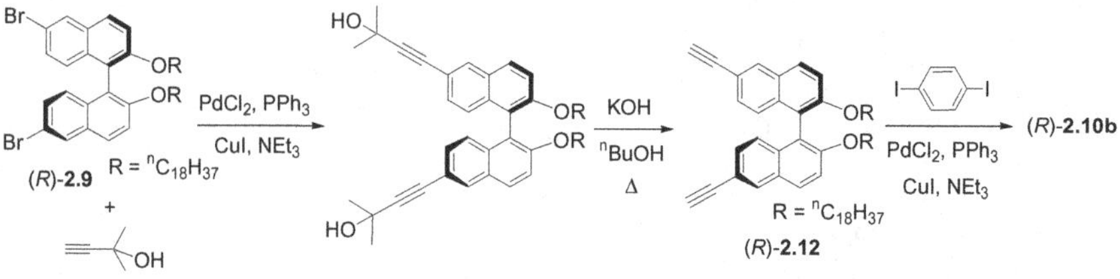

Scheme 2.14. Synthesis of monomer (R)-**2.12** and polymer (R)-**2.10b**.

Table 2.3 summarizes the polymerization conditions, molecular weights (measured by GPC) and specific optical rotations for polymers **2.10a–c** and **2.11a–c**. All these polymers are soluble in organic solvents. The best conditions for the synthesis of these polymers involve the use of Pd(PPh$_3$)$_4$ and CuI for the coupling of monomer **2.12** with aryl iodides, which produces polymers **2.10c** and **2.11c**. As shown by polymers (R)-**2.11a–c**, the optical rotation of these chiral polymers increases with increasing molecular weight of the polymers. The optical rotations of these polymers also have the sign opposite to that of the 6,6′-dibromo monomer (R)-**2.9**. Replacement of the 6,6′-bromine atoms of (R)-**2.9** with the two ethynyl groups of (R)-**2.12** inverts the optical rotation.

TGA shows that polymer (R)-**2.10c** is stable up to 396°C. From 396 to 486°C, this polymer loses its alkyl groups. After the loss of the alkyl groups, it becomes extremely stable with no significant weight loss up to 800°C. The thermal stability of polymer (R)-**2.11c** is similar to that of (R)-**2.10c**. It starts to lose its alkyl groups at 395°C until 540°C. After that, the polymer is stable up to 800°C.

In collaboration with Wu, we have compared the GPC molecular weights of polymers (R)-**2.11a,b** and rac-**2.11c** with those obtained by laser light scattering (LLS) in THF solution. As shown in Table 2.4, the GPC data relative to polystyrene standards underestimate the molecular weight of the polymers by 1.4–2.5 times.

Compounds **2.13** and **2.14** are prepared as the repeating units of polymers (R)-**2.10** and (R)-**2.11** from the coupling of 6-neopentyloxy-naphthyl-2-boronic acid with p-diethynylbenzene and 4,4′-diethynylbiphenyl, respectively. The NMR and IR spectra of these compounds are very similar to those of their corresponding polymers (R)-**2.10** and (R)-**2.11**, which support the structures of these polymers. Table 2.5 gives the UV absorption data of the polymers and the repeating units. The UV absorptions of repeating units **2.13** and **2.14** are similar to those of

Table 2.3. Conditions to prepare **2.10a–c** and **2.11a–c** and their molecular weights based on GPC analysis.

Polymers	Monomers	Catalysts	M_w	M_n	PDI	$[\alpha]_D$	$[\Phi]_D$[a]
(R)-**2.10a**	(R)-**2.9** + p-diethynylbenzene	PdCl$_2$, CuI	7400	3800	2.0	−154.9	−1415
rac-**2.10a**	rac-**2.9** + p-diethynylbenzene	PdCl$_2$, CuI	8500	3200	2.7		
(R)-**2.10b**	(R)-**2.12** + 1,4-diiodobenzene	PdCl$_2$, CuI	21,000	15,000	1.4	−278.2	−2541
rac-**2.10b**	rac-**2.12** + 1,4-diiodobenzene	PdCl$_2$, CuI	13,000	8300	1.6		
(R)-**2.10c**	(R)-**2.12** + 1,4-diiodobenzene	Pd(PPh$_3$)$_4$, CuI	29,000	12,000	2.4	−272.2	−2486
rac-**2.10c**	rac-**2.12** + 1,4-diiodobenzene	Pd(PPh$_3$)$_4$, CuI	47,000	25,000	1.9		
(R)-**2.11a**	(R)-**2.9** + 4,4′-diethynylbiphenyl	PdCl$_2$, CuI	9300	4800	1.9	−158.9	−1572
rac-**2.11a**	Rac-**8** + 4,4′-diethynylbiphenyl	PdCl$_2$, CuI	6600	3900	1.7		
(R)-**2.11b**	(R)-**2.12** + 4,4′-diiodobiphenyl	PdCl$_2$, CuI	29,000	15,000	1.9	−250.6	−2480
rac-**2.11b**	rac-**2.12** + 4,4′-diiodobiphenyl	PdCl$_2$, CuI	20,000	12,000	1.7		
(R)-**2.11c**	(R)-**2.12** + 4,4′-diiodobiphenyl	Pd(PPh$_3$)$_4$, CuI	37,000	18,000	2.1	−281.7	−2787
rac-**2.11c**	rac-**2.12** + 4,4′-diiodobiphenyl	Pd(PPh$_3$)$_4$, CuI	54,000	25,000	2.2		

a. Molar rotation $[\Phi] = [\alpha]M_w/100$ (M_w: molecular weight of the polymer repeating unit).

Table 2.4. Comparison of the GPC molecular weights of (R)-**2.11a,b** and rac-**2.11c** with those by the LLS study.

Polymers	M_w (GPC)	M_w/g/mol (LLS data)	M_w/M_n
(R)-**2.11a**	9300	23,000	~3.7
(R)-**2.11b**	20,000	32,000	~3.5
rac-**2.11c**	54,000	73,000	~2.1

Table 2.5. UV absorptions of chiral polymers (R)-**2.10c** and (R)-**2.11c** and repeating units **2.13** and **2.14**.

Polymer or repeating unit	(R)-**2.10c**	(R)-**2.11c**	**2.13**	**2.14**
UV absorption (CH$_2$Cl$_2$) λ_{max} (nm)	238, 300, 354, 384 (sh)	234, 306 (sh), 352	236, 290, 350, 374 (sh)	236, 296 (sh), 348

polymers (R)-**2.10c** and (R)-**2.11c**, respectively, indicating no significant extension of conjugation from the repeating units to the polymers. It is also found that the UV absorptions of (R)-**2.10a–c**/rac-**2.10a–c** and (R)-**2.11a–c**/rac-**2.11a–c** are mostly independent of the molecular weight and steric structure. Comparison of repeating unit **2.14** with **2.13** shows that even though compound **2.14** has one extra phenyl ring, its UV absorption appears at a shorter wavelength than that of compound **2.13**. This can be attributed to the interaction of the ortho hydrogens in the biphenyl units of **2.14**, which can disturb the planarity of the molecule and reduce its effective conjugation. The fluorescence spectrum of (R)-**2.10c** shows emissions at 401 (sh), 419 and 461 (sh) nm, and that of (R)-**2.11c** at 407 (sh), 420 and 460 (sh) nm. Intense CD signals are observed for polymers (R)-**2.10c** and (R)-**2.11c**.

2.13

2.14

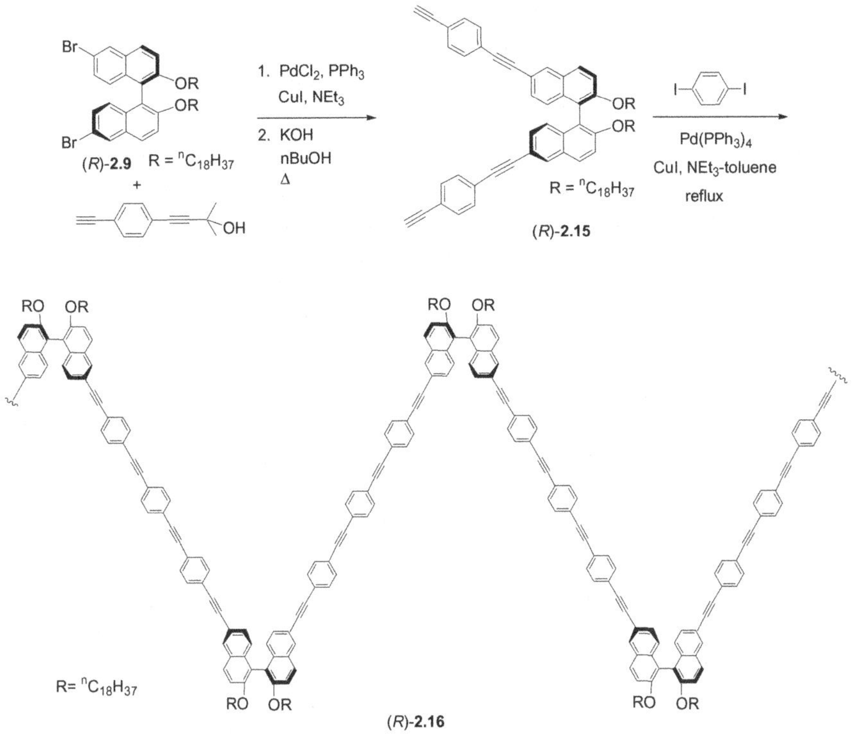

Scheme 2.15. Synthesis of polymer (R)-**2.16**.

The chiral binaphthyl monomer (R)-**2.15** that contains two p-ethynylphenylethynyl substitutents at the 6,6′-positions is synthesized from the coupling of (R)-**2.9** with a monoprotected p-diethynylbenzene followed by the base-promoted deprotection (Scheme 2.15). Its specific optical rotation $[\alpha]_D$ is -176.3 ($c = 1.0$, CH_2Cl_2) and the molar rotation $[\Phi]_D = -1833$. Monomer (R)-**2.15** has more extended conjugation than monomer (R)-**2.12**, and it also has a much greater optical rotation value with the same sign. Polymerization of (R)-**2.15** with p-diiodobenzene generates polymer (R)-**2.16** in 98% yield. This polymer contains significantly extended phenyleneethynylenes in the repeating unit and is found to be still soluble in common organic solvents. GPC shows its molecular weight as $M_w = 38{,}200$ and $M_n = 29{,}000$ (PDI = 1.3). Its specific optical rotation $[\alpha]_D$ is -288.6 ($c = 0.5$, CH_2Cl_2). This polymer gives fairly well-resolved NMR spectra that support its structure. In its ^{13}C NMR spectrum, four different alkyne carbon signals are observed.

In methylene chloride solution, polymer (R)-**2.16** gives UV absorptions at λ_{max} = 238 and 366 nm and shows strong emissions at 420, 464 and 500 (sh) nm. From the racemic monomer, polymer rac-**2.16** is obtained with M_w = 42,000 and M_n = 20,000 (PDI = 2.1). This polymer exhibits almost the same spectral data as (R)-**2.16**. The CD spectrum of (R)-**2.16** in methylene chloride solution shows the following strong Cotton effects: $[\theta]$ = $+4.46 \times 10^5$ (235 nm), -7.74×10^4 (262 nm), -1.21×10^5 (284 nm), $+3.94 \times 10^4$ (338 nm) and -2.03×10^5 (385 nm).

Crown ether functions are incorporated into the binaphthyl-based chiral polyaryleneethynylenes. As shown in Scheme 2.16, polymerization of the binaphthyl crown ether monomer (S)-**2.17** with a p-diethynylbenzene linker gives polymer (S)-**2.18** in 86% yield. This polymer is soluble in common organic solvents. GPC shows its molecular weight as M_w = 10,500 and M_n = 5,300 (PDI = 2.0). The specific optical rotation $[\alpha]_D$ of (S)-**2.18** is 120.9 (c = 0.95, THF), which is opposite to that of monomer (S)-**2.17** $\{[\alpha]_D = -8.3\ (c = 0.95, \text{THF})\}$. The UV spectrum of (S)-**2.18** in methylene chloride solution displays λ_{max} at 240, 296, 334 and 378 nm. When its solution is excited at 378 nm, the polymer shows strong emissions at 420, 445 and 455 nm. The CD spectrum of (S)-**2.18** gives the following strong Cotton effects: $[\theta]_\lambda$ = -3.72×10^5 (233 nm), 1.88×10^5 (251 nm), 5.21×10^4 (285 nm), -8.86×10^3 (349 nm) and 3.46×10^4 (398 nm).

The cross-coupling of the racemic monomer rac-**2.17** with the racemic monomer rac-**2.15** gives polymer rac-**2.19** (Scheme 2.17). GPC shows the molecular weight of rac-**2.19** as M_w = 27,900 and M_n = 11,500 (PDI = 2.4). This higher molecular weight crown ether polymer is also soluble in common organic solvents. The UV absorptions of rac-**2.19** are observed at λ_{max} = 246, 300 and 354 nm, which are of shorter wavelengths than the corresponding absorptions of polymer (S)-**2.18**. This is attributed to the two extra electron-donating alkoxy groups in the repeating units of (S)-**2.18**.

2.5. Binaphthyl–Thiophene Copolymers[18]

2.5.1. *Copolymerization of Binaphthyl and Oligothiophene Monomers*

We have incorporated oligothiophene units into the main chain of the polybinaphthyls in order to systematically tune their absorption and emission properties. Scheme 2.18 shows the synthesis of the α, α'-dibromooligothiophene monomers **2.20**–**2.25**. In compounds **2.24** and **2.25**, an alkyl group R^1 is introduced to render them soluble in organic solvents. A nine thiophene

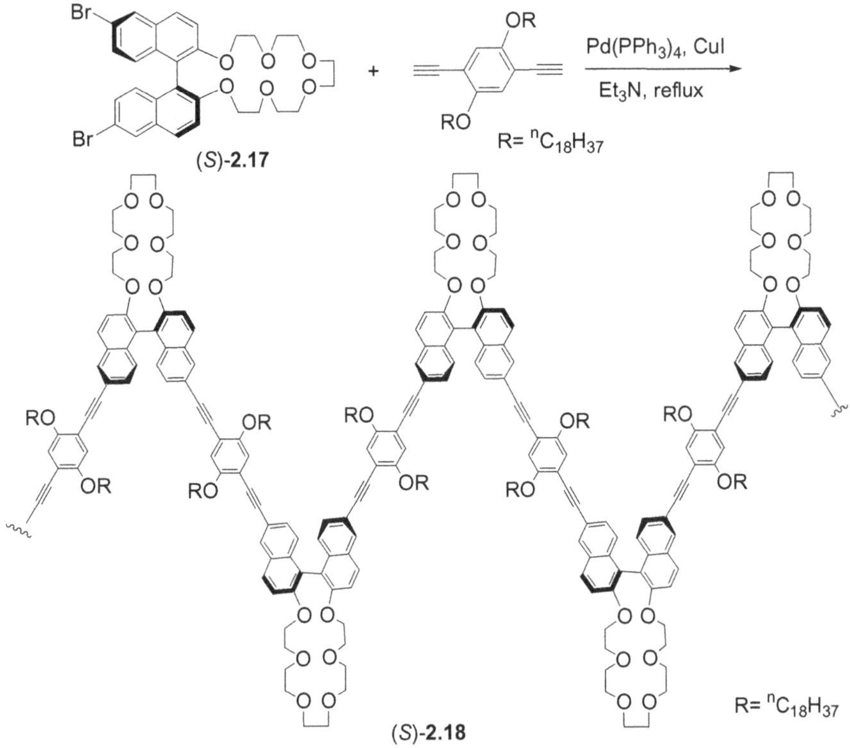

Scheme 2.16. Synthesis of the main chain chiral-conjugated polyaryleneethynylene containing crown ether functions.

oligomer prepared from the coupling of the α,α'-dibromopentathiophene **2.24** with a 2,2'-bithiophen-5-yl Grignard reagent is not soluble in organic solvents with R^1 being either $^nC_6H_{13}$ or $^nC_{18}H_{37}$.

Coupling of the above α,α'-dibromooligothiophene monomers with the 6,6'-diboronic acid binaphthyl monomer *rac*-**2.1** generates a series of binaphthyl-thiophene copolymers **2.26–2.30** (Scheme 2.19). The binaphthyl-monothiophene copolymer **2.26** is obtained as a greenish-yellow solid in 93% yield. GPC shows its molecular weight as $M_w = 35{,}500$ and $M_n = 13{,}900$ (PDI = 2.6). The binaphthyl-bithiophene copolymer **2.27** is obtained as a yellow solid in 97% yield. GPC shows its molecular weight as $M_w = 30{,}300$ and $M_n = 18{,}100$ (PDI = 1.7). Because of the low solubility of the tetrathiophene monomer **2.22**, its coupling with *rac*-**2.1** gives a lower molecular weight polymer **2.28** as an orange solid in 88% yield. GPC shows its molecular weight as $M_w = 5500$ and $M_n = 5100$ (PDI = 1.1).

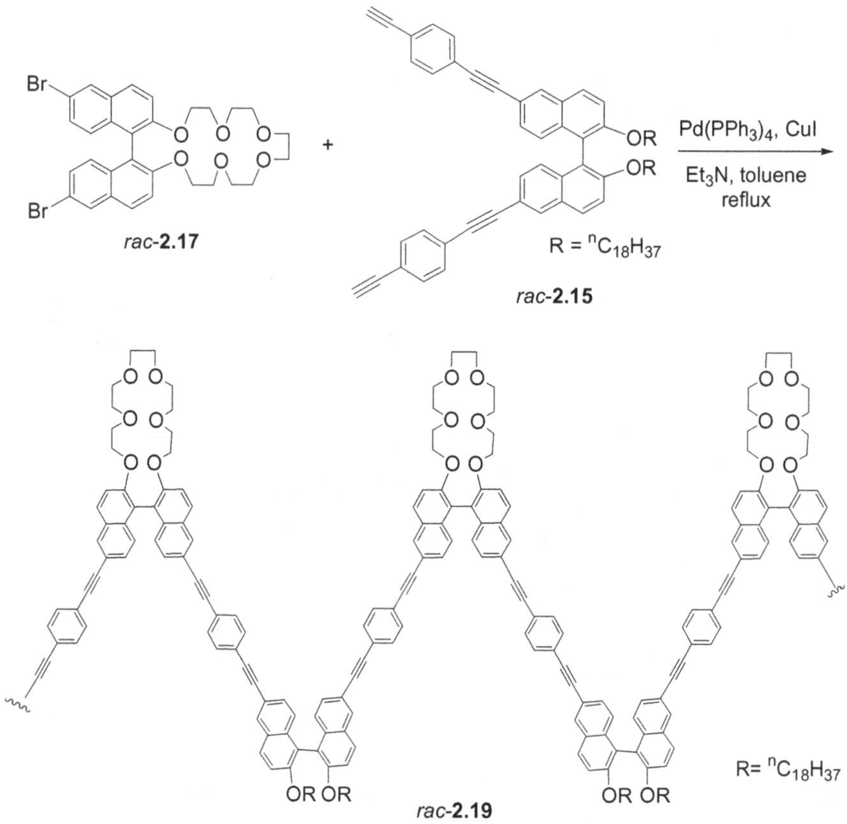

Scheme 2.17. Synthesis of polymer *rac*-**2.19**.

The alkylated pentathiophene monomer **2.24** is very soluble. Its coupling with *rac*-**2.1** gives polymer **2.29** as a red solid in 94% yield. GPC shows its molecular weight as $M_w = 28{,}000$ and $M_n = 17{,}400$ (PDI = 1.6). The alkylated heptathiophene monomer **2.25** has low solubility in organic solvents. Its coupling with *rac*-**2.1** gives **2.30** as a dark red solid in 54% yield. The lower yield is attributed to the formation of insoluble and probably higher molecular weight materials. GPC shows the molecular weight of the soluble portion of polymer **2.30** as $M_w = 6000$ and $M_n = 2300$ (PDI = 2.6). All the polymers **2.26–2.30** are soluble in common organic solvents. They give well-resolved ^{1}H and ^{13}C NMR spectra consistent with the expected structures.

Table 2.6 gives the UV absorptions of the oligothiophenes. As the number of thiophene units increases from thiophene to heptathiophene,

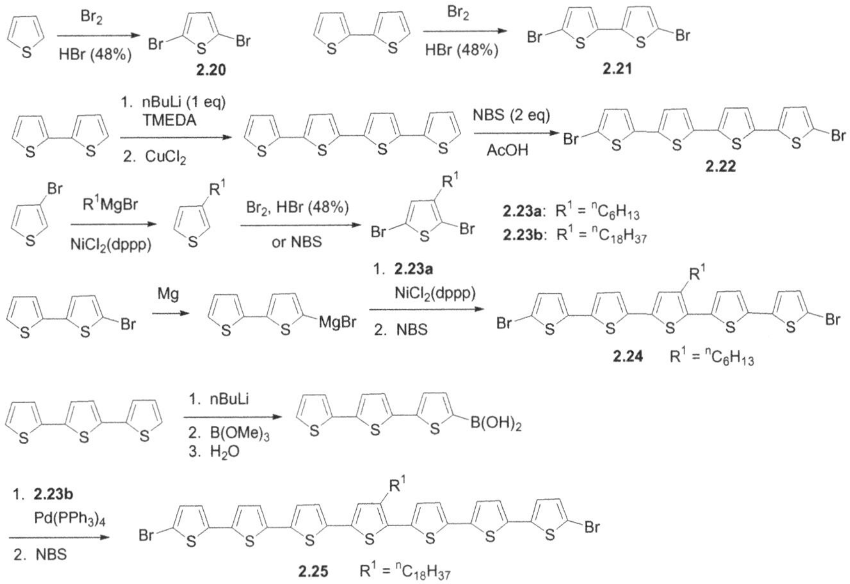

Scheme 2.18. Synthesis of oligothiophene monomers.

there is a very large red shift for the UV absorptions with the longest wavelength absorption increased by about 200 nm. Table 2.7 gives the UV absorption and fluorescence spectral data of the binaphthyl-thiophene copolymers. As the repeating unit length of these polymers increases, there is a systematic increase in both absorption and emission wavelengths. The pentathiophene copolymer **2.29** gives an abnormal absorption at 434 nm that is slightly shorter than that of the tetrathiophene copolymer **2.28** at 440 nm. This is probably because the alkyl group in the middle of the repeating unit of the pentathiophene copolymer **2.29** has reduced the ground state conjugation. Using 9-anthracene carboxylic acid as the reference, we have estimated the fluorescence quantum yields of polymers **2.26**–**2.30** in methylene chloride solution as 0.54, 0.26, 0.19, 0.23 and 0.072, respectively. Polymer **2.30** containing heptathiophene units shows a much lower quantum yield than other polymers.

TGA shows that polymers **2.26**, **2.27** and **2.29** have similar thermal decomposition patterns. As shown in Table 2.8, these three polymers lose their hexyl groups in the temperature range of 380–480°C under nitrogen. Further heating to 800°C leads to only a small additional mass loss (∼10%). However, the two low-molecular-weight polymers **2.28** and **2.30** are much

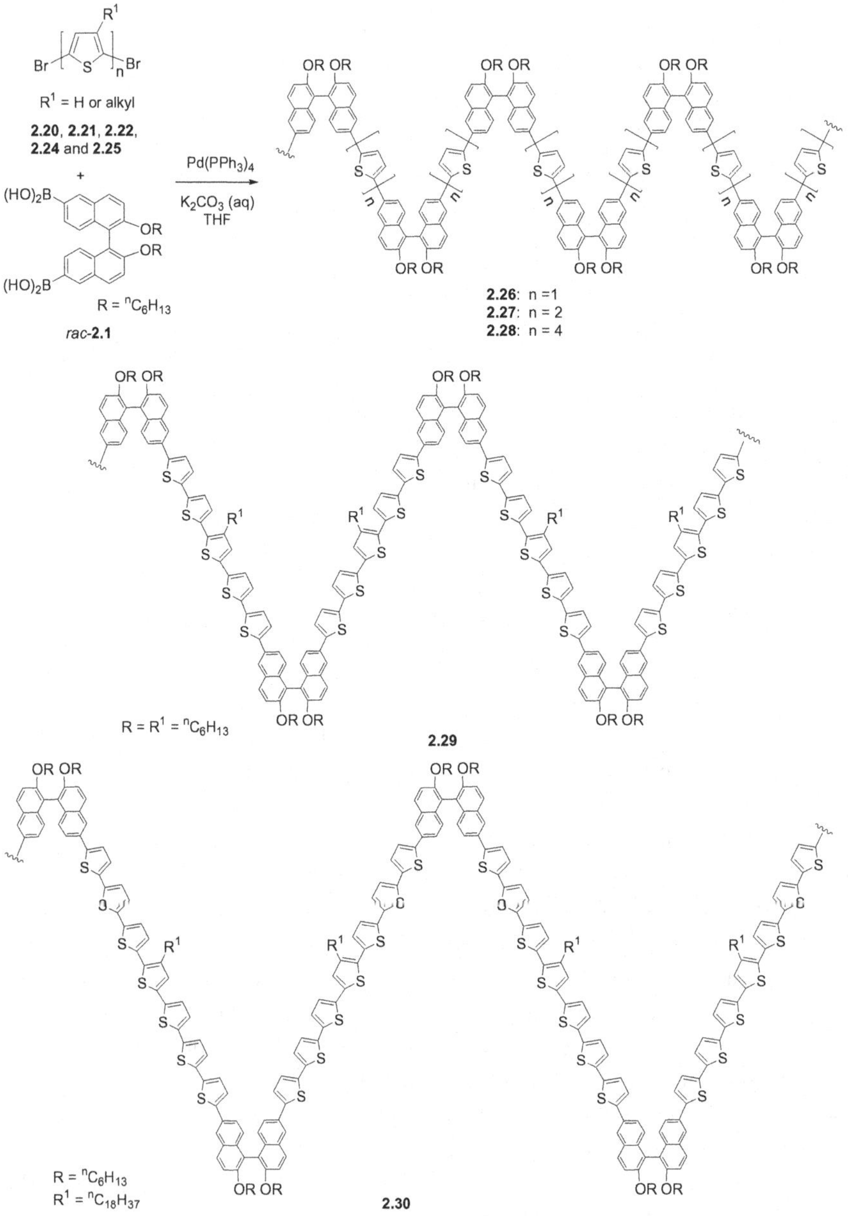

Scheme 2.19. Synthesis of binaphthyl-thiophene copolymers.

Table 2.6. The UV absorption wavelengths of the oligothiophenes.

oligothiophene	λ_{max} (nm)
(thiophene)	246
(bithiophene)	248 306
(quaterthiophene)	254 394
(R¹-substituted quinquethiophene)	254 406
(R¹-substituted septithiophene)	210 218 234 336 438

Table 2.7. The UV and fluorescence spectral data of the binaphthyl-thiophene copolymers.

Polymer	Number of thiophenes per repeat unit	Color of the polymer	UV λ_{max} (nm)	Fluorescence λ_{emi} (nm)
2.26	1	Yellow-green	236 298 368	421 446 475 (sh)
2.27	2	Yellow	246 406	463 498
2.28	4	Orange	232 270 (sh) 440	515 549 (sh)
2.29	5	Bright red	236 348 (sh) 434	530 568 (sh)
2.30	7	Dark red	234 264 (sh) 454	545 583 631 (sh)

Table 2.8. TGA data for polymers **2.26**–**2.30**.

Polymer	T (°C) at 5% mass loss	T (°C) at loss of hexyl	Additional mass loss before 800°C (%)
2.26	383	475	10
2.27	393	460	12
2.28	240		54
2.29	389	485	10
2.30	339		63

less stable. They begin to decompose at much lower temperatures and continuously lose significant amount of mass while heated to 800°C.

Further study on binaphthyl-bithiophene copolymer **2.27** is conducted in collaboration with Jen.[19] The cyclic voltammetry of **2.27** shows two quasi-reversible waves (formal potentials $E_{1/2\text{ox},\text{I}} = 0.80$ V, $E_{1/2\text{ox},\text{II}} = 1.04$ V versus Ag/Ag$^+$) under an anodic sweep, and a single quasi-reversible wave (formal potential $E_{1/2\text{red},\text{I}} = \pm 2.21$ V versus Ag/Ag$^+$) under a cathodic sweep. On the basis of the CV measurement, the energy values for HOMO and LUMO of **2.27** and the HOMO–LUMO gap ($E_g E_C$) are calculated to be ± 5.42, ± 2.50 and 2.92 eV, respectively.

A thin film of polymer **2.27** is obtained by spin-coating its chloroform solution onto a quartz plate. The UV spectrum of this thin film gives signals at $\lambda_{\max} = 246$ and 407 nm almost identical to those observed in solution. When the polymer film is excited at 366 nm, it shows emissions at 568 and 590 (sh) nm. Thus, there is about 100 nm red shift for the emission of **2.27** going from its solution to the solid state. As no difference in the UV absorptions of **2.27** between the solution phase and the solid state is observed, the large difference observed in the fluorescence spectra is attributed to the formation of excimer in the solid state. The excimer emission occurs at much longer wavelengths.

2.5.2. Electroluminescence Study

A light-emitting diode made of ITO/polymer **2.27**/Al is prepared in collaboration with Jen.[19] This device emits orange light when positive voltage is applied to the ITO electrode. The turn-on voltage and rectification ratio are 5.7 V and 4.36×10^3 at ± 8.0 V, respectively. This device exhibits electroluminescence at 568 and 600 (sh) nm, which closely match with the fluorescence of the polymer thin film. Thus, both the photoluminescence and electroluminescence should originate from the same excited state of **2.27**. This light-emitting diode shows a luminance of 25 cd/m^2 at a voltage of 8.0 V with an external quantum efficiency of 0.005%.

An electron transporting material FCuPc [=copper(II)-1,2,3,4,8, 9,10,11,15,16,17,18,22,23,24,25-hexadecafluoro-29H,31H-phthalocyanine] is incorporated to obtain a double-layer device with a configuration of ITO/polymer **2.27**/FCuPc/Al. This double-layer device does not show improved electroluminescence efficiency over the single-layer device. A device using polymer **2.27** as the hole transporter and Alq3 as the emitting material with a configuration of ITO/polymer **2.27**/Alq3/Al is prepared. The electroluminescence spectrum of this double-layer device gives the characteristic emission of Alq3 at 525 nm. At an applied voltage of 24.5 V, this device has a brightness of 1100 cd/m^2 and a luminous efficiency of 0.57 lm/W. No emission is observed without polymer **2.27** as the hole transporter in this device.

2.6. Copolymers of BINAM and Thiophene-Containing Conjugated Linkers[20]

1,1'-Binaphthyl-2,2'-diamine (BINAM)-derived monomers are synthesized as shown in Scheme 2.20 (see Scheme 2.39 for more details). Compounds (R)-**2.31a,b** and (R)-**2.32** contain 6,6'-dihalogen atoms that are introduced for the subsequent cross-coupling polymerization. The two alkyl groups on each of the nitrogens make these compounds more air stable and they also contribute to the solubility of the corresponding chiral-conjugated polymers. The synthesis of thiophene-containing conjugated linker molecules **2.33** and **2.34** is shown in Scheme 2.21. Compound **2.34** is found to be not very stable in air over time. Therefore, this compound is normally freshly prepared before polymerization.

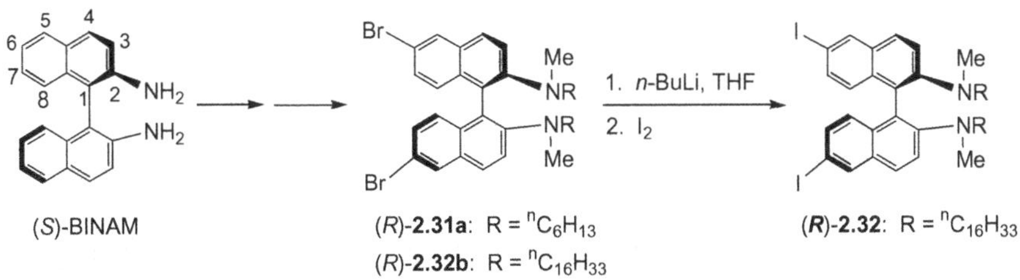

Scheme 2.20. Synthesis of chiral BINAM monomers.

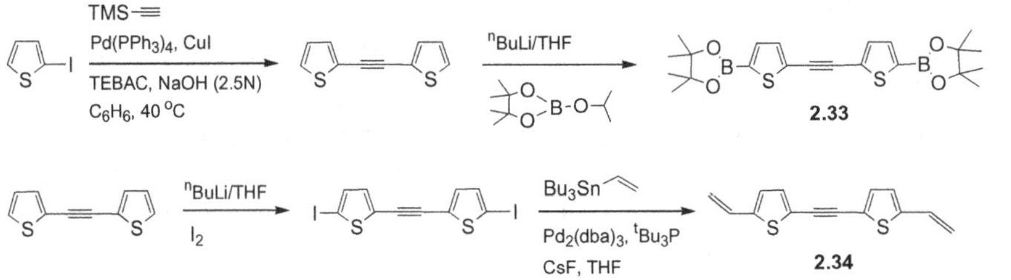

Scheme 2.21. Synthesis of conjugated linker monomers.

Polymerization of (R)-**2.31a** with **2.33** is conducted in the presence of Pd$_2$(dba)$_3$, tri-*tert*-butylphosphine and cesium carbonate (Scheme 2.22). This produces polymer (R)-**2.35** as a dark red solid in 85% yield. This polymer is soluble in common organic solvents. GPC shows its molecular weight as $M_w = 7500$ and $M_n = 3000$ (DPI = 2.5).

The Heck coupling of the binaphthyl monomers (R)-**2.31a** and (R)-**2.32** with the divinyl monomer **2.34** is conducted to generate polymers (R)-**2.36a** and (R)-**2.36b**, respectively (Scheme 2.23). The polymerizations involve the use of Pd(OAc)$_2$, CuI and tri-*o*-tolylphosphine in DMF and triethylamine. 2,6-Di-*tert*-butyl-4-methylphenol (BHT) is added to inhibit a possible radical polymerization of **2.34**. Both polymers (R)-**2.36a,b** are dark red solids and soluble in common organic solvents. GPC shows that the molecular weights of (R)-**2.36a** and (R)-**2.36b** are $M_w = 12{,}000$ (PDI = 2.4) and $M_w = 11{,}800$ (PDI = 2.5), respectively. Thus, both the halogen atoms and the size of the alkyl groups on these monomers do not significantly influence the polymerization.

The UV and CD spectral data of polymers (R)-**2.35** and (R)-**2.36a,b** in chloroform solution are summarized in Table 2.9. The more extended conjugation in the repeating units of polymers (R)-**2.36a,b** leads to their longer wavelength absorptions than that of (R)-**2.35**. Polymer (R)-**2.36b** has good solubility in chloroform and methylene chloride, but it is only partially soluble in benzene and poorly in hexane. Its UV absorption undergoes about 20–30 nm red shift as the solvent polarity as well as the solubility of the polymer increases from hexane to benzene and methylene chloride. In hexane, polymer (R)-**2.36b** shows a strong emission at 490 nm, which also undergoes about 20–30 nm red shift in benzene and methylene chloride. The fluorescence intensity of this polymer in hexane is much stronger than that in benzene and methylene chloride. This indicates that in the poor solvent, the dilute solution of the polymer probably

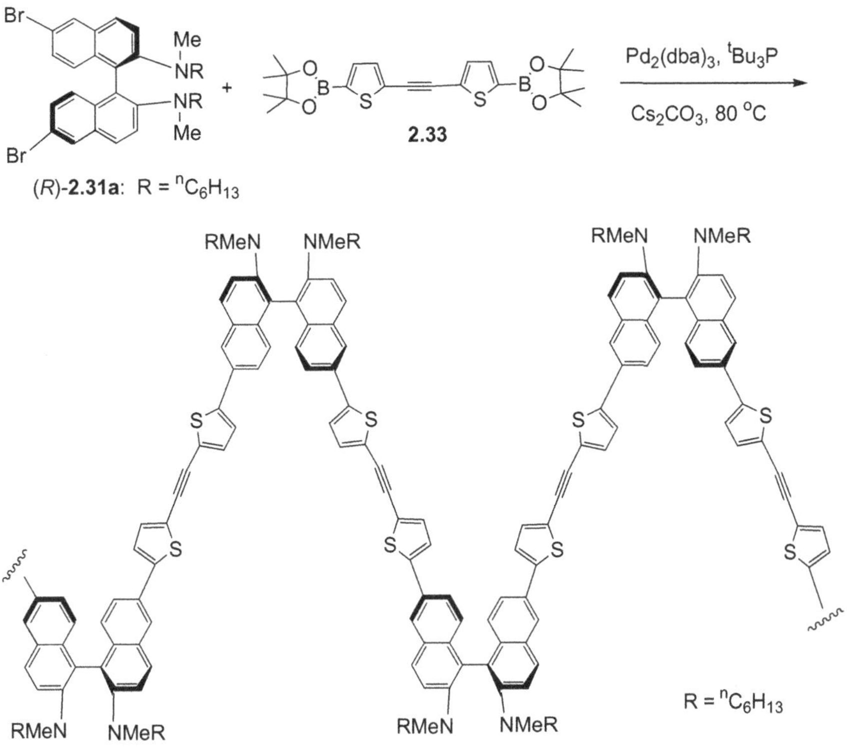

Scheme 2.22. Synthesis of chiral conjugated polymer (*R*)-**2.35**.

adopts a much tighter and more rigid conformation, which could be much more emissive than the more flexible and extended conformations in good solvents. The more extended conformations in better solvents could lead to better conjugation, giving the observed longer wavelength absorptions.

2.7. Polybinaphthyls Without Conjugated Linkers

2.7.1. *Using Nickel Complexes to Promote Polymerization*[21]

Polybinaphthyls that do not contain any of the phenylene, phenyleneviny-lene, phenyleneethynylene and thiophene linkers as described in the previous sections are synthesized. Polymerization of (*R*)-DBDH $\{[\alpha]_D = 25.9\ (c = 0.52,\ \text{THF})\}$ in the presence of a stoichiometric amount

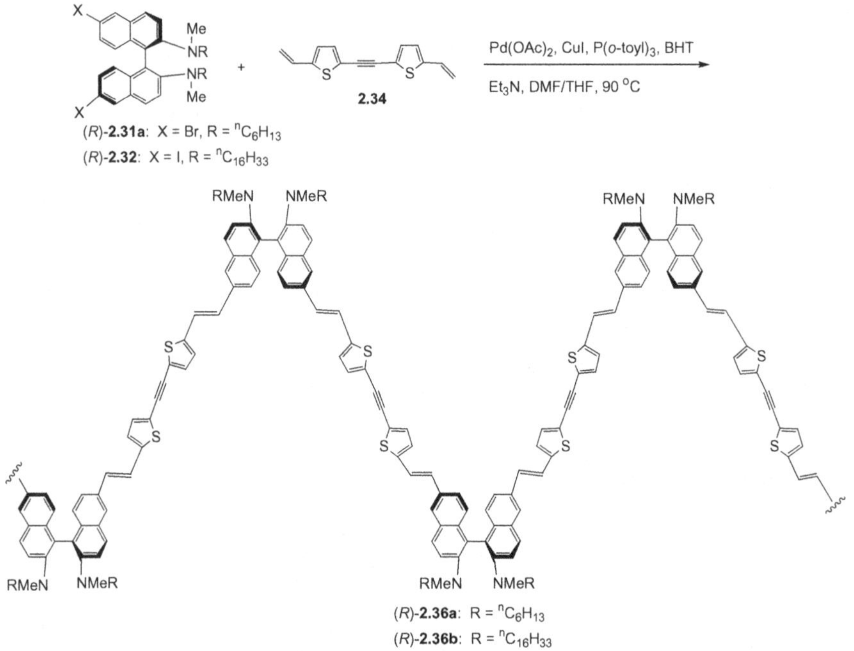

Scheme 2.23. Synthesis of chiral conjugated polymers (R)-**2.36a,b**.

Table 2.9. Optical spectral data of the chiral conjugated polymers.

Polymers	(R)-**2.35**	(R)-**2.36a**	(R)-**2.36b**
UV–vis λ_{max}, nm (ε) CHCl$_3$	376 (2.88 × 10^4) 263 (4.73 × 10^4)	397 (2.83 × 10^4) 263 (4.25 × 10^4)	401 (5.25 × 10^4) 260 (6.34 × 10^4)
CD [θ] (nm), CHCl$_3$	−6.78 × 10^4 (275)	−1.48 × 10^5 (275)	−1.28 × 10^5 (277)

of bis(1,5-cyclooctadiene)nickel(0), 2,2′-bipyridine, 1,5-cyclooctadiene and DMF gives the polybinaphthyl (R)-**2.37** (Scheme 2.24). GPC shows the molecular weight of (R)-**2.37** as M_w = 15,000 and M_n = 6800 (PDI = 2.2). Its specific optical rotation [α]$_D$ is −215 (c = 0.52, CH$_2$Cl$_2$). This polymer gives well-resolved ^{1}H and ^{13}C NMR spectra that are consistent with its structure. TGA shows that polymer (R)-**2.37** is stable up to 392°C before it starts to lose mass. Its T_g is 119°C as measured by DSC analysis. Polymerization of the racemic monomer gives polymer rac-**2.37** with a molecular weight of M_w = 11,000 and M_n = 4700 (PDI = 2.3) as measured

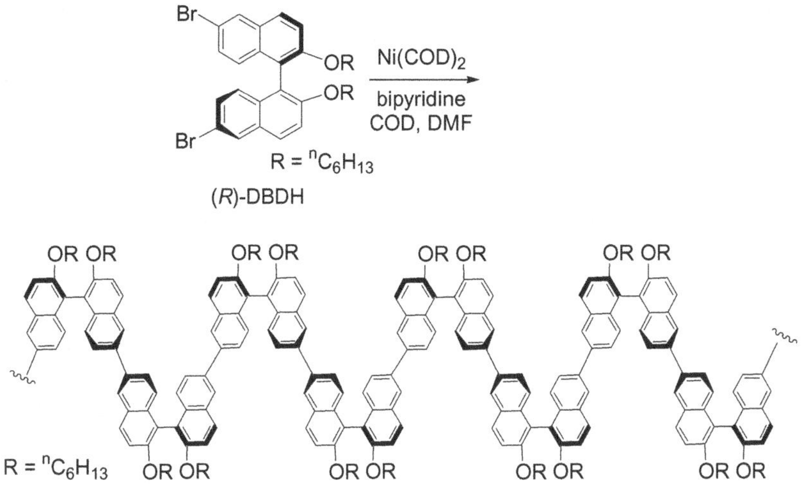

Scheme 2.24. Synthesis of optically active poly(binaphthyl) (R)-**2.37**.

by GPC. The NMR spectra of rac-**2.37** are almost identical to those of (R)-**2.37**.

The racemic monomer DBDH is also polymerized by using a catalytic amount of NiCl$_2$ and an excess amount of Zn. This gives a polymer with M_w = 11,300 (PDI = 2.1) close to that prepared with the use of Ni(COD)$_2$. The cheaper NiCl$_2$ and Zn makes this polymerization method more economical.

The methylene chloride solution of polymer (R)-**2.37** displays UV absorptions at $\lambda_{\max}$ = 240 (sh), 262, 284 and 316 nm. Polymer rac-**2.37** gives an almost identical UV spectrum. Table 2.10 compares the UV absorptions of polymer (R)-**2.37** with those of 2,2'-bishexyloxy-1,1'-binaphthyl and 2-hexyloxy naphthalene. It shows that polymer (R)-**2.37** does not have more extended conjugation than the small molecules. That is, the naphthalene units in the polymer probably have mutually orthogonal conformation with respect to both the 1,1'- and 6,6'-bonds.

Strong Cotton effects are observed in the CD spectrum of (R)-**2.37**. When this polymer is heated at reflux in toluene for 24 h, there is only 6% loss of its optical rotation. Thus, the chiral configuration of this polymer is very stable.

Polymerization of the neopentyl substituted racemic monomer in the presence of NiCl$_2$ and excess Zn is conducted to give polymer rac-**2.38**

Table 2.10. Comparison of the UV absorptions of polymer (*R*)-**2.37** with those of 2,2′-bishexyloxy-1,1′-binaphthyl and 2-hexyloxy naphthalene.

Polymer or compound	(*R*)-**2.37**	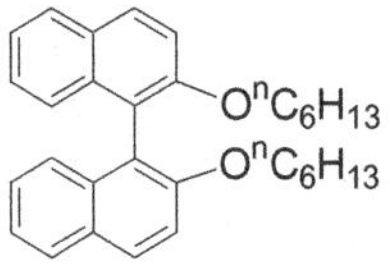 2,2′-bishexyloxy-1,1′-binaphthyl	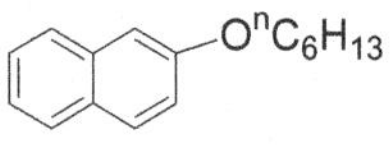 2-hexyloxy naphthalene
λ_{max} (nm)	240 (sh), 262, 284, 316	252, 273 (sh), 284, 293 (sh), 328 (sh), 340	240, 266, 274, 318, 330

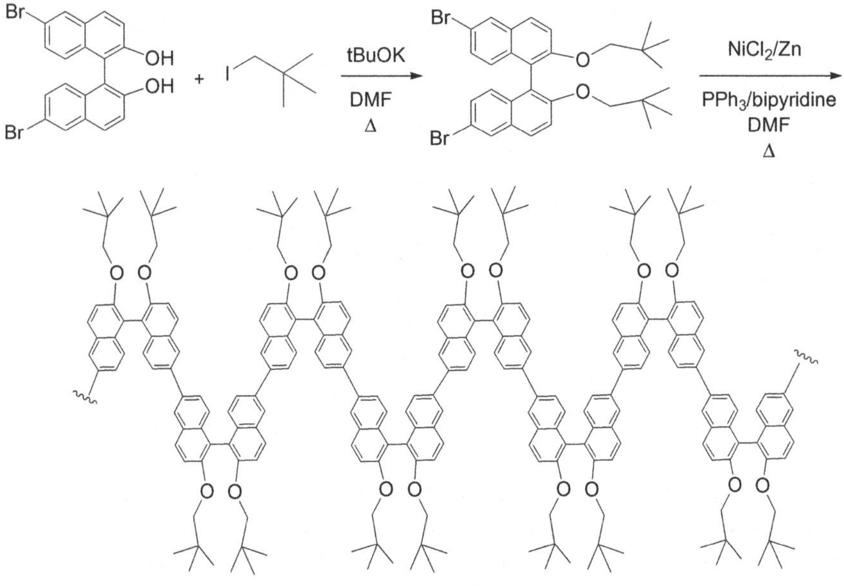

Scheme 2.25. Synthesis of neopentyl-substituted polymer *rac*-**2.38**.

(Scheme 2.25).[22] GPC shows the molecular weight of *rac*-**2.38** as $M_w = 7100$ (PDI = 1.6). This GPC molecular weight is almost the same as that calculated by a ^{1}H NMR analysis of the end groups. It indicates that the polystyrene standard used in the GPC analysis is suitable to measure the molecular weight of this polymer. Polymer *rac*-**2.38** has good solubility in common organic solvents and can be easily cast into thin films from solution.

The UV spectrum of rac-**2.38** in solution displays absorptions at λ_{max} = 268, 316 and 341 (sh) nm. When this polymer is excited at 319 nm, it gives intense emissions at λ_{emi} = 385 and 405 nm. Both the absorption and emission spectra of rac-**2.38** are similar to those of rac-**2.37**.

TGA shows that rac-**2.38** loses 5% weight at 415°C under nitrogen (heating rate: 10°C/min). At 497°C, it loses 42% of its weight, which is equivalent to the loss of its neopentyloxy groups. After the loss of the alkoxy groups, rac-**2.38** becomes very stable. There is only ca. 5% additional mass loss when heated up to 800°C. TGA of polymer rac-**2.37** shows a 5% weight loss at 366°C under the same conditions. At 478°C, it loses 44% of its weight corresponding to the loss of the hexyloxy groups. From 478 to 800°C, ca. 5% additional weight loss was observed.

DSC shows that rac-**2.38** undergoes irreversible changes above 310°C. In the temperature range of 0–300°C, an irreversible phase transition is observed at 110°C during the first heating cycle, which disappears in the subsequent heating cycles. A glass transition temperature at ca. 256°C is observed for rac-**2.38** in all the heating cycles. The DSC plot of polymer rac-**2.37** showed a glass transition temperature at ca. 98°C during multiple heating cycles from 0 to 275°C. Above 300°C, rac-**2.37** undergoes irreversible thermal transitions probably due to structural changes.

As shown by the TGA and DSC studies, polymer rac-**2.38** with the more bulky neopentyl groups exhibits much higher thermal stability than polymer rac-**2.37** with the linear hexyl groups even though the molecular weight of rac-**2.38** is lower.

2.7.2. *Synthesis of the Binaphthyl-Based Polydendrimers by Using Ni Complexes to Promote Polymerization*[23]

Dendritic arms are introduced to the binaphthyl units of the polybinaphthyls. The reactions of (R)-6,6'-Br$_2$BINOL with benzyl bromide and its dendritic derivatives give compounds (R)-**2.39** (generation 0), (R)-**2.40** (generation 1) and (R)-**2.41** (generation 2), respectively (Scheme 2.26). These binaphthyl compounds contain the Fréchet-type dendrons.[24] Polymerization of these compounds in the presence of a catalytic amount of NiCl$_2$ and excess zinc powder gives the optically active polydendrimers (R)-**2.42** (generation 0), (R)-**2.43** (generation 1) and (R)-**2.44a** (generation 2), respectively. Monomers (R)-**2.41** and (S)-**2.41** are also polymerized by using a stoichiometric amount of Ni(COD)$_2$ (COD = 1,5-cyclooctadiene) to give the polydendrimers (*R*)-**2.44b** and (*S*)-**2.44**, respectively. All the polydendrimers are soluble in common organic solvents.

The GPC molecular weights relative to polystyrene standards and those obtained by using the multiangle LLS (MALLS) detector are listed in Table 2.11 for these polydendrimers. It shows that using Ni(COD)$_2$ gives much higher molecular weight polymers than using NiCl$_2$/Zn. The MALLS molecular weights of the generation 2 polydendrimers **2.44** are

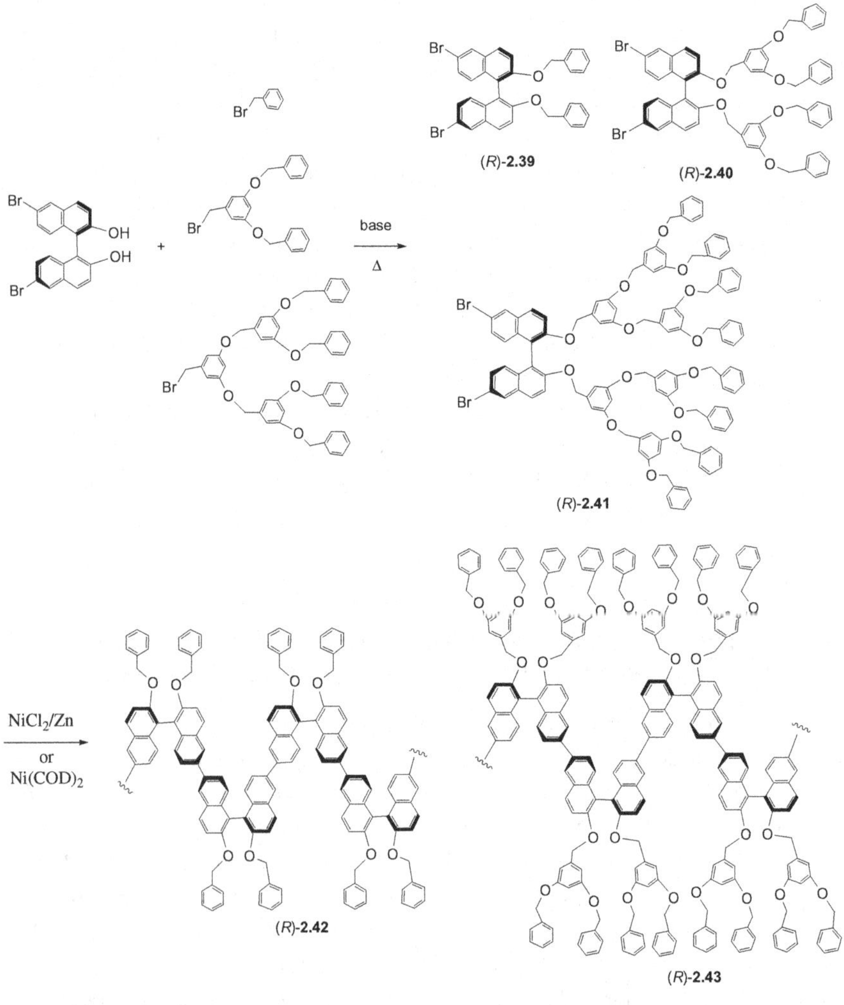

Scheme 2.26. Synthesis of polydendrimers (*R*)-**2.42**–(*R*)-**2.44**.

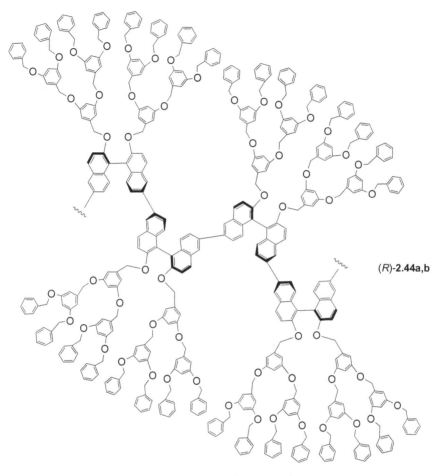

(R)-**2.44a,b**

Scheme 2.26. (*Continued*)

about twice as much as the GPC molecular weights. Thus, because of the significant difference between the structure of these polydendrimers and that of the polystyrene standards, the GPC analysis greatly underestimates the molecular weights of these materials. This is quite different from that observed for polymer *rac*-**2.38** containing neopentyl groups. For *rac*-**2.38**, the GPC molecular weight data relative to polystyrene standards matches that obtained from the end group analysis of its NMR spectrum.

The model compounds of the binaphthyl-based polydendrimers are synthesized from the Suzuki coupling of 6-methoxynaphth-2-ylboronic acid with monomers (*R*)-**2.39**, (*R*)-**2.40** and (*S*)-**2.41** (Scheme 2.27). Both

Table 2.11. Molecular weights of the polydendrimers.

Polymer	Polymerization method	GPC			MALLS			Isolated yield (%)
		M_w	M_n	PDI	M_w	M_n	PDI	
(R)-**2.42**	Ni(II)/Zn	6600	3600	1.8				62
(R)-**2.43**	Ni(II)/Zn	8200	5200	1.6				71
(R)-**2.44a**	Ni(II)/Zn	6800	5100	1.3	14,100	11,100	1.3	76
(R)-**2.44b**	Ni(COD)$_2$	11,200	7200	1.6	26,200	19,300	1.4	86
(S)-**2.44**	Ni(COD)$_2$	16,200	10,500	1.5	35,300	23,000	1.5	82

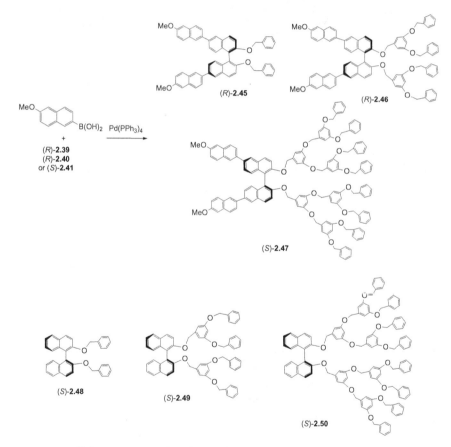

Scheme 2.27. Synthesis of dendritic model compounds.

the specific and molar optical rotations of these compounds decrease as the dendritic generation increases (Table 2.12). Thus, the flexible dendritic units in these compounds have a diluting effect on the optical rotation of the binaphthyl core. Previously, Meijer reported the binaphthyl dendrimers (S)-**2.48**–(S)-**2.50**.[25] In Table 2.12, the optical rotations of our polydendrimers are compared with those of monomers (R)-**2.39**–(R)-**2.41**, the model compounds (R)-**2.45**–(R)-**2.47** and Meijer's compounds (S)-**2.48**–(S)-**2.50**. Although the molar optical rotation increases significantly as the molecular weight of the polydendrimers increases from (R)-**2.44a** to (R)-**2.44b** and (S)-**2.44**, the specific and molar optical rotations of the high-molecular-weight polydendrimer (S)-**2.44** are very close to those of the model compound (S)-**2.47**. This indicates that polydendrimer (S)-**2.44** does not form a helical chain conformation beyond its individual binaphthyl unit and there is no amplification of the optical rotation by the polymer chain. The lower optical rotations for the lower molecular weight materials such as (R)-**2.44a** can be attributed to the higher proportion of the end groups such as the brominated or debrominated binaphthyl units. The optical rotations of the brominated monomers such as (R)-**2.41** and the debrominated compounds such as (S)-**2.50** are opposite to those of the corresponding model compound (R)- or (S)-**2.47** of the same chiral binaphthyl configuration. Therefore, higher percentage of the end groups in the polydendrimers in the lower molecular weight materials should reduce their optical rotations.

The UV spectra of polydendrimers (R)-**2.42**, (R)-**2.43** and (S)-**2.44** show two similar absorption bands at λ_{max} = 265–270 and 308–317 nm. However, there is a large increase in the absorption at about λ_{max} = 210 nm going from the zero generation polydendrimer (R)-**2.42** to the first generation (R)-**2.43** and the second generation (S)-**2.44**. This is due to the absorption of the greatly increased number of phenylene units at the higher generation dendrimers.

When the THF solutions of (R)-**2.42**, (R)-**2.43** and (S)-**2.44** are excited at 320 nm, they display almost identical strong emissions at λ = 385 and 405 nm. The fluorescence quantum yields are 46.0, 54.1 and 55.8% for (R)-**2.42**, (R)-**2.43** and (S)-**2.44**, respectively, as measured by using quinine sulfate as the standard. The fluorescence quantum yield of the model compound (S)-**2.47** is 48.6% (λ_{exc} = 270 nm), which is similar to that of the polydendrimers. When the polydendrimers are excited at 208 nm where the phenylene dendrons absorb mainly, only emissions at 385 and 405 nm are observed, but no emission at 312 nm where the phenylene dendrons alone

Table 2.12. Comparison of the optical rotations of the polydendrimers with the monobinaphthyl dendrimers.

Compound		Specific rotation $[\alpha]_D$	Molar rotation $[\phi]$
Precursor dendrimer	(R)-**2.39**	+30.3	+189
	(R)-**2.40**	+21.5	+225
	(R)-**2.41**	+20.5	+389
Polydendrimer[a]	(R)-**2.42**	−174.9	−812
	(R)-**2.43**	−89.7	−797
	(R)-**2.44a**	−14.6	−254
	(R)-**2.44b**	−30.5	−530
	(S)-**2.44**	+34.6	+601
Meijer's dendrimer	(S)-**2.48**	−45.5	−212
	(S)-**2.49**	−22.8	−203
	(S)-**2.50**	−15.6	−271
Model dendrimer	(R)-**2.45**	−142.2	−1107
	(R)-**2.46**	−73.5	−885
	(S)-**2.47**	+34.5	+716

a. The molar optical rotations of the polydendrimers are calculated based on their repeating binaphthyl units.

emit. This indicates that the light absorbed by the phenylene dendrons is transferred to the more conjugated naphthalene units.

TGA shows that the polydendrimers are stable up to around 300°C before they begin to lose the dendritic substituents. They are much less stable than polymer rac-**2.37** containing hexyl groups and polymer rac-**2.38** containing neopentyl groups. That is, it is easier to lose the benzyl groups in the polydendrimers than the alkyl groups in other polybinaphthyls. DSC shows that the T_g of these polydendrimers decreases significantly as the dendritic generation increases. The T_g's of (R)-**2.42**, (R)-**2.43** and (S)-**2.44** are 162, 87 and 55°C, respectively. They are much lower than the glass transition temperature of the neopentyl containing polymer rac-**2.38**. The hexyl containing polybinaphthyl rac-**2.37** has a glass transition temperature lower than (R)-**2.42** (G0) but higher than (R)-**2.43** (G1).

2.7.3. Using the Suzuki Coupling Reaction for Polymerization[21b]

The Suzuki coupling of (R)-DBDH with (R)-**2.1** is conducted to generate the polybinaphthyl (R)-**2.51**, which has the same structure as (R)-**2.37** (Scheme 2.28). GPC shows the molecular weight of (R)-**2.51** as

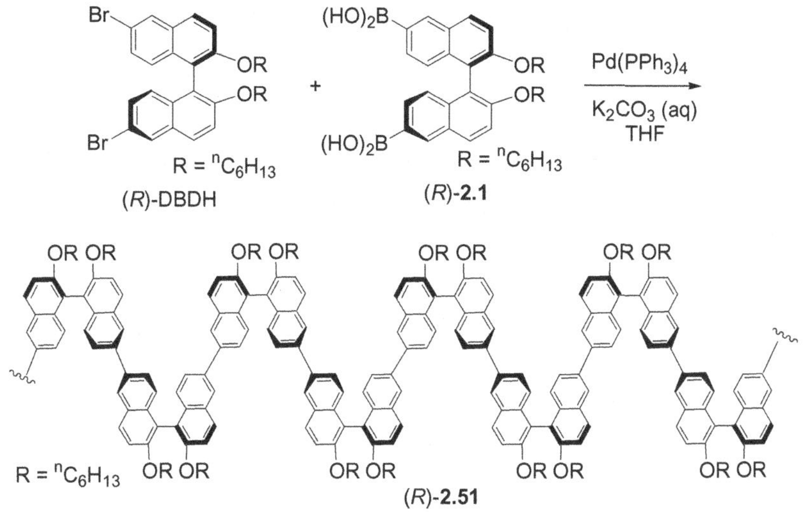

Scheme 2.28. The Suzuki polymerization to synthesize polybinaphthyl (R)-**2.51**.

$M_w = 17{,}600$ and $M_n = 7600$ (PDI = 2.3). Its specific optical rotation $[\alpha]_D$ is -281 ($c = 0.50$, THF). The NMR and UV spectra of (R)-**2.51** are almost the same as those of (R)-**2.37**. The coupling of the racemic monomer rac-DBDH with rac-**2.1** gives rac-**2.51** with $M_w = 21{,}500$ and $M_n = 8100$ (PDI = 2.6). Using the Suzuki coupling produces higher molecular weight polymers than using the Ni(0) complex.

The coupling of two equivalents of rac-DBDH with one equivalent of (R)-**2.1** is conducted, which produces an oligomer with $M_w = 3600$ and $M_n = 2300$ (PDI = 1.5). The specific optical rotation of this oligomer $[\alpha]_D$ is -97.7 ($c = 1.0$, THF). At the end of the coupling reaction, about 20% of the unreacted rac-DBDH is recovered whose specific optical rotation $[\alpha]_D$ is -0.7. This demonstrates that (R)-**2.1** couples with both enantiomers of rac-DBDH equally to form the oligomer with no enrichment of either enantiomer in the unreacted rac-DBDH. Therefore, polymer rac-**2.51** generated from the polymerization of rac-DBDH with rac-**2.1** should contain randomly distributed R and S binaphthyl units in the polymer chain. There should be no stereoselectivity in this polymerization. The oligomer from the coupling of rac-DBDH and (R)-**2.1** is still optically active because using (R)-**2.1** allows the oligomer to contain excess (R)-binaphthyl units.

2.7.4. Electroluminescence Study of the Polybinaphthyls[26]

Further study of the polybinaphthyl rac-**2.51** ($M_w = 21{,}500$) is conducted in collaboration with Jen. The pinhole-free homogeneous thin films of this polymer can be obtained by spin-coating its chloroform solution. The CV of the polymer film is studied under nitrogen in acetonitrile containing 0.1 M tetrabutylammonium perchlorate as the electrolyte. Three electrodes are used including ITO glass coated with the polymer film ($\sim$3 cm^2), Pt gauze and Ag/Ag$^+$. The polymer film shows a quasi-reversible wave at $E_{\text{onset}} = 1.03$ V versus Ag/Ag$^+$ under an anodic sweep. No wave is observed under a cathodic sweep up to -3 V. This demonstrates that this polymer resists to reduction much more than the bithiophene containing polymer **2.27** described earlier. The HOMO energy of ferrocene is -4.8 eV below the vacuum level. Since $E_{\text{ferrocene}} = 0.12$ V vs. Ag/Ag$^+$, the HOMO energy of polymer rac-**2.51** is calculated to be -5.71 eV.

The UV spectrum of the thin films of polymer rac-**2.51** coated on quartz plate shows absorptions at $\lambda_{\max} = 218$, 264 and 320 (sh) nm, which are similar to those observed in solution. The π–π^* band gap of rac-**2.51** is determined to be 3.33 eV from the extrapolation of the UV spectrum. Thus, the LUMO energy level of rac-**2.51** is -2.38 eV. The fluorescence spectrum of the polymer film displays signals at $\lambda_{\text{emi}} = 390$, 408 and 450 (sh) nm. A broad secondary peak between 500 and 560 nm is also observed, which are attributed to the excimer emission.

A single-layer light-emitting diode with a configuration of ITO/polymer rac-**2.51**/Al is prepared by spin-coating the polymer film ($\sim$50 nm) on ITO glass followed by vacuum deposition of Al. This device shows light emission at $\lambda = 554$ nm with a shoulder at 500 nm when a positive voltage is applied on the ITO glass of the single-layer device. Thus, the electroluminescence of this device should be predominately from the excimer emission. The turn-on voltage for the light emission is at 15.1 V and for the current at 9.0 V. The external quantum efficiency is about 0.15% relatively independent of the current. The luminescence of the device is 3200 cd/m^2 at 8 mA, corresponding to a current density of 114 mA/cm^2.

A double-layer device with a configuration of ITO/polymer rac-**2.51**/FCuPc/Al is prepared. In this device, FCuPc is incorporated. This compound is known to be an air-stable n-type semiconductor with high field-effect mobility. The HOMO and LUMO energy levels of FCuPc are -5.94 and -4.22 eV, as determined by CV measurements using vacuum sublimed film of FCuPc under the same experimental conditions as rac-**2.51**. Since there is no energy barrier between the interface of FCuPc and Al, the introduction of the FCuPc layer should enhance the electron injection and transporting. Interestingly, although the energy barrier between the LUMO of FCuPc and rac-**2.51** (-1.84 eV) is much higher than that of the HOMO (-0.23 eV), a strong blue–green emission with maximum at 540 nm is observed from the device. This might be attributed to the emission of an exciplex formed at the interface between FCuPc and rac-**2.51**. The turn-on voltages both for the current and for the light are approximately the same at 21.9 V, indicative of a balanced injection of holes and electrons. The double-layer device exhibits an external quantum efficiency of 2.0%, 14 times higher than the single-layer device. A luminous of 9400 cd/m^2 is observed at a current of 1.7 mA, corresponding to a current density of 24.3 mA/cm^2. The luminous efficiency is calculated to be 4.9 lm/W. The light intensity increases with increasing of the forward bias. The high performance for the double-layer device may be attributed to (a) the steric effect of the twisted binaphthyl unit in the polymer backbone, which prevents polymer emissive segments from closely packing, reducing the self-quenching processes, and (b) the incorporation of FCuPc, which provides balanced charge injection/transport.

2.8. Propeller-Like Polybinaphthyls

2.8.1. *Synthesis of the Propeller-Like Polymers Derived from BINOL*[27,28]

We have designed and synthesized the propeller-like polybinaphthyls such as **2.52** shown in Fig. 2.2. Polymer **2.52** is generated by linking binaphthyl units at the 6,6'-positions with *ortho*-bisethynylphenylene units. In this polymer, if the aromatic rings in each repeating dipole unit are assumed to be coplanar, the homochirality of the binaphthyl units will produce a helical polymer chain structure. When viewed against the polymer axis, the repeating dipole units of the polymer should progressively revolve around the polymer chain axis similar to the blades of a propeller. The collective dipole moments of the repeating dipole units containing electron donors (D) and electron acceptors (A) will be diminished along the polymer chain because of the propeller-like structure. If the dihedral angle of the binaphthyl unit is 120°, the arrangement of the dipole units in the polymer is similar to that of the octopole molecules such as **2.53**.[29] These materials are potentially useful for non-linear optics.

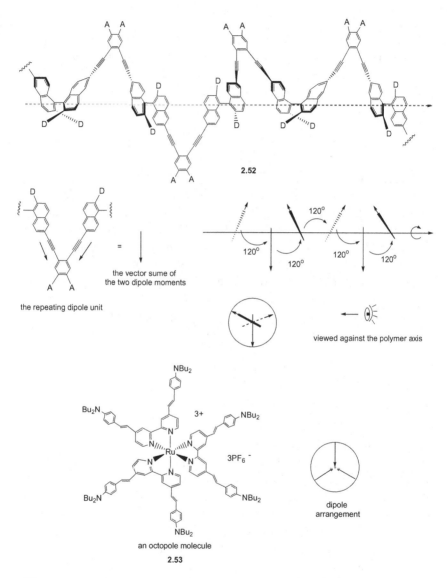

Figure 2.2. A propeller-like polymer **2.52** and an octopole molecule **2.53**.

Polymerization of (R)-**2.12** with 3,4-dibromonitrobenzene in the presence of Pd(PPh$_3$)$_4$, CuI and Et$_3$N in toluene at reflux gives polymer (R)-**2.54** in 92% yield (Scheme 2.29). Since the nitro groups are expected to be randomly distributed at position 1 or 2 in the repeating unit of polymer (R)-**2.54**, the overall dipole arrangement of this polymer should

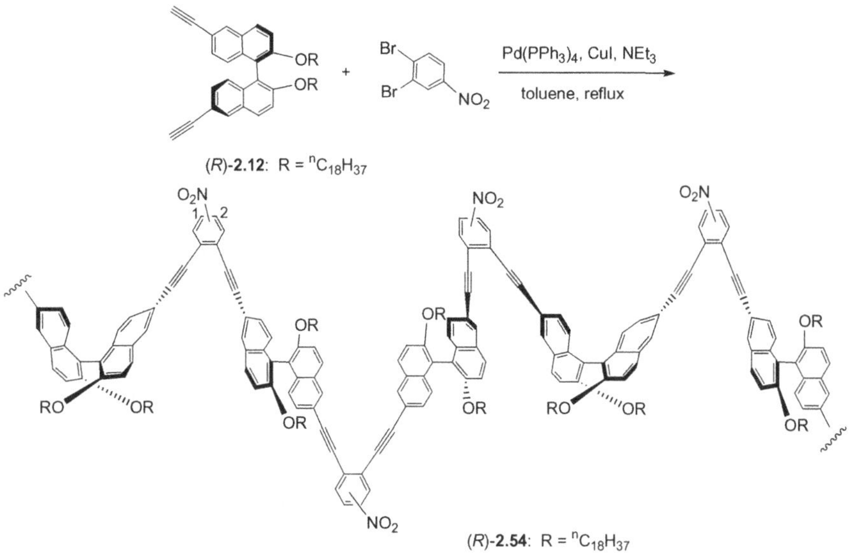

Scheme 2.29. Synthesis of propeller-like polymer (*R*)-**2.54**.

be similar to that of **2.52**. The donors in (*R*)-**2.54** are the alkoxy groups and the acceptors are the nitro groups. This polymer is soluble in common organic solvents. GPC shows its molecular weight as $M_w = 7200$ and $M_n = 5100$ (PDI = 1.4). Its specific optical rotation $[\alpha]_D$ is -160.0 ($c = 0.15$, CH$_2$Cl$_2$). From the racemic monomer *rac*-**2.12**, polymer *rac*-**2.54** is obtained with $M_w = 7400$ and $M_n = 5200$ (PDI = 1.4). The chiral configuration of the monomer has little effect on the polymerization.

Reaction of (*R*)-**2.12**′ ($R = {}^nC_6H_{13}$) with 3,4-dibromonitrobenzene at room temperature gives compound (*R*)-**2.55** in high yield (Scheme 2.30). This demonstrates that the bromine atom *para* to the nitro group in 3,4-dibromonitrobenzene is much more reactive than the bromine at the *meta* position. When (*R*)-**2.55** is treated with a monoprotected bisethynyl monomer (*R*)-**2.56** (see Scheme 2.43 for preparation), a significant self-coupling of (*R*)-**2.56** occurs to give (*R*)-**2.57** in ∼50% yield. This indicates that polymer (*R*)-**2.54** should contain a mixture of structures with a significant amount of alkyne–alkyne homo-coupling. The NMR spectra of (*R*)-**2.54** show broad and complicated signals.

Polymerization of (*R*)-**2.12** with 1,2-dibromo-4,5-dinitrobenzene is studied (Scheme 2.31). In 1,2-dibromo-4,5-dinitrobenzene, both of the bromine atoms are activated by their *para* nitro groups and they should

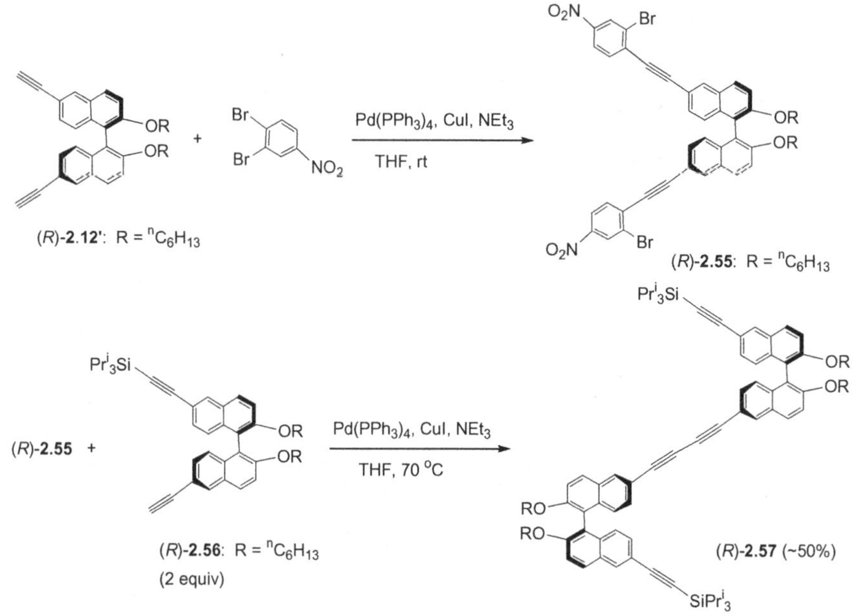

Scheme 2.30. A step-wise coupling study.

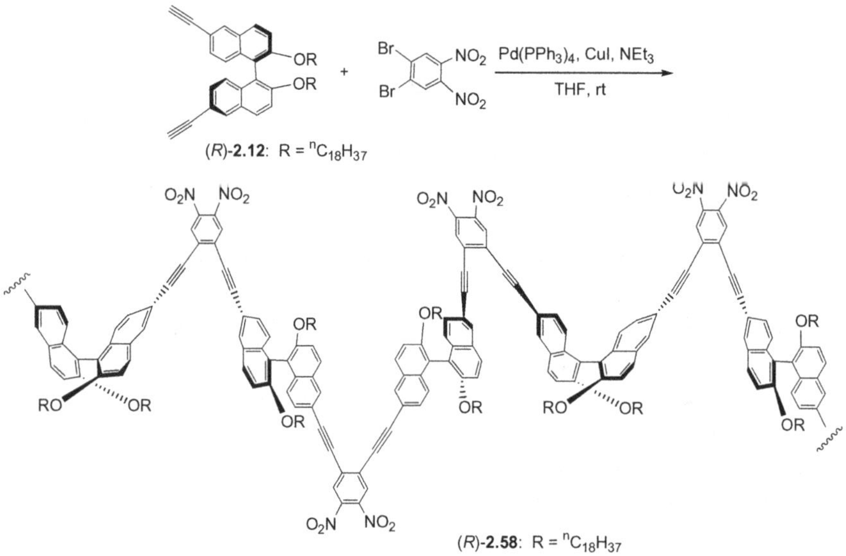

Scheme 2.31. Synthesis of *ortho*-phenyleneethynylene-based chiral polymer (*R*)-**2.58**.

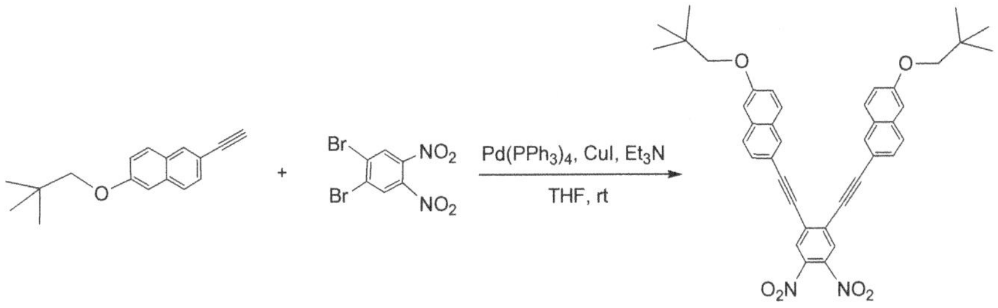

Scheme 2.32. Synthesis of repeating unit **2.59**.

react readily with (R)-**2.12**. In fact, the polymerization of (R)-**2.12** with this dibromobenzene proceeds smoothly at room temperature to give polymer (R)-**2.58**. The reaction can even be conducted at 0°C. The structure of (R)-**2.58** is much better defined than that of (R)-**2.54**. The ^{13}C NMR spectrum of polymer (R)-**2.58** is consistent with the expected structure. GPC shows the molecular weight of this polymer as $M_w = 18{,}000$ and $M_n = 12{,}000$ (PDI = 1.5), which is much higher than that of (R)-**2.54**. This polymer is soluble in common organic solvents. Its specific optical rotation $[\alpha]_D$ is -163.9 ($c = 0.16$, CH_2Cl_2). Powder X-ray diffraction study of (R)-**2.58** shows that this polymer is non-crystalline. Polymerization of the racemic monomer rac-**2.12** gives polymer rac-**2.58** with $M_w = 14{,}200$ and $M_n = 10{,}000$ (PDI = 1.4).

The repeating unit **2.59** is synthesized as shown in Scheme 2.32. The coupling of 2-neopentoxy-6-ethynylnaphthalene with 1,2-dibromo-4,5-dinitrobenzene in the presence of Pd(PPh$_3$)$_4$ and CuI at room temperature gives **2.59** in over 95% yield without any side product. The NMR and IR data of **2.59** are similar to those of (R)-**2.58**. Thus, the clean formation of **2.59** and its spectroscopic analysis support the structure of polymer (R)-**2.58**.

The fluorine-containing monomer o-diiodotetrafluorobenzene is polymerized with (R)-**2.12** in the presence of Pd(PPh$_3$)$_4$/CuI in a refluxing triethylamine/toluene (1:4) solution (Scheme 2.33). After 2 days, polymer (R)-**2.60** is obtained in 93% yield. GPC shows its molecular weight as $M_w = 7200$ and $M_n = 5100$ (PDI = 1.4). Its specific optical rotation $[\alpha]_D$ is -103 ($c = 0.29$, CH_2Cl_2). The racemic monomer rac-**2.12** is used to synthesize polymer rac-**2.60** with $M_w = 7600$ and $M_n = 5300$ (PDI = 1.4). Polymerization of rac-**2.12** with o-diiodobenzene gives polymer rac-**2.61**

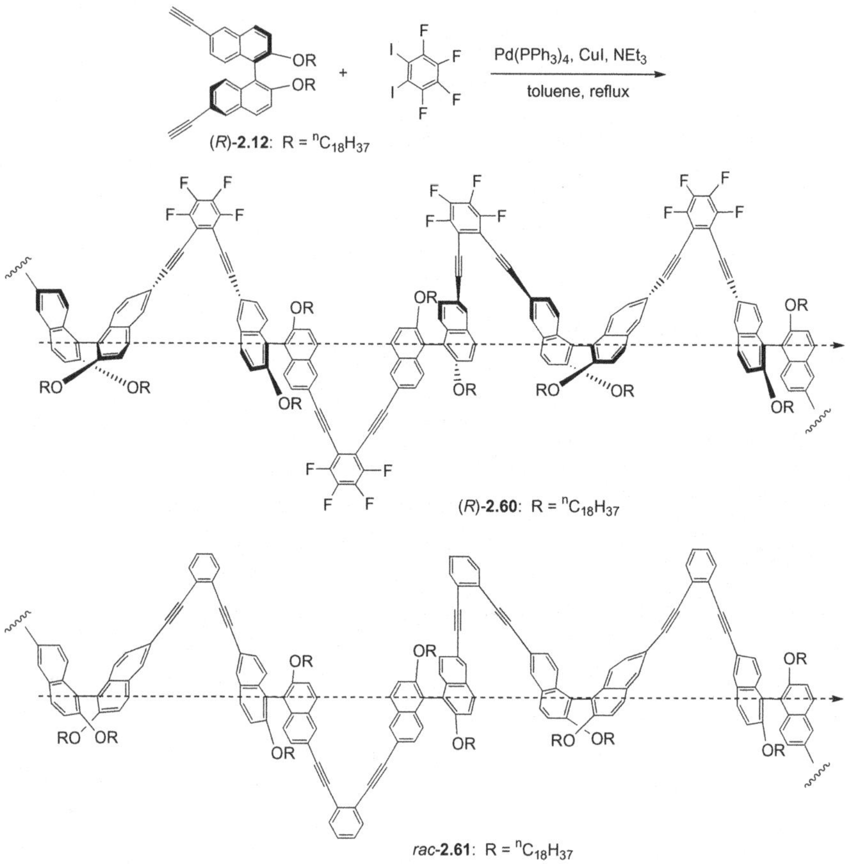

Scheme 2.33. Synthesis of fluorine-containing chiral polymer (R)-**2.60**.

with $M_w = 8100$ and $M_n = 5400$ (PDI = 1.5). The ^{13}C NMR spectrum of this polymer indicates that it contains a mixture of structures probably resulted from both the alkyne–aryl iodide coupling and the alkyne–alkyne homo-coupling.

Polymerization of (R)-**2.15** with 1,2-dibromo-4,5-dinitrobenzene generates polymer (R)-**2.62** that contains more extended conjugation in the repeating unit (Scheme 2.34). GPC shows its molecular weight as $M_w = 10,700$ and $M_n = 6000$ (PDI = 1.80). Its specific optical rotation $[\alpha]_D$ is -228 ($c = 0.13$, CH$_2$Cl$_2$). Monomer (R)-**2.15** is also polymerized with 3,4-dibromonitrobenzene to generate polymer (R)-**2.13** with $M_w = 14,200$ and $M_n = 6900$ (PDI = 2.1). Its specific optical rotation $[\alpha]_D$ is -284

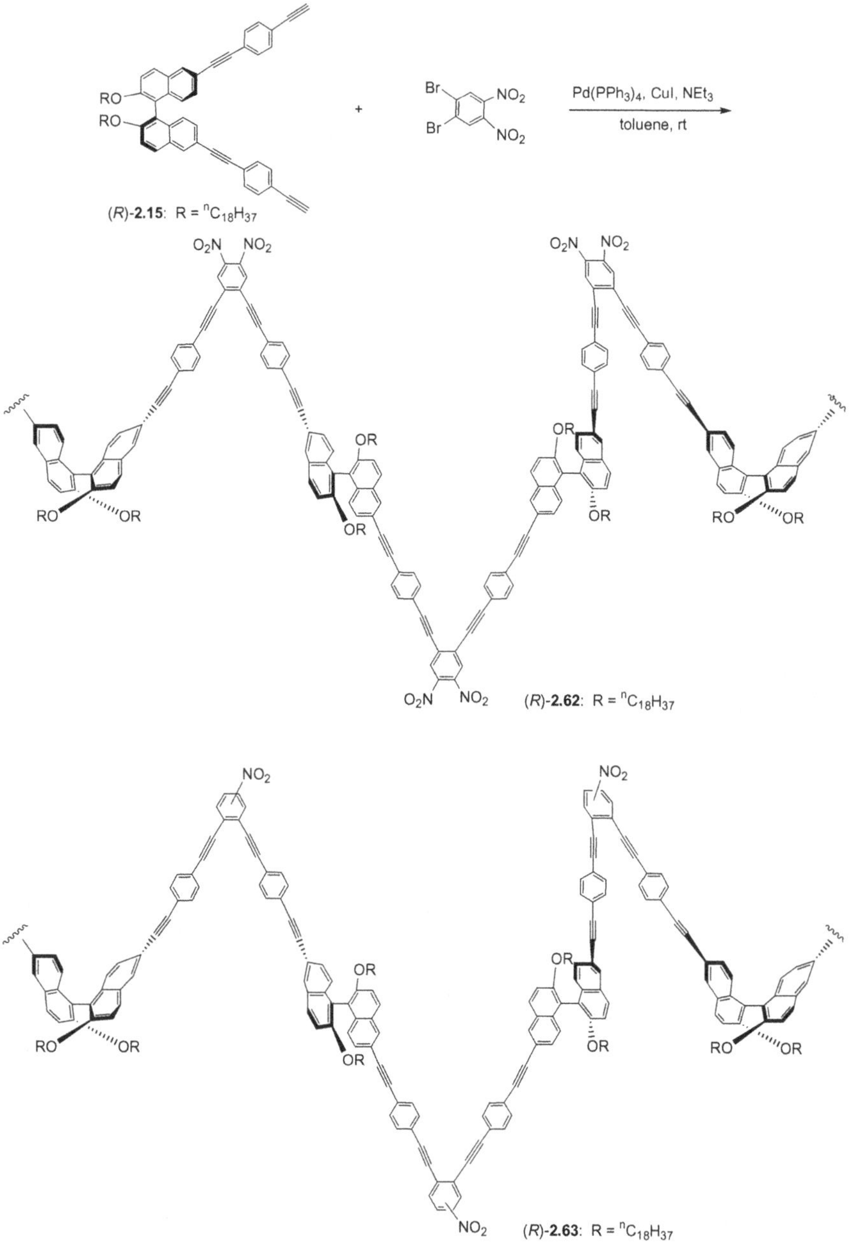

Scheme 2.34. Synthesis of chiral polymer (*R*)-**2.62**.

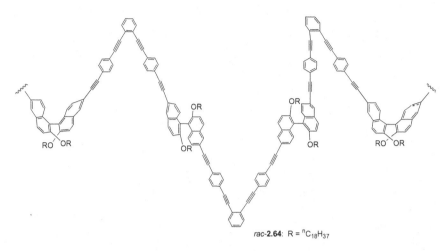

rac-**2.64**: R = $^nC_{18}H_{37}$

Scheme 2.34. (*Continued*)

($c = 0.14$, CH_2Cl_2). The corresponding optically inactive polymer *rac*-**2.63** is obtained from *rac*-**2.15** and 1,2-dibromo-4,5-dinitrobenzene with $M_w = 10{,}500$ and $M_n = 6000$ (PDI = 1.8). Polymerization of *rac*-**2.15** with *o*-diiodobenzene gives polymer *rac*-**2.64** with $M_w = 11{,}600$ and $M_n = 6400$ (PDI = 1.8).

Besides the Sonogashira couplings, the Suzuki coupling is also used to prepare the propeller-like polymers. As shown in Scheme 2.35, polymerization of the diboronic acid monomer (*R*)-**2.1** with 3,4-dibromonitrobenzene in the presence of $Pd(PPh_3)_4$ gives polymer (*R*)-**2.65**. GPC shows the molecular weight of this polymer as $M_w = 10{,}100$ and $M_n = 4900$ (PDI = 2.1). Its specific optical rotation $[\alpha]_D$ is -70.9 ($c = 0.5$, CH_2Cl_2). Polymerization of the racemic monomer *rac*-**2.1** with 3,4-dibromonitrobenzene gives *rac*-**2.65** with $M_w = 6300$ and $M_n = 3700$ (PDI = 1.7). For the propeller-like polymers that contain ethynylene linkers, the NMR spectra of the polymers made of the enantiomerically pure binaphthyl monomers are almost the same as those made of the racemic monomers. This indicates that there is little interference between the adjacent binaphthyl units regardless of their chiral configuration. However, the ^1H NMR spectrum of *rac*-**2.65** displays much more complicated aromatic signals than (*R*)-**2.65**. The shorter *o*-phenylene linkers of these polymers should allow the adjacent binaphthyl units to interact with each other, which should lead to different NMR signals for the binaphthyl units of

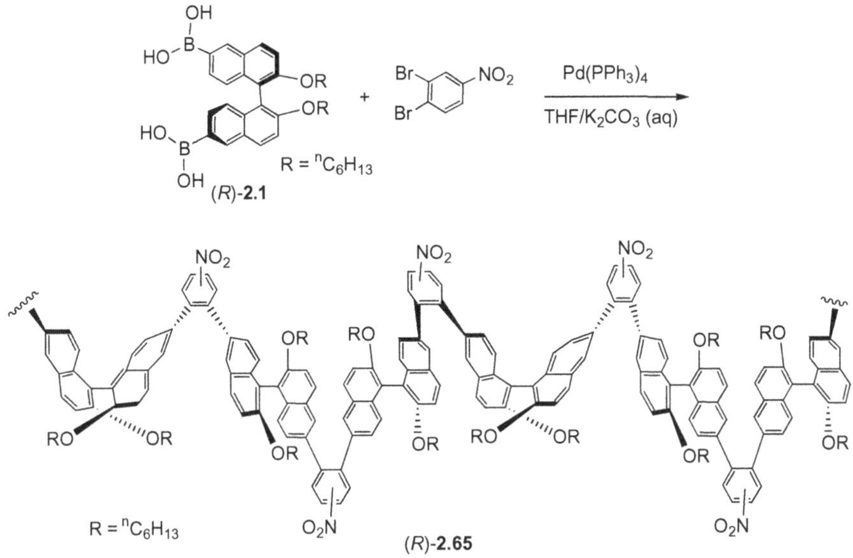

Scheme 2.35. Synthesis of chiral polymer (R)-**2.65** by the Suzuki coupling.

different configurations in the polymer chain. Therefore, the randomly distributed R and S binaphthyl units in rac-**2.65** should have more complicated NMR signals than the homochiral polymer (R)-**2.65**. The significantly lower molecular weight of rac-**2.65** than that of (R)-**2.65** also indicates that the steric interaction between the adjacent binaphthyl units in these polymers can strongly influence their polymerization process.

The binaphthyls are also linked with meta-phenylene units. From the coupling of (R)-**2.12** with 1,5-dibromo-2,4-dinitrobenzene, polymer (R)-**2.66** is obtained in 92% yield (Scheme 2.36). GPC shows the molecular weight of (R)-**2.66** as $M_w = 30{,}500$ and $M_n = 16{,}300$ (PDI = 1.9). The high molecular weight of this polymer is attributed to the doubly activated aryl bromides by the *ortho* and *para* nitro groups in monomer 1,5-dibromo-2,4-dinitrobenzene. This polymer is soluble in common organic solvents. From monomer rac-**2.12**, polymer rac-**2.66** is obtained with $M_w = 20{,}000$ and $M_n = 12{,}000$ (PDI = 1.61).

Compound **2.67** is obtained from the reaction of 1,5-dibromo-2,4-dinitrobenzene with two equivalents of 2-neopentyl-6-ethynylnaphthalene in the presence of Pd(PPh$_3$)$_4$ and CuI in high yield. The NMR and IR

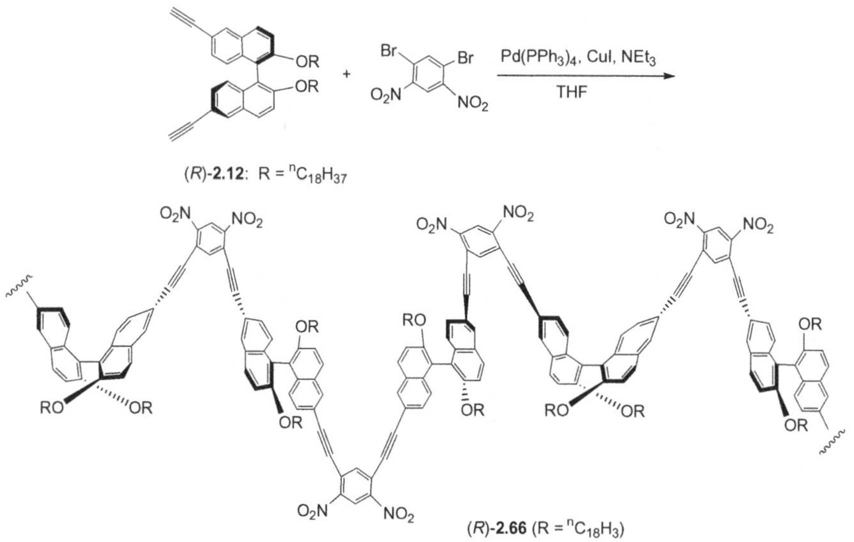

Scheme 2.36. Polymerization of (*R*)-**2.12** with 1,5-dibromo-2,4-dinitrobenzene.

spectral data of **2.67** are very similar to those of (*R*)-**2.66**. This supports the structural characterization of polymer (*R*)-**2.66**.

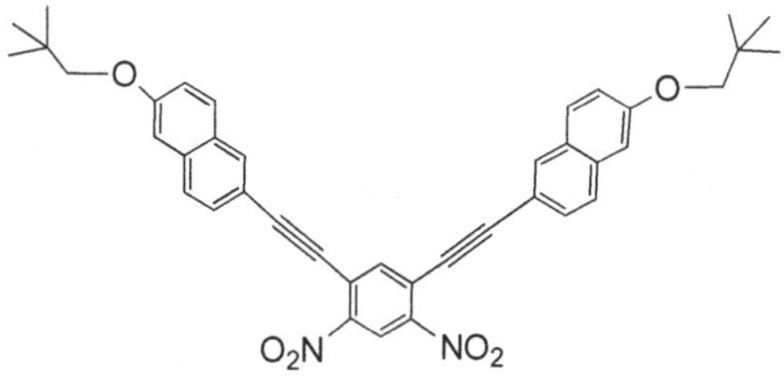

The UV spectral data of the propeller-like polymers and the repeating units are summarized in Table 2.13. The absorptions of polymers (*R*)-**2.58** and (*R*)-**2.66** are very similar to those of their repeating units **2.59** and **2.67**, respectively. This indicates that there is little conjugation between the repeating units in these polymers. The increased distance between the electron donors and acceptors in the repeating unit of polymer (*R*)-**2.62** leads to a blue shift in the absorptions in comparison with those of (*R*)-**2.58** in spite of the longer conjugation length of (*R*)-**2.62**. In the repeating unit **2.67**, because the electron donors and acceptors are *para* and *ortho* with respect to each other, its UV absorptions are shifted to the red of

Table 2.13. The UV–vis absorption wavelengths of the propeller-like polymers and the repeating units.

Polymer or repeating unit	λ_{max} (nm)
(R)-**2.54**	234, 295 (sh), 346, 380
(R)-**2.58**	238, 277 (sh), 309 (sh), 358, 418
2.59	236, 300, 354, 406
(R)-**2.60**	262, 278, 344
rac-**2.61**	236, 276, 340 (sh)
(R)-**2.62**	234, 296, 346
(R)-**2.63**	232, 296, 362
rac-**2.64**	234, 302, 342
(R)-**2.65**	246, 350, 400 (sh)
(R)-**2.66**	256, 284 (sh), 359 (sh), 442
2.67	244, 282, 354, 430

those of **2.59** where the donors and acceptors are *para* and *meta* with respect to each other. There is better conjugation in **2.67** than in **2.59**. The polymers containing nitro groups show very weak or no fluorescence at all. Polymers **2.60** and **2.61** that do not have nitro groups exhibit strong fluorescence at 420–460 nm. These polymers display two strong positive and negative CD signals at around 235 and 270 nm, respectively. The CD data of these polymers are given in Table 2.14. The LLS study of the propeller-like polymers shows that the molecular weights of most of the polymers are about 20% higher than the GPC data relative to polystyrene standards.

The propeller-like polymers that contain triple bonds show an irreversible exothermic peak at about 200°C in their DSC plots, indicating a chemical decomposition. Table 2.15 summarizes the decomposition temperatures and the heat released at this temperature. As shown in Table 2.15, polymers containing nitro groups give much larger exothermic peaks than the polymers that do not contain nitro groups, and the *meta*-dinitrophenylene-based polymers **2.66** and the repeating unit **2.67** give much larger exothermic peak than the *ortho*-phenylene-based polymers. In the *ortho*-phenylene-based polymers, their repeating units contain enediyne fragments that are known to undergo the Bergman cyclization at *ca.* 200°C (Scheme 2.37).[30] Thus, the enediyne cyclization might have occurred for the *ortho*-phenylene-based polymers. Structure **2.68** is a proposed intermediate generated from the Bergman cyclization of polymer (R)-**2.58**. Polymers **2.60**, **2.61** and **2.64** do not contain nitro groups and they either show no decomposition or release much less heat at the decomposition temperature. This is probably because these polymers might contain a small number of the enediyne fragment because of a competitive

Table 2.14. The CD spectral data of the propeller-like polymers.

Polymer	λ_{max}(nm)	$[\theta]$
(R)-**2.54**	235	2.19×10^5
	264	-5.42×10^4
	351	4.82×10^3
	396	-5.05×10^3
(R)-**2.58**	234	2.68×10^5
	277	-1.40×10^5
	356	9.14×10^3
	419	-9.71×10^3
(R)-**2.60**	237	2.38×10^5
	284	-2.14×10^5
	381	-2.38×10^4
(R)-**2.62**	234	2.29×10^5
	264	-4.63×10^4
	296	-6.56×10^4
	383	-2.78×10^4
(R)-**2.65**	240	3.34×10^5
	275	-1.80×10^5
	306 (sh)	-6.18×10^4
	348	2.12×10^4
(R)-**2.66**	238	2.53×10^5
	270	-1.16×10^5
	470	-3.11×10^4

Table 2.15. The DSC data of the propeller-like polymers and the repeating units.

Polymer or repeating unit	Temperature (°C)	ΔH (J/g)	ΔH (kcal/mol)
(R)-**2.54**	200	−79.5	−18.8
rac-**2.54**	193	−107.3	−25.3
(R)-**2.58**	202	−94.2	−23.3
rac-**2.58**	202	−113.3	−28.0
2.59	196	−142	−21.7
(R)-**2.60**	231	−34.7	−8.4
rac-**2.60**	231	−35.7	−8.7
rac-**2.61**	*		
(R)-**2.62**	190	−57.9	−15.5
(R)-**2.63**	176	−61.0	−15.7
rac-**2.64**	183	−15.8	−3.9
(R)-**2.66**	154	−359.6	−88.8
rac-**2.66**	154	−447.9	−110.3
2.67	226	−466.4	−71.4

*denotes no decomposition up to 300°C.

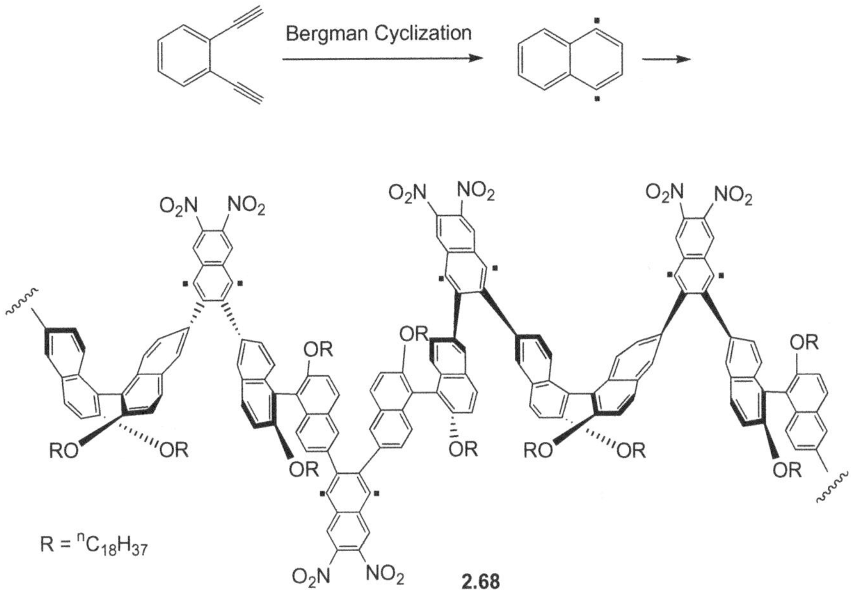

Scheme 2.37. The Bergman cyclization of *ortho*-diethynylbenzene.

homo-coupling of the binaphthyl alkyne monomers during the preparation. When the linker molecules do not have the strong electron-withdrawing nitro groups, they become less reactive toward the cross-coupling reaction. This leads to a significant amount of the alkyne–alkyne coupling structure as shown for compound (R)-**2.54** and thus much less enediyne fragment.

Polymers **2.66** do not have *ortho*-bisalkynylbenzene unit and cannot undergo the Bergman cyclization. However, these polymers and their repeating unit **2.67** still show large irreversible exothermic peaks at 154 and 226°C, respectively, in their DSC plots. It is proposed that the two nitro groups in the repeating unit of these compounds could react with the *ortho*-triple bonds on the benzene ring, leading to the exothermic transition. Scheme 2.38 shows one of the possible chemical transformations for the repeating unit of polymers **2.66** at high temperature.

Polymers **2.65** do not contain triple bonds and show very different thermal transitions. In the DSC plots, (R)-**2.65** shows a T_g at 148°C (T_g) and *rac*-**2.65** shows a T_g at 142°C. Although an irreversible exothermic peak at *ca.* 270°C is observed, the heat released ($\Delta H = -7$ to $9\,\text{J/g}$) is very small, which is more like a phase transition rather than a chemical reaction.

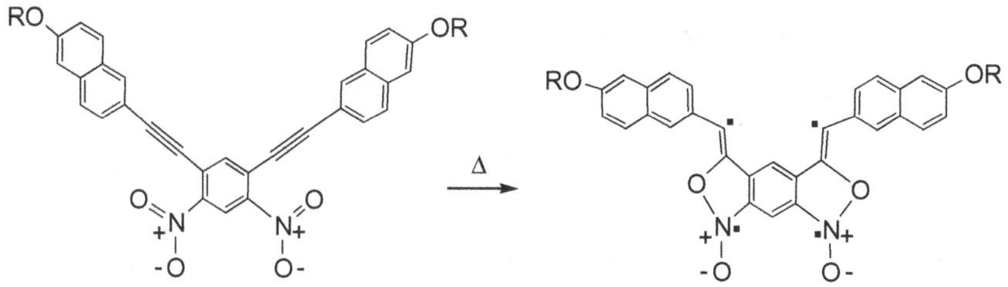

Scheme 2.38. Thermal transformation of the repeating unit of polymers **2.66**.

TGA shows that all the propeller-like polymers start to lose 5% of their mass at temperatures in the range of 310–380°C with the polymers containing nitro groups at lower temperatures and those without nitro groups at higher temperatures. These polymers probably lose their alkyl groups first and then become thermally more stable up to 800°C, which is similar to other binaphthyl polymers. The nitro-containing polymers may slowly lose their nitro groups after the lose of the alkyl groups.

2.8.2. Synthesis of the Propeller-Like Polymers Derived from BINAM[31]

BINAM-derived propeller-like polymers are synthesized. Scheme 2.39 shows the preparation of the BINAM-based monomers. From the dimerization of 2-napthtol in the presence of hydrazine followed by optical resolution, alkylation and bromination, compound (R)-**2.69** is obtained. It is necessary to use ^{n}Bu$_4$NBr$_3$ for the bromination, since the use of Br$_2$ leads to unknown products. Compound (R)-**2.69** is then converted to the 6,6'-bisethynyl monomer (R)-**2.70**. The specific optical rotation $[\alpha]_D$ of (R)-**2.70** is −170 (c = 1.0, THF).

Polymerization of (R)-**2.70** with 1,2-dibromo-4,5-dinitrobenzene in the presence of Pd(PPh$_3$)$_4$ and CuI generates polymer (R)-**2.71** as a dark red solid in 89% yield (Scheme 2.40). This polymer is soluble in common organic solvents. GPC shows its molecular weight as $M_w = 23{,}500$ (PDI = 1.9).

A diiodo monomer (R)-**2.72** is prepared from the dibromo compound (R)-**2.69** (Scheme 2.41). Polymerization of (R)-**2.72** with 1,2-dinitro-4,5-divinylbenzene in the presence of Pd(OAc)$_2$ and CuI is conducted. BHT is added to inhibit a radical polymerization of the divinyl monomer. An oligomeric product (R)-**2.73** is obtained as a shiny black solid in 74% yield with $M_w = 2800$ (PDI = 1.45). The repeating unit of (R)-**2.73** is

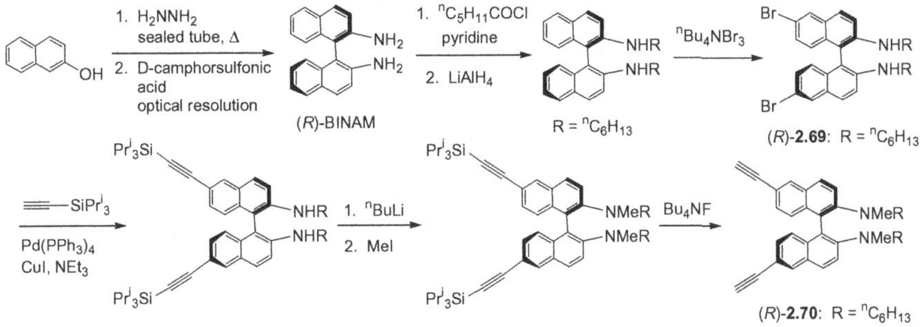

Scheme 2.39. Synthesis of BINAM-derived monomers.

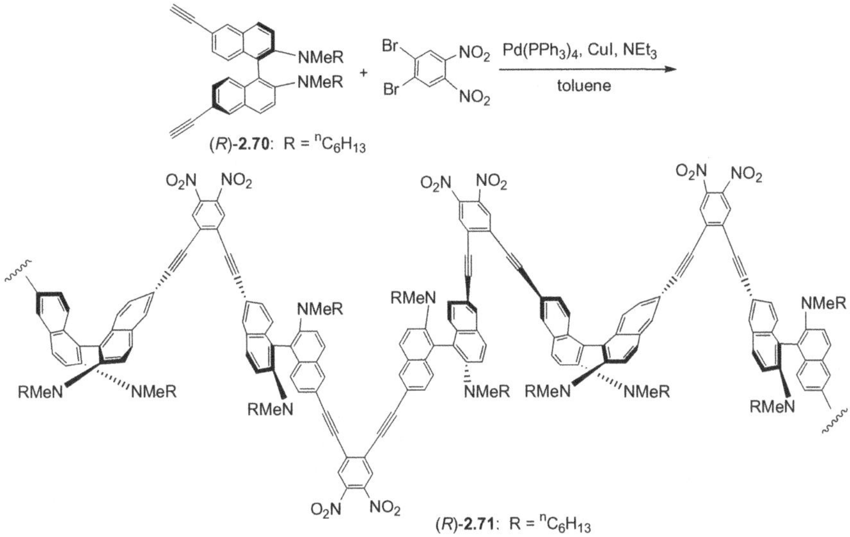

Scheme 2.40. Synthesis of BINAM-based propeller-like polymer (R)-**2.71**.

synthesized as shown in Scheme 2.42 where compound **2.74** is obtained as a dark red solid. The structure of **2.74** is determined on the basis of its NMR and mass spectroscopic data. Since the oligomer (R)-**2.73** gives broad and complicated ^{1}H NMR signals that are very different from those of **2.74**, it probably contains a mixture of structures.

The UV absorption data of (R)-**2.71**, (R)-**2.73** and **2.74** in methylene chloride solution are given in Table 2.16. There is red shift in the absorptions when the triple bonds of (R)-**2.71** are replaced with the double bonds

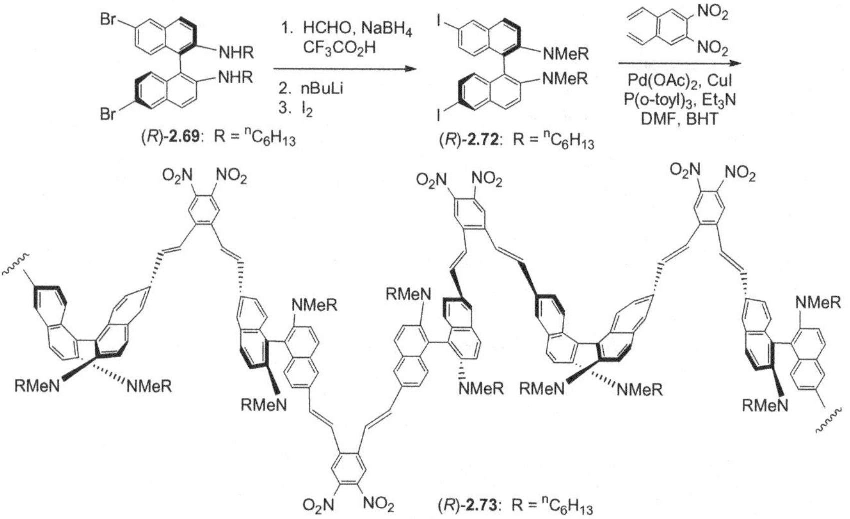

Scheme 2.41. Polymerization of (*R*)-**2.72** with 1,2-dinitro-4,5-divinylbenzene.

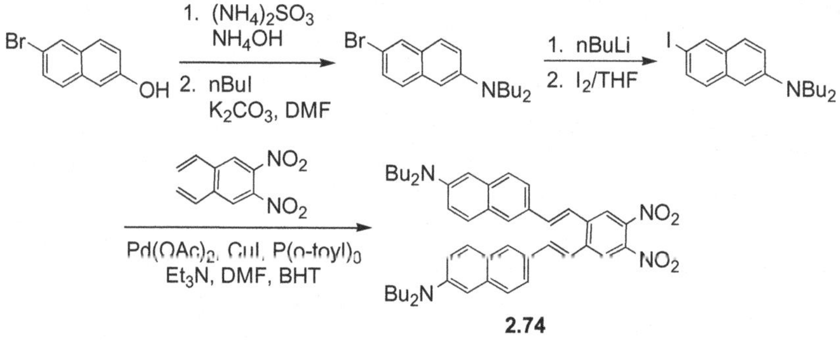

Scheme 2.42. Synthesis of repeating unit **2.74**.

of (*R*)-**2.73**. This could be attributed to the more polarizable π electrons in the vinylene units than those in the ethynylene units, leading to a better conjugation of (*R*)-**2.71**. The UV absorptions of (*R*)-**2.73** are similar to those of **2.74**. This indicates that in spite of the probably more complicated structure of (*R*)-**2.73**, it should still contain units similar to **2.74**. Both (*R*)-**2.71** and (*R*)-**2.73** exhibit intense positive and negative CD signals.

Table 2.16. UV and CD spectral data of (R)-**2.71**, (R)-**2.73** and **2.74**.

Polymer or repeating unit	(R)-**2.71**	(R)-**2.73**	**2.74**
λ_{max} (nm)	272	254	250
	324	308	342
	383	400	414
	470	486	494
$[\theta]$ (λ_{max} nm)	2.30×10^5 (220, sh)	1.86×10^5 (223)	
	-1.18×10^5 (286)	-5.68×10^4 (278)	
	1.50×10^3 (377)	-1.49×10^4 (356)	
	-1.70×10^4 (409, sh)	4.74×10^3 (415)	
	-1.40×10^4 (469, sh)	-3.34×10^4 (503)	

A BINOL-derived vinyl polymer (R)-**2.75** is also synthesized from the coupling of (R)-DBDH with 1,2-dinitro-4,5-divinylbenzene.

(R)-**2.75**: R = $^nC_6H_{13}$

2.8.3. Study of the Non-linear Optical Properties of the Propeller-Like Polymers[32]

In collaboration with Persoons, we have studied the second-order non-linear optical properties of the propeller-like polymers. The Langmuir–Blodgett (LB) films of the optically active polymers (R)-**2.58**, (R)-**2.71**, (R)-**2.73** and (R)-**2.75** are prepared. The dilute chloroform solutions of these polymers are spread on water. After evaporation of the solvent, the monolayer is compressed to form LB films with thickness between 1 and 8 monomolecular layers. The second-harmonic response (SHG) of the films is measured using a Q-switched Nd:YAG laser (10 ns, 50 Hz, 1064 nm).

The films of polymers (R)-**2.58**, (R)-**2.73** and (R)-**2.75** are of very good optical quality, but those of polymer (R)-**2.71** are poor with large holes visible. Thus, the long chain alkyls of (R)-**2.58** and the double bonds of (R)-**2.73** and (R)-**2.75** are very good for generating high-quality films because they introduce more structural flexibility.

It is found that the maximum SHG intensity from both polymers (R)-**2.58** and (R)-**2.75** is very weak. In addition, the SHG signals of polymer (R)-**2.75** decrease to 75% of its initial value within minutes because of the instability of this polymer upon irradiation by the infrared laser light. Polymer (R)-**2.71** shows a factor of 50 higher maximum SHG intensity than that of polymers (R)-**2.58** and (R)-**2.75** even though the LB films of (R)-**2.71** are of poor optical quality. This is attributed to the stronger electron donors of the amine groups in (R)-**2.71**. Replacement of the triple bonds of (R)-**2.71** with double bonds led to the high-quality films of (R)-**2.73**. This polymer shows four times more efficient SHG signals than (R)-**2.71** and 200 times more efficient than (R)-**2.58** and (R)-**2.75**. By analyzing the polarization dependence of the SHG, the tensor components of the nonlinear susceptibility $\chi^{(2)}_{ijk}$ are determined for polymer (R)-**2.73** (Table 2.17). Although the second-order susceptibility values in Table 2.17 are resonantly enhanced, they demonstrate that this optically active polybinaphthyl is a new class of promising NLO material. The chirality of the LB films is reflected by the xyz component of 2 pm/V.

These polymers also show different second-harmonic efficiency when they are irradiated with left- and right-hand circularly polarized light.[33] The experiments are conducted on the LB films of (R)-**2.58** by spreading the chloroform solution of the polymer on water. The thickness of the films is ten X-type layers with an absorption maximum at 430 nm. The fundamental beam of a Q-switched and injection-seeded Na:YAG laser (1064 nm) is used to pump the chiral LB films. It is in a 45° angle of incidence with respect to the sample. The fundamental beam is initially p-polarized with respect to

Table 2.17. The second-order susceptibility [$\chi^{(2)}_{ijk}$] of polymer (R)-**2.73**.

Second-order susceptibility component	Absolute value (pm/V)	Chirality clasification
xyz	2	Chiral
xxz	7	Achiral
zxx	5	Achiral
zzz	35	Achiral

the films and the polarization is varied by a quarter-wave plate. The signals of the *s*- and *p*-components of the transmitted second-harmonic fields are detected, which show significant CD responses. These CD effects can be represented by the following equation: $\Delta I/I = 2(I_l - I_r)/(I_l + I_r)$ where I_l and I_r are the intensities of the second-harmonic light generated when the fundamental beam is left- or right-hand circularly polarized. The second-harmonic CD effect for (*R*)-**2.58** is 65%, which are orders of magnitude higher than the linear circular dichroism effect ($\Delta\varepsilon/\varepsilon$) in solution. For other chiral polymers of this class, the second-harmonic CD effects are also observed between 25 and 60%.

2.9. Dipole-Oriented Propeller-Like Polymers[34]

The propeller-like polymers described in Section 2.7 would have cancelled dipole moment along the polymer axis because of their designated arrangement of the dipolar repeating units. We have also designed a chiral polybinaphthyl that could have increased dipole moment as the polymer chain extends by orienting the dipole units in one direction along the polymer axis. Scheme 2.43 shows the synthesis of monomer (*R*)-**2.75** for

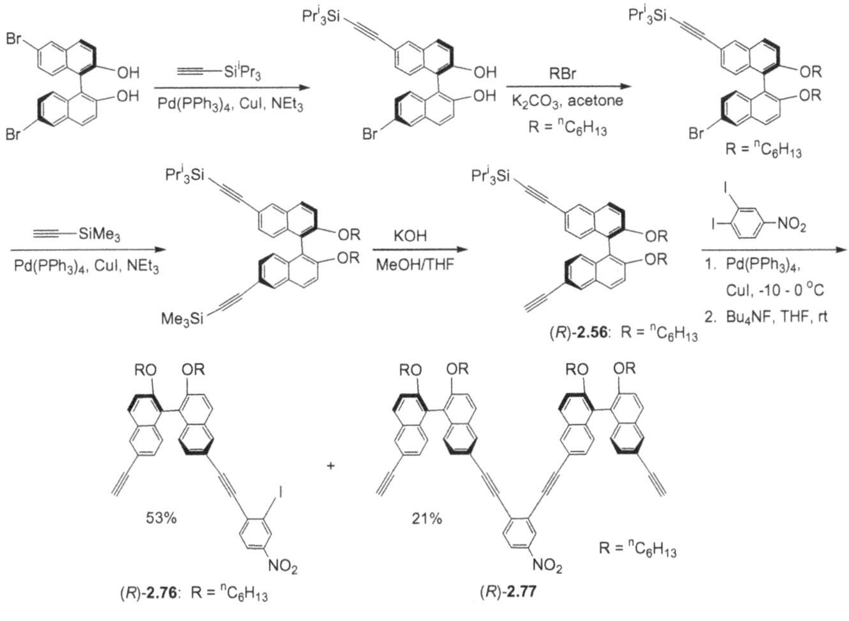

Scheme 2.43. Synthesis of AB monomer (*R*)-**2.75a**.

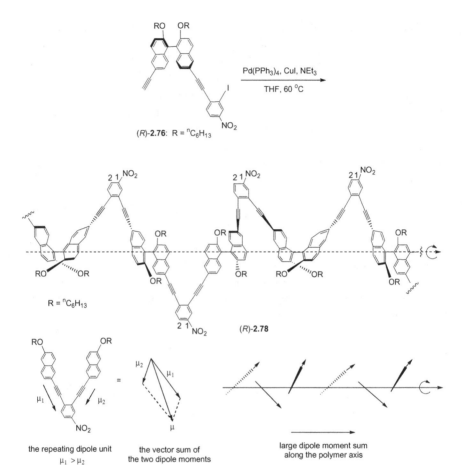

Scheme 2.44. Homopolymerization of AB monomer (*R*)-**2.76** to generate the dipole-oriented propeller-like polymer (*R*)-**2.78**.

the dipole-oriented polymer from the reaction of the mono-protected 6,6′-diethynylbinaphthyl compound (*R*)-**2.56** with 1,2-diiodo-5-nitrobenzene. The mono coupling product (*R*)-**2.76** is obtained in 53% yield. In the mean time, the double coupling product (*R*)-**2.77** is obtained in 21% yield.

Compound (*R*)-**2.76** is an AB-type monomer. Its self-coupling polymerization is conducted at 60°C in the presence of Pd(PPh$_3$)$_4$ and CuI at 0.24 M in an Et$_3$N-THF solution (Scheme 2.44). This produces polymer (*R*)-**2.78** in 80% yield as a light yellow solid. This polymer is soluble in common organic solvents. GPC shows its molecular weight as $M_w = 34{,}000$

and $M_n = 17,000$ (PDI = 1.97). Unlike polymer (R)-**2.54** in which the nitro groups are randomly distributed at position 1 or 2 in the polymer chain, polymer (R)-**2.78** has a nitro group specifically located at the position 1 in the repeating unit. In the repeating dipole unit of the polymer as shown in Scheme 2.44, because $\mu 1 > \mu 2$, the sum of the dipole moment of each repeating unit should tilt along the polymer chain in one direction. Thus, the dipole moment along the polymer chain should increase with the chain length. The CD spectrum of (R)-**2.78** gives strong Cotton effects at $\lambda_{\max}$ ($[\theta]$) = 232(2.7×10^5), 271 (sh) (-1.6×10^5), 282 (-1.8×10^5), 350 (1.9×10^4), 387 (-1.8×10^4).

DSC shows that polymer (R)-**2.78** has a large irreversible exothermic peak at 209°C ($\Delta H = -90.2$ J/g). This could be attributed to the Bergman cyclization of the enediyne moiety in the repeating unit of (R)-**2.78**, which is similar to that observed for polymer (R)-**2.58** and its analogs. TGA shows that (R)-**2.78** begins to lose mass at about 300°C. At 475°C, this polymer probably loses its hexyl groups (25% weight loss). From 500 to 800°C, the polymer does not lose significant amount of its mass.

When the homo-coupling of (R)-**2.76** is conduced at a lower concentration of 0.064 M, several macrocyclic products are obtained with 34% of the cyclic dimer (R)-**2.79**, 18% of the cyclic trimer (R)-**2.80** and 12% of the cyclic tetramer (R)-**2.81** after purification by flash chromatograph on silica gel (Scheme 2.45). These molecules are characterized by NMR and FAB mass spectroscopic methods.

Table 2.18 compares the UV spectral data and optical rotations of the monomer, dimer, polymer and macrocycles. All these compounds show similar absorption maxima as they all contain the same repeating units as the chromophore. The macrocycles have much greater specific optical rotations than the polymer and monomers.

2.10. Binaphthyl-Based Polysalophens[35]

When the homochiral binaphthyls are linked by a planar metallosalophen (salophen = N,N-disalicylidene-1,2-phenylenediaminato) unit, the resulting polymer could have a rigid helical structure as shown in Fig. 2.3.

In order to explore the possibility of constructing a rigid helical polysalophen, the 3,3'-diformylBINOL (S)-**2.82** is synthesized from (S)-BINOL (Scheme 2.46). Condensation of (S)-**2.82** with o-phenylenediamine, however, gives a macrocyclic salophen (S)-**2.83** in high yield. No polymer formation is observed. The macrocycle (S)-**2.83** has a very large optical

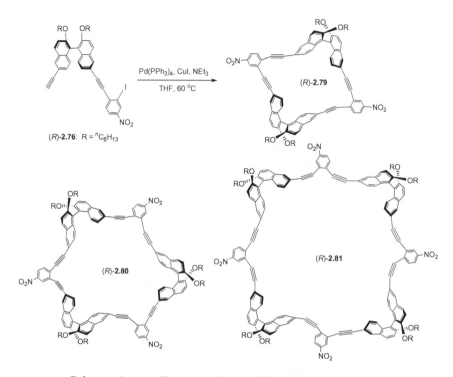

Scheme 2.45. Homo-coupling of (R)-**2.76** to macrocycles.

Table 2.18. The UV spectroscopic data and the optical rotations of the monomer, dimer, polymer and macrocycles.

Compound	(R)-**2.76**	(R)-**2.77**	(R)-**2.78**	(R)-**2.79**	(R)-**2.80**	(R)-**2.81**
UV–vis, (CH$_2$Cl$_2$)	250	254	236	254	236	234
λ_{max}/nm	300	288	284	286	286	286
	354	350	346	342	346	344
	390	393	380	390	388	384
$[\alpha]_D$ (CH$_2$Cl$_2$)	−102.6	−122.7	−197.6	−301.3	−282.3	−297.0

rotation with $[\alpha]_D = +1561$ ($c = 0.3$, CH$_2$Cl$_2$). Treatment of the macrocycle with two equivalents of Ni(OAc)$_2 \cdot$ 4H$_2$O gives a paramagnetic complex (S)-**2.84**. The mass spectroscopic analysis supports the proposed structure for (S)-**2.84**. This complex probably contains a diamagnetic square planar Ni(II)-salophen unit and a paramagnetic tetrahedral Ni(II) unit. Treatment of macrocycle (S)-**2.83** with one equivalent of Ni(OAc)$_2 \cdot$ 4H$_2$O cannot generate a mononuclear diamagnetic complex.

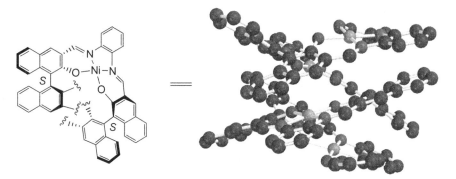

Figure 2.3. The proposed structure of a binaphthyl-based polysalophen.

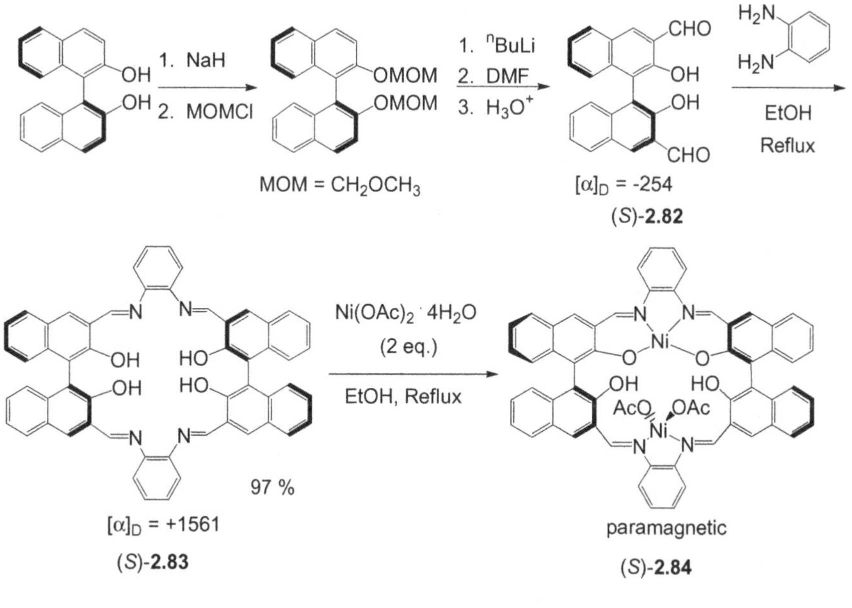

Scheme 2.46. Condensation of (S)-**2.82** with o-phenylenediamine.

The alkylated o-phenylenediamines **2.85a** and **2.85b** are synthesized as shown in Scheme 2.47. Coupling of 1,2-dibromo-4,5-dinitrobenzene with 1-octyne and 1-octadecyne followed by hydrogenation gives **2.85a** and **2.85b**, respectively. Condensation of (S)-**2.82** with **2.85b** in the presence of Ni(OAc)$_2 \cdot$4H$_2$O in refluxing THF produces an oligomer (S)-**2.86** in 66% yield (Scheme 2.48). The alkyl groups on the phenylene units make (S)-**2.86** soluble in common organic solvents. GPC shows its molecular weight as

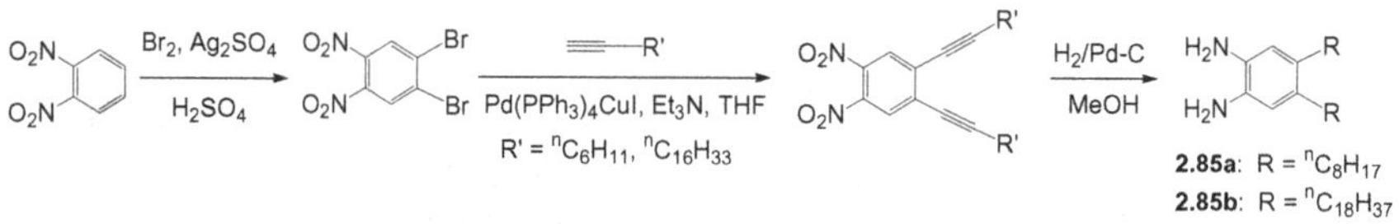

Scheme 2.47. Synthesis of alkylated *o*-phenylenediamines.

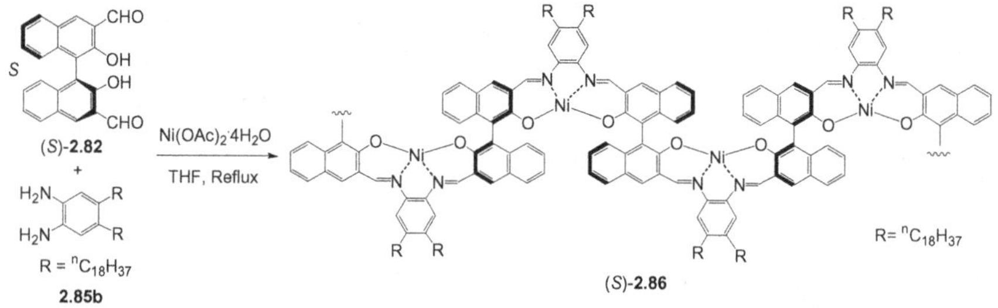

Scheme 2.48. Polymerization of (*S*)-**2.82** and **2.85b** in the presence of Ni(OAc)$_2 \cdot$ 4H$_2$O.

$M_w = 3600$ (PDI $= 1.62$), which corresponds to a degree of polymerization of 7. Polymer (*S*)-**2.86** is found to be a paramagnetic material. This indicates that (*S*)-**2.86** contains non-planar Ni(II) coordination. The non-planar Ni(II) coordination in (*S*)-**2.86** should make it deviate from the proposed helical structure as shown in Fig. 2.3.

The condensation of the achiral biphenyl dialdehyde **2.87** with the diamine **2.85b** is also conducted (Scheme 2.49). Unlike the reaction of (*S*)-**2.82** with *o*-phenylenediamine, the condensation of **2.87** with **2.85b** cannot generate the corresponding macrocycle in high yield. In the presence of Ni(OAc)$_2 \cdot$ 4H$_2$O, a low-molecular-weight polymer **2.88** is produced in 68% yield. The longer chain alkyl groups of **2.85b** than those of **2.85a** give a much more soluble material. GPC shows the molecular weight of **2.88** as $M_w = 4900$ (PDI $= 1.2$) corresponding to a degree of polymerization of 10. This polymer is also paramagnetic, indicating non-planar Ni(II) coordinations in the structure. Comparison of the UV spectrum of polymer **2.88** with that of the diamagnetic square planar compound **2.89** shows that the polymer has significantly decreased absorptions in the range of 450–550 nm. This indicates that the non-planar Ni(II) coordination in **2.88** should lead to a less-effective conjugation. Polymer **2.88** also shows broad near-visible-infrared absorptions in the range of 700–1200 nm ($\varepsilon = 32$–100)

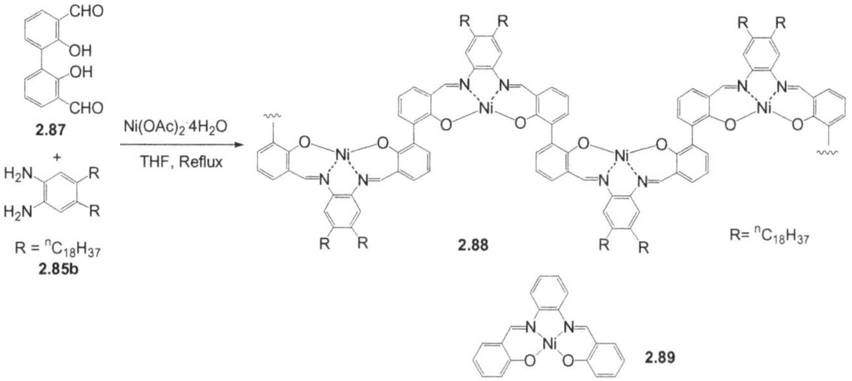

Scheme 2.49. Condensation of **2.87** with **2.85b** in the presence of Ni(OAc)$_2 \cdot$ 4H$_2$O.

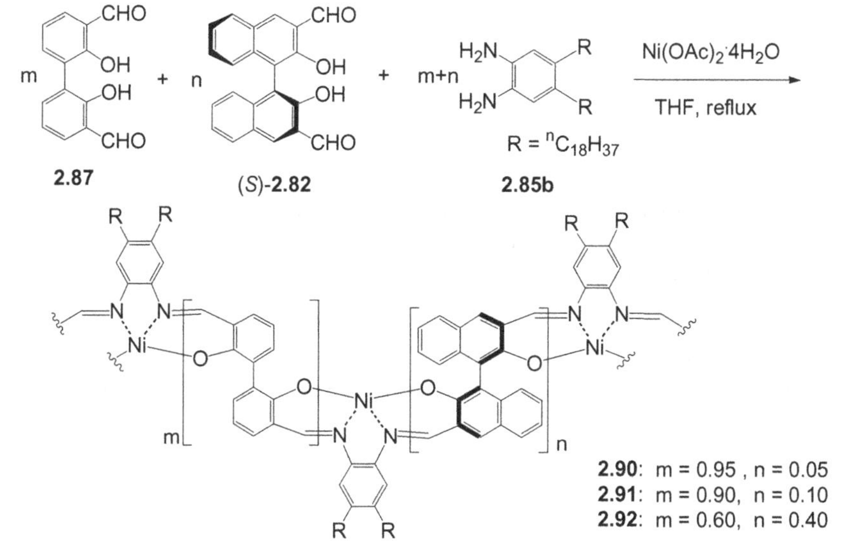

Scheme 2.50. Copolymerization of **2.87**, (S)-**2.82** and **2.85b** in the presence of Ni(OAc)$_2 \cdot$ 4H$_2$O.

due to the d–d transition of the non-planar Ni(II) center. The planar diamagnetic compound **2.89** does not have absorption in this region.

Copolymerization of the achiral biphenyl monomer **2.87** and the chiral BINOL monomer (S)-**2.82** with the diamine **2.85b** at various monomer ratios is conducted to produce a series of copolymers **2.90**–**2.92** (Scheme 2.50). The molecular weights of these polymers are obtained by GPC analysis. Polymer **2.90** contains 5% of the BINOL unit with $M_w = 5400$ (PDI = 1.6); Polymer **2.91** contains 10% of the BINOL unit with $M_w = 4400$ (PDI = 1.1); and polymer **2.92** contains 40% of the BINOL unit with $M_w = 3400$ (PDI = 1.3). The CD spectra of these polymers and (S)-**2.86** show a linear increase in the signal intensity at the longest wavelength maximum as the BINOL component increases. This indicates that the homochiral BINOL units in these polymers do not induce a preferred helicity for the biphenol units and the chiral optical properties are determined by individual BINOL units. No cooperation between the monomer units is observed to form a propagating helical chain. The non-planar Ni(II) coordination in these materials should be one of the factors contributing to the non-helical chain structure.

2.11. Helical Ladder Polybinaphthyls[36]

A high-yield room temperature conversion of arylalkynes to fused polyaromatic compounds was reported by Swager (Scheme 2.51).[37] We have used this reaction to construct helical ladder polybinaphthyls.

From the Suzuki coupling of the binaphthyl diborate (R)-**2.93** (see Scheme 3.31 in Chapter 3 for preparation) with the aryldibromide **2.94**, polymer (R)-**2.95** is obtained (Scheme 2.52). This polymer is soluble in common organic solvents and GPC shows its molecular weight as $M_w = 11{,}200$ and $M_n = 7300$ (PDI = 1.54). Its specific optical rotation $[\alpha]_D$ is $+36.8$ ($c = 0.1$, CH_2Cl_2). When this polymer is treated with CF_3CO_2H in methylene chloride at room temperature, a yellow solution of polymer (R)-**2.95** is converted to a dark green solution within an hour. After 12 h, polymer (R)-**2.96** is obtained in 95% yield. This polymer is also soluble in

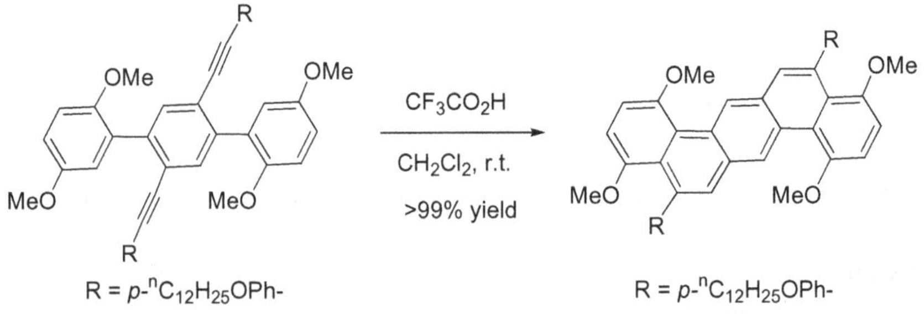

Scheme 2.51. Swager's acid-catalyzed cyclization of arylalkynes to polyaromatic compounds.

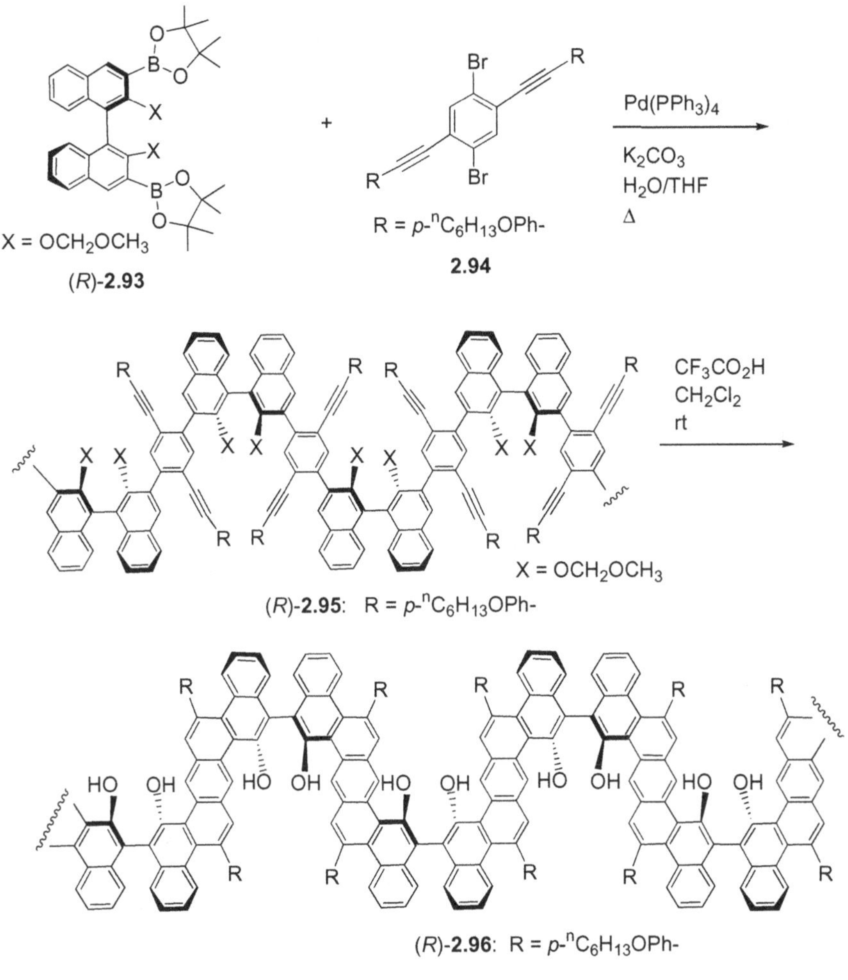

Scheme 2.52. Synthesis of helical ladder polybinaphthyl (R)-**2.96**.

common organic solvents. GPC shows its molecular weight as $M_w = 17{,}900$ and $M_n = 10{,}500$ (PDI = 1.70). The dark color of the polymer solution makes its optical rotation fluctuate at around -531 ($c = 0.1$, CH$_2$Cl$_2$). This optical rotation of (R)-**2.96** has a greatly increased value and an opposite sign in comparison with that of the precursor polymer (R)-**2.95**. The NMR spectra of (R)-**2.96** show the disappearance of the signals for the CH$_2$OCH$_3$ groups as well as the alkyne carbons. The UV spectrum of (R)-**2.96** shows a large red shift ($\Delta\lambda_{\max} = 87$ nm) for the longest

Table 2.19. Spectral data of polymers (R)-**2.95**, (R)-**2.96**, (R)-**2.98**, (R)-**2.99**, (R)-**2.101** and (R)-**2.102** in methylene chloride solution.

Polymer	UV λ_{max} (ε) nm	Fluorescence λ_{emi} (nm)	CD [θ] (λ_{max}, nm, 1.0×10^{-5} M)
(R)-**2.95**	366(sh, 3.58×10^4) $341(4.86 \times 10^4)$	398	5.86×10^4 (365) -1.17×10^5 (292) 1.90×10^5 (258)
(R)-**2.96**	$453(1.26 \times 10^4)$ $344(7.33 \times 10^4)$	521	-6.98×10^4 (360) 7.49×10^4 (330) 1.23×10^5 (268)
(R)-**2.98**	$320(6.28 \times 10^4)$ $281(5.53 \times 10^4)$	408	-4.06×10^3 (359) 8.07×10^3 (341) -1.64×10^5 (284)
(R)-**2.99**	$411(1.03 \times 10^4)$ $339(6.06 \times 10^4)$	450 476(sh)	-2.86×10^5 (351) 1.62×10^5 (320) -3.54×10^3 (273)
(R)-**2.101**	372(sh, 3.47×10^4) $315(7.23 \times 10^4)$ $268(5.28 \times 10^4)$	408	2.80×10^4 (337) -1.75×10^5 (289) 3.90×10^4 (252)
(R)-**2.102**	$410(1.28 \times 10^4)$ $330(7.14 \times 10^4)$	455 484(sh)	-2.49×10^5 (350) 6.05×10^4 (324) -4.76×10^4 (303) -8.19×10^4 (277) 1.56×10^5 (241)

wavelength absorption from that of (R)-**2.95** (Table 2.19). The fluorescence spectrum of (R)-**2.96** also gives a large red shift ($\Delta\lambda = 123$ nm) for the emission maximum in comparison with that of (R)-**2.95**. These spectral data support the formation of a more conjugated planar fused polyaromatic system in polymer (R)-**2.96**. The fluorescence quenching of (R)-**2.96** by an amino alcohol is found to be much more efficient than that of BINOL. For the interaction of (R)-BINOL and (R)-**2.96**, both at 1.0×10^{-5} M in CH$_2$Cl$_2$/hexane (1:1), with (1R,2S)-(−)-N-methylephedrine (3.0×10^{-2} M), it is observed that the I_0/I is 1.3 for (R)-BINOL, but 3.4 for polymer (R)-**2.96**.

Figure 2.4 gives an energy minimized molecular modeling structure (PCSpartan-Pro, semiemperical AM1) of a segment of (R)-**2.96** where the R groups of (R)-**2.96** are replaced with H atoms. Because of the planar fused polyaromatic repeating units and the homochiral biaryl linkage, this polymer is expected to have a rigid helical chain structure.

Polymerization of the binaphthyl monomer (R)-**2.97** with the aryldibromide **2.94** gives polymer (R)-**2.98** (Scheme 2.53). GPC shows

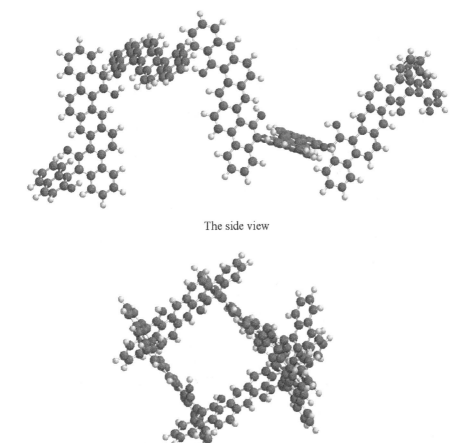

The side view

Viewed along the helical axis

Figure 2.4. The molecular modeling structure of polymer (R)-**2.96**. (The R groups of (R)-**2.96** are replaced with hydrogen atoms.)

the molecular weight of this polymer as $M_w = 14{,}300$ and $M_n = 8700$ (PDI = 1.64). Its specific optical rotation $[\alpha]_D$ is -115.7 ($c = 0.1$, CH_2Cl_2). Polymer (R)-**2.98** is treated with CF_3CO_2H in methylene chloride at room temperature, which produces polymer (R)-**2.99** as a brown solid in 97% yield. The MOM groups are removed similar to that observed for (R)-**2.96**. In the acid promoted cyclization of the repeating unit of (R)-**2.98**, the reaction could occur at either the α- or β-positions of the naphthalene rings. It is proposed that the reaction probably occurs at the generally more

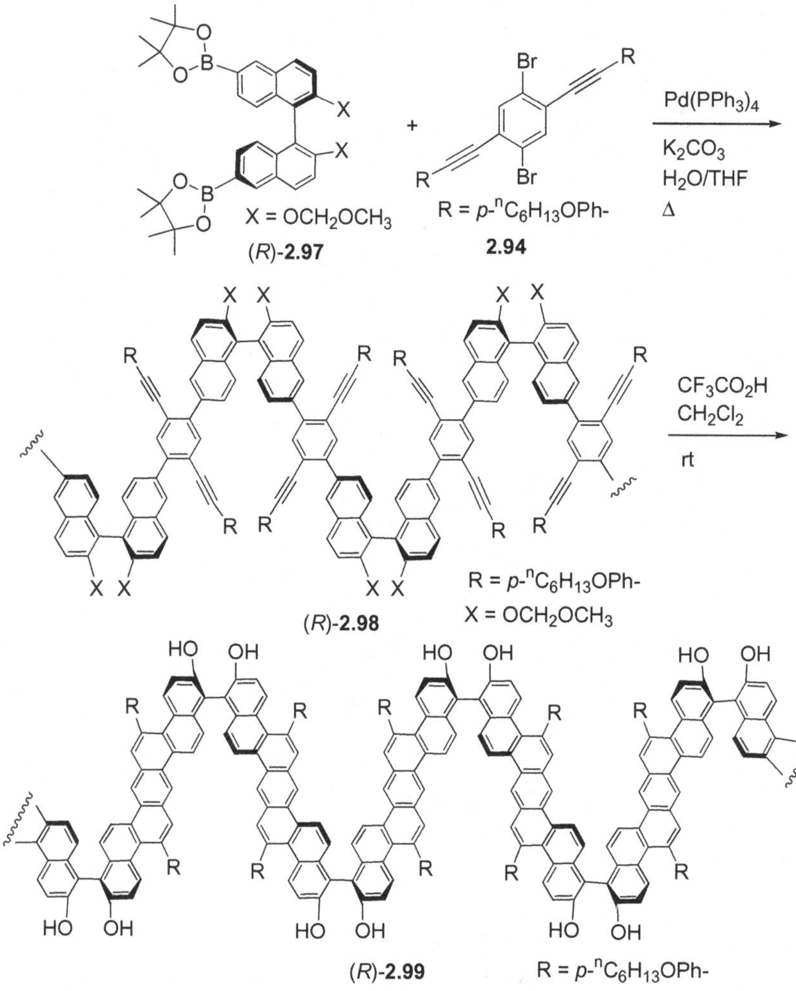

Scheme 2.53. Synthesis of helical ladder polybinaphthyl (R)-**2.99**.

reactive α-positions of the naphthalene rings to form the ladder polymer. The structure of polymer (R)-**2.99** is proposed with the reaction occuring at the α-position of the naphthalene rings. Both (R)-**2.98** and (R)-**2.99** are soluble in common organic solvents. GPC shows the molecular weight of (R)-**2.99** as $M_w = 17,400$ and $M_n = 10,200$ (PDI = 1.71). The dark color of the polymer solution gives a fluctuating optical rotation at around $[\alpha]_D \approx -708$ ($c = 0.1$, CH_2Cl_2), which is a much greater value than that

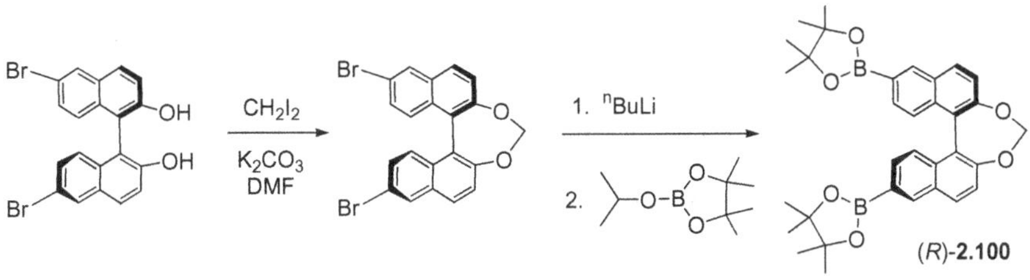

Scheme 2.54. Synthesis of monomer (R)-**2.100**.

of (R)-**2.98**. NMR analysis shows that both the removal of the CH$_2$OCH$_3$ groups and the cyclization take place in the presence of CF$_3$CO$_2$H. The UV spectrum of (R)-**2.99** shows a large red shift ($\Delta\lambda_{max}$ = 91 nm) for the longest wavelength absorption compared with that of (R)-**2.98**. The fluorescence spectrum of (R)-**2.99** also shows a large red shift ($\Delta\lambda_{max}$ = 42–68 nm) for the emission maximum compared with that of (R)-**2.98**. These data are consistent with the formation of the more conjugated fused polyaromatic repeating units in polymer (R)-**2.99**.

The methylene bridged binaphthyl diborate monomer (R)-**2.100** is synthesized from 6,6'-Br$_2$BINOL by protection with CH$_2$I$_2$ followed by borate synthesis (Scheme 2.54). Polymerization of monomer (R)-**2.100** with **2.94** is conducted to give polymer (R)-**2.101** as a light yellow solid in 90% yield (Scheme 2.55). GPC shows the molecular weight of this polymer as M_w = 38,800 and M_n = 12,100 (PDI = 3.22). Its specific optical rotation $[\alpha]_D$ is −466.7 (c = 0.1, CH$_2$Cl$_2$). From the reaction of polymer (R)-**2.101** with CF$_3$CO$_2$H, polymer (R)-**2.102** is isolated as a dark brown solid in 95% yield. GPC shows the molecular weight of (R)-**2.102** as M_w = 29,800 and M_n = 11,400 (PDI = 2.61). The ^{1}H NMR spectrum of (R)-**2.102** indicates that this polymer still contains the bridging methylene groups without hydrolysis. Large red shifts are observed for the absorptions ($\Delta\lambda_{max}$ = 38 nm) and emission signals ($\Delta\lambda$ = 47–76 nm) of (R)-**2.102** in comparison with those of (R)-**2.101**, which is similar to that observed in the conversion of (R)-**2.95** and (R)-**2.98** to (R)-**2.96** and (R)-**2.99**, respectively (Table 2.19).

Conversions of polymers (R)-**2.95**, (R)-**2.98** and (R)-**2.101** to polymers (R)-**2.96**, (R)-**2.99** and (R)-**2.102** also lead to dramatic changes in their CD signals (Table 2.19). As an example, the CD spectra of polymers (R)-**2.98** and (R)-**2.99** are compared in Fig. 2.5. New positive and negative CD signals centered at $ca.$ 340 nm are observed in the CD spectrum of

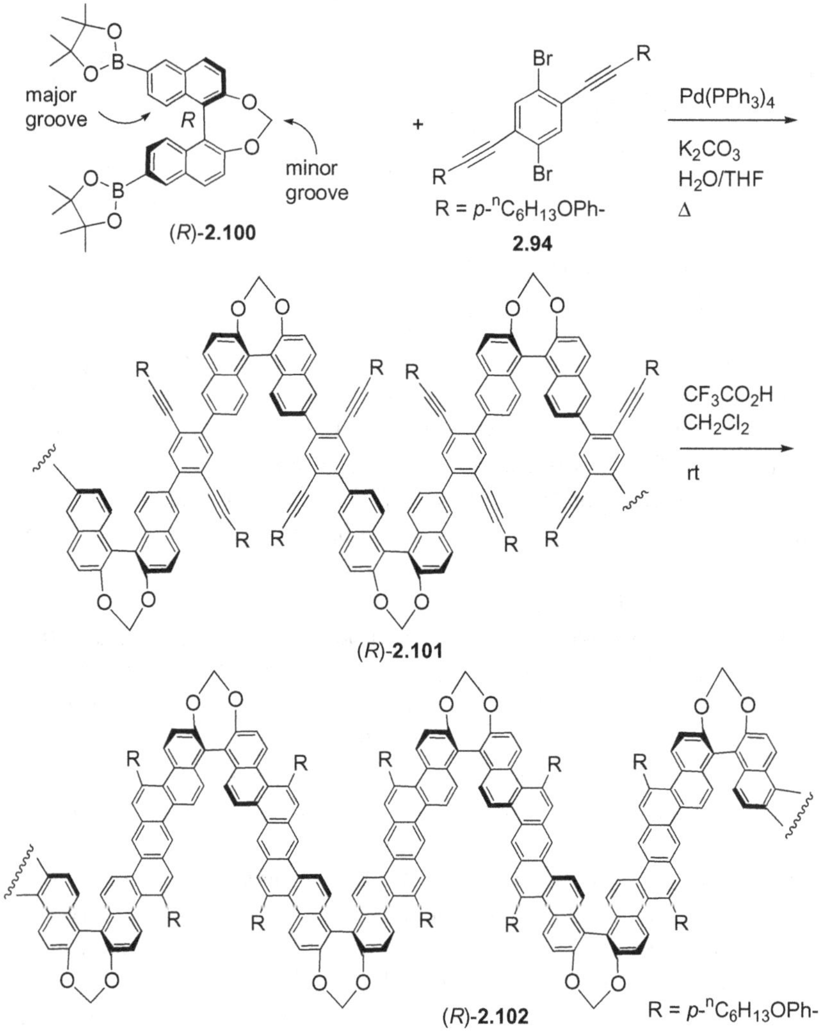

Scheme 2.55. Synthesis of helical ladder polybinaphthyl (*R*)-**2.102**.

(*R*)-**2.99**, which are probably due to the exciton coupling of the helical aromatic units.

Polymerization of (*R*)-**2.97** with (*R*)-**2.94** is quenched at 4, 6, 10 and 24 hours, respectively to generate polymer (*R*)-**2.98** of various molecular weights. Treatment of these polymers with CF_3CO_2H produces polymers (*R*)-**2.99a** ($M_n = 3300$, PDI = 1.27, DP ≈ 4.3), (*R*)-**2.99b** ($M_n = 4200$,

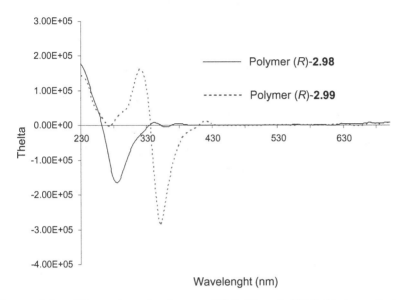

Figure 2.5. CD spectra of polymers (R)-**2.98** and (R)-**2.99** in methylene chloride.

PDI = 1.31, DP ≈ 5.5), (R)-**2.99c** (M_n = 5600, PDI = 1.38, DP ≈ 7.3) and (R)-**2.99d** (M_n = 10,200, PDI = 1.71, DP ≈ 13.4), respectively.

The UV spectra of polymers (R)-**2.99a–d** are compared in Fig. 2.6. As the molecular weight increases from (R)-**2.99a** to (R)-**2.99b** and (R)-**2.99c**, there is significant increase in the long wavelength absorptions of these polymers. However, when the molecular weight further increases (doubles) from (R)-**2.99c** to (R)-**2.99d**, the difference in their UV absorptions is much smaller. We attribute this to the end group effect, which is much greater at the low molecular weights but diminishes at higher molecular weights. The end groups of these materials have shorter conjugation length than their internal repeating units. The fluorescence spectra of these polymers show no difference in the shape and position of the emission signals (Fig. 2.7). This is probably because there should be efficient energy migration from the less conjugated end units to the more conjugated internal units.

Unlike the observed changes in the UV spectra from the lower molecular ones to the higher ones, the CD spectra of these polymers show continuous increase in the molar ellipticity ([θ] based on the repeating unit) at the exciton coupling around 340 nm as the molecular weight increases from (R)-**2.99a** to (R)-**2.99d** (Fig. 2.8). This demonstrates that there is a

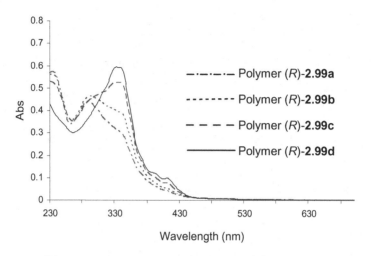

Figure 2.6. UV spectra of polymers (R)-**2.99a–d**.

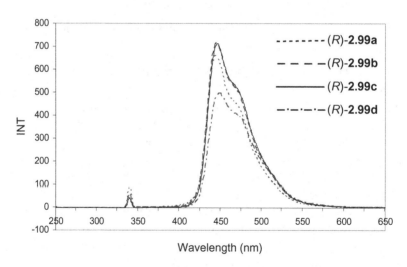

Figure 2.7. Fluorescence spectra of polymers (R)-**2.99a–d**.

cooperative effect between the repeating units in the polymer chain and the CD intensity contributed by each repeating unit increases as the number of the repeating unit increases. This supports a helical structure for the polymer chain. The helical chain amplifies the chiral optical properties of the individual chiral biaryl units.

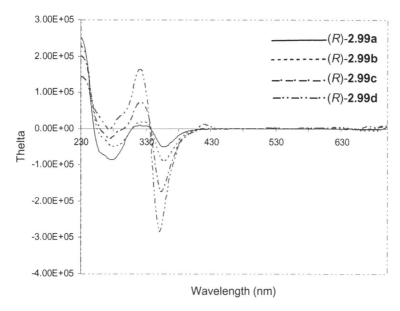

Figure 2.8. CD spectra of polymers (R)-**2.99a–d**.

References

1. Shirakawa H, Louis EJ, MacDiarmid AG, Chiang CK, Heeger J, *J Chem Soc Chem Commun* 578, 1977.
2. Skotheim TA (ed.), *Handbook of Conducting Polymers*, Vols. 1 and 2, Marcel Dekker, New York, 1986.
3. (a) Bredas JL, Silbey R, *Conjugated Polymers: The Novel Science and Technology of Highly Conducting and Nonlinear Optically Active Materials*, Kluwer Academic Publishers, Boston, 1991. (b) Kiess, HG (ed.), *Conjugated Conducting Polymers*, Springer-Verlag, New York, 1992.
4. Komori T, Nonaka T, *J Am Chem Soc* **106**:2656, 1984.
5. Moore JS, Gorman CB, Grubbs RH, *J Am Chem Soc* **113**:1704, 1991.
6. Yashima E, Huang S, Matsushima T, Okamoto Y, *Macromolecules* **28**:4184, 1995.
7. Lemaire M, Delabouglise D, Garreau R, Guy A, Roncali J, *J Chem Soc Chem Commun* 658, 1988.
8. Pickett CJ, Ryder KS, Moutet J-C, *J Chem Soc Chem Commun* 694, 1992.
9. Bross PA, Schöberl U, Daub J, *Adv Mater* **3**:198, 1991.
10. (a) Green MM, Peterson NC, Sato T, Teramoto A, Cook R, Lifson S, *Science* **268**:1860, 1995. (b) Green MM, Andreola C, Muñoz B, Reidy MP, Zero K, *J Am Chem Soc* **110**:4063–4065, 1988.
11. Hu Q-S, Vitharana D, Liu G, Jain V, Wagaman MW, Zhang L, Lee TR, Pu L, *Macromolecules* **29**:1082–1084, 1996.

12. Hu Q-S, Vitharana DR, Pu L, *Electrical, Optical, and Magnetic Properties of Organic Solid State Materials III,* Jen AK-Y, Lee CY-C, Dalton LR, Rubner MF, Wnek GE, Chiang LY (eds.), MRS, Pittsburgh, p. 621, 1996.
13. (a) Zhang JZ, Kreger MA, Hu Q-S, Vitharana D, Pu L, Brock PJ, Scott JC, *J Chem Phys* **106**:3710–3720, 1997; (b) Zhang JZ, Kreger MA, Goka EH, Pu L, Hu Q-S, Vitharana D, Rothberg LJ, *Springer Ser Chem Phys* **62**:284, 1996.
14. Hu Q-S, Vitharana D, Liu G, Jain V, Pu L, *Macromolecules* **29**:5075–5082, 1996.
15. Ma L, Hu Q-S, Musick K, Vitharana D, Wu C, Kwan CMS, Pu L, *Macromolecules* **29**:5083–5090, 1996.
16. Ma L, Hu Q-S, Pu L, *Tetrahedron: Asymmetry* **7**:3103–3106, 1996.
17. Cheng H, Ma L, Hu Q-S, Zheng X-F, Anderson J, Pu L, *Tetrahedron: Asymmetry* **7**:3083–3086, 1996.
18. Musick KY, Hu Q-S, Pu L, *Macromolecules* **31**:2933–2942, 1998.
19. Liu YQ, Yu G, Jen AKY, Hu QS, Pu L, *Macromol Chem Phys* **203**(1):37–40, 2002.
20. Zhang H-C, Pu L, *Tetrahedron* **59**:1703–1709, 2003.
21. (a) Hu Q-S, Zheng X-F, Pu L, *J Org Chem* **61**:5200–5201, 1996. (b) Hu Q-S, Vitharana D, Zheng X-F, Wu C, Kwan CMS, Pu L, *J Org Chem* **61**:8370–8377, 1996.
22. Wyatt SR, Pu L, *Polymer Preprints* **41**(1):190–191, 2000.
23. Wyatt SR, Hu Q-S, Yan X-L, Bare W, Pu L, *Macromolecules* **34**:7983–7988, 2001.
24. Hawker CJ, Fréchet JMJ, *J Am Chem Soc* **112**:7638, 1990.
25. (a) Peerlings HW, Meijer EW, *Eur J Org Chem* 573–577, 1998; (b) Rosini C, Superchi S, Peerlings HWI, Meijer EW, *Eur J Org Chem* 61–71, 2000.
26. Jen AK-Y, Liu Y, Hu Q-S, Pu L, *Appl Phys Lett* **75**(24):3745–3747, 1999.
27. Ma L, Hu Q-S, Vitharana D, Wu C, Kwan CMS, Pu L, *Macromolecules* **30**:204, 1997.
28. Ma L, Hu Q-S, Vitharana D, Pu L, *Polymer Preprints* **37**(2):462, 1996.
29. (a) Zyss J, Ledoux I, *Chem Rev* **94**:77, 1994; (b) Dhenaut C, Ledoux I, Samuel IDW, Zyss J, Bourgault M, Bozec HL, *Nature* **374**:339, 1995.
30. (a) Bergman RG, *Acc Chem Res* **6**:25, 1973; (b) Grissom JW, Calkins TL, McMillen HA, Jiang Y, *J Org Chem* **59**:5833, 1994.
31. Cheng H, Pu L, *Macromol Chem Phys* **200**:1274–1283, 1999.
32. Elshocht SV, Verbiest T, Kauranen M, Ma L, Cheng H, Musick KY, Pu L, Persoons A, *Chem Phys Lett* **309**:315–320, 1999.
33. Persoons A, Kauranen M, Van Elshocht S, Verbiest T, Ma L, Pu L, Langeveld-Voss BMW, Meijer EW, *Mol Cryst Liq Cryst* **315**:93–98, 1998.
34. Ma L, Pu L, *Macromol Chem Phys* **199**:2395–2401, 1998.
35. Zhang H-C, Huang W-S, Pu L, *J Org Chem* **66**:481–487, 2001.
36. Zhang H-C, Pu L, *Macromolecules* **37**:2695–2702, 2004.
37. Goldfinger MB, Crawford KB, Swager TM, *J Am Chem Soc* **119**:4578–4593, 1997.

Chapter 3

Polybinaphthyls in Asymmetric Catalysis

3.1. Introduction about Chiral Polymers in Asymmetric Catalysis

In the previous chapter, we have described the synthesis of various main chain chiral-conjugated polybinaphthyls for potential electrical and optical applications. In this chapter we will show the use of main chain chiral polybinaphthyls to develop a new class of polymeric chiral catalysts.

There are important advantages using polymer-supported catalysts in industrial processes.[1] Owing to the solubility difference between macromolecules and small molecules, polymeric reagents or catalysts can be easily separated from the reaction mixture by simple filtration or by precipitation with the addition of a non-solvent. The recovery of chiral catalysts is particularly important since it is often quite expensive to prepare these optically pure materials. Using polymeric catalysts makes it possible to carry out reactions in flow reactors or in flow membrane reactors for continuous production. Polymer-supported catalysts may also have durable catalytic activity than the monomeric metal complexes. The traditional polymeric chiral catalysts are synthesized by covalently bonding a chiral metal complex to an achiral and sterically irregular polymer backbone. As shown by the polystyrene-supported chiral rhodium catalyst **3.1**,[2a] each of the catalytic sites in this polymer has a very different microenvironment

3.1

because of the flexibility and steric irregularity of the polymer backbone. The polymer-supported chiral catalysts such as **3.1** were often found to be less efficient than their monomeric version. This indicates that the microenvironment of the catalytic sites in the polymers might have contributed to the different behavior between the polymer-supported catalysts and the monomeric ones. The randomly oriented catalytic sites in the polymer-supported catalysts as shown by **3.1** make it difficult to systematically modify their microenvironment. Nevertheless, some highly enantioselective polymeric chiral catalysts have been obtained in this way when the catalytic centers can overcome the unfavorable effect of the polymers.[1c] For example, a polyethyleneglycol-supported alkaloid ligand in combination with OsO_4 and K_3FeCN_6 has been prepared to catalyze the dihydroxylation of *trans*-stilbene with up to 99% ee (Scheme 3.1).[2b]

Synthetic main chain chiral polymers have also been used to conduct asymmetric catalysis.[3] For example, a series of chiral polypeptides have been prepared from the amine-initiated ring-opening polymerization of N-carboxyanhydrides (Scheme 3.2). These polypeptides are found to catalyze the asymmetric epoxidation of α,β-unsaturated ketones in the

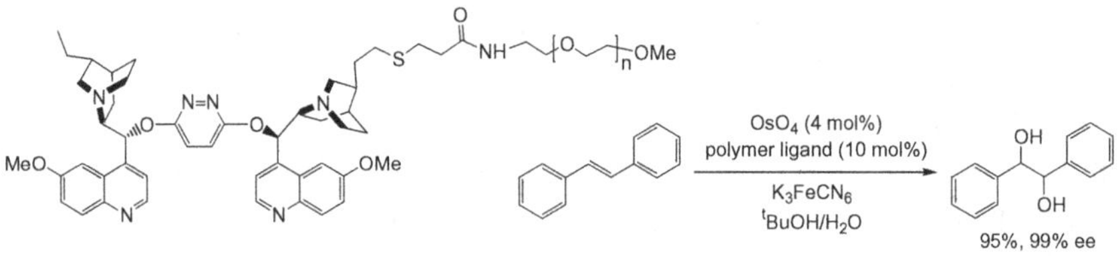

Scheme 3.1. Asymmetric dihydroxylation catalyzed by a chiral polymer.

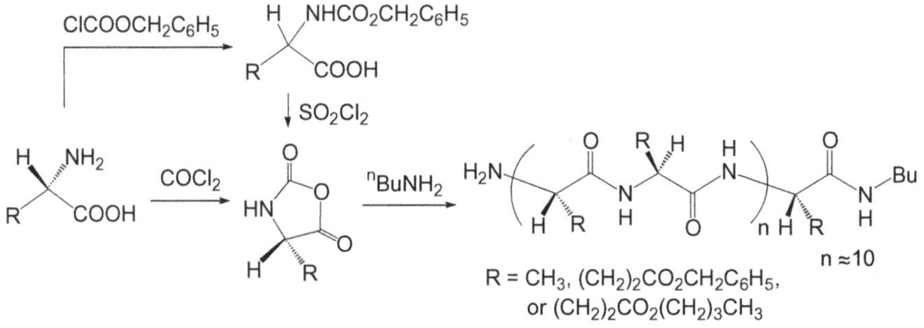

Scheme 3.2. Synthesis of chiral polypeptides.

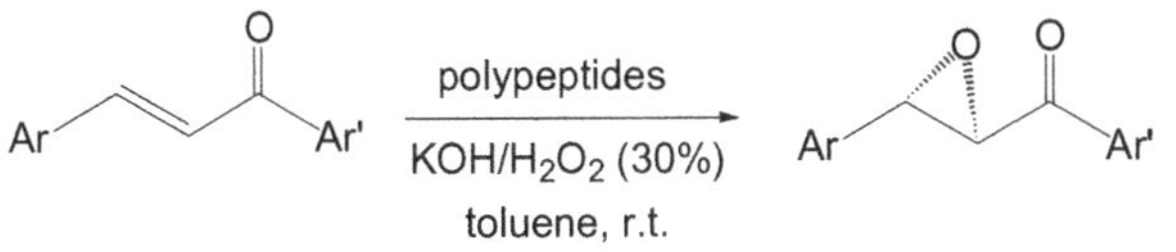

Scheme 3.3. Asymmetric epoxidation of α,β-unsaturated ketones catalyzed by chiral polypeptides.

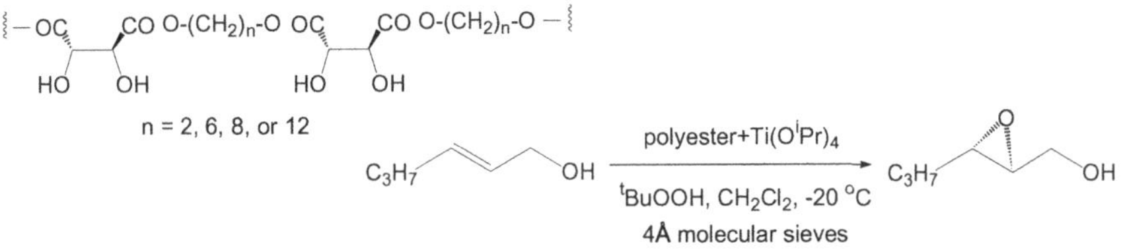

Scheme 3.4. Asymmetric epoxidation of an allylic alcohol catalyzed by chiral polyesters.

presence of H_2O_2 with up to 86% *ee* (Scheme 3.3).[4a] Scheme 3.4 shows another example in which the main chain chiral polyesters prepared from the condensation of diols with L-(+)-tartaric acids are used in combination with $Ti(O^iPr)_4$ and ^tBuOOH to catalyze the asymmetric epoxidation of an allylic alcohol with 79% *ee* at $-20°C$ (Scheme 3.4).[4b]

We consider that when the rigid and sterically regular main chain chiral polybinaphthyls are used to construct the polymeric chiral catalysts, they should be able to produce structurally better defined catalytic sites than many other chiral polymers. This makes it possible to systematically design and modify the microenvironment of the catalytic centers in the polybinaphthyls to achieve the optimized activity and stereoselectivity. Highly enantioselective polybinaphthyl catalysts for various organic reactions have been developed.

3.2. Synthesis of Major-Groove Poly(BINOL)s

In the helical structure of a 1,1'-binaphthyl molecule, there exists a major groove and a minor groove divided by the 1,1'-bond as indicated in the structure of (S)-BINOL shown below. We have conducted the polymerization of BINOL at either the 6,6'-positions or 3,3'-positions. Polymers obtained from the polymerization at the 6,6'-positions are named as the major-groove poly(BINOL)s and those obtained from the polymerization at the 3,3'-positions are named as the minor-groove poly(BINOL)s. These two types of polymers produce very different chiral cavities, which lead to very different catalytic properties.

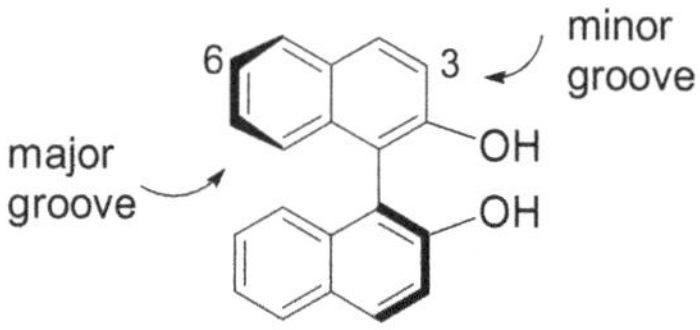

(S)-BINOL

In order to synthesize the major-groove poly(BINOL)s, we have attempted a direct polymerization of (R)-6,6'-Br$_2$BINOL, prepared from the bromination of (R)-BINOL in over 95% yield. However, in the presence of a catalytic amount of nickel(II) chloride and excess zinc (R)-6,6'-Br$_2$BINOL is mostly converted back to (R)-BINOL by debromination. Only

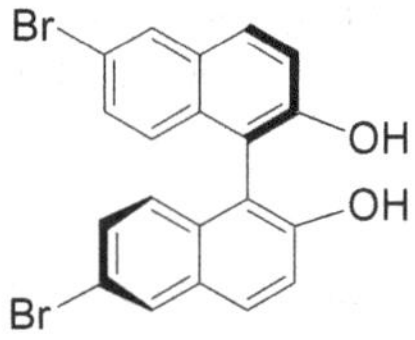

(R)-6,6'-Br$_2$BINOL

less than 5% of oligomers is produced. The hydroxyl groups of (R)-6,6'-Br$_2$BINOL are therefore protected with various protecting groups to give monomers that can be polymerized to make chiral polybinaphthyls. Removal of the protecting groups in the polymers gives the major-groove poly(BINOL)s that can be used to bind various Lewis acidic metals to catalyze diverse organic reactions.

3.2.1. Synthesis of Polybinaphthyls with Various Protecting Groups[5]

In Section 2.7 of Chapter 2, preparation of the alkyl-protected polybinaphthyl (R)-**3.2** [=(R)-**2.37**] is described. We have also prepared polybinaphthyls with protecting groups other than the alkyl in (R)-**3.2**. Scheme 3.5 shows the synthesis of monomer (R)-**3.3** that contains two methoxy methyl protecting groups. The specific optical rotation of (R)-**3.3** is $[\alpha]_D = +23.1$ ($c = 1.0$, THF). This monomer is polymerized in the presence of a catalytic amount of $NiCl_2$ and excess Zn. It is found that increasing the amount of $NiCl_2$ from 2 to 10 mol% leads to increased molecular weights for the resulting polymer (R)-**3.4**. However, further increasing the amount of $NiCl_2$ cannot further increase the molecular weight. Table 3.1 summarizes the molecular weights of the polymers measured by GPC when various amounts of nickel(II) chloride are used. In the presence of 10 mol% of nickel(II) chloride and excess zinc, polymer

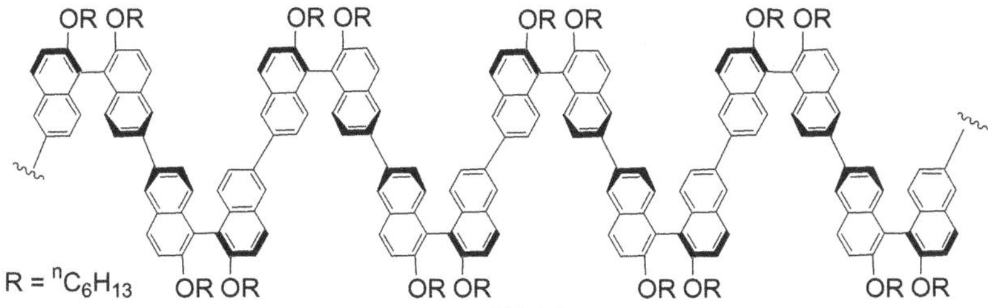

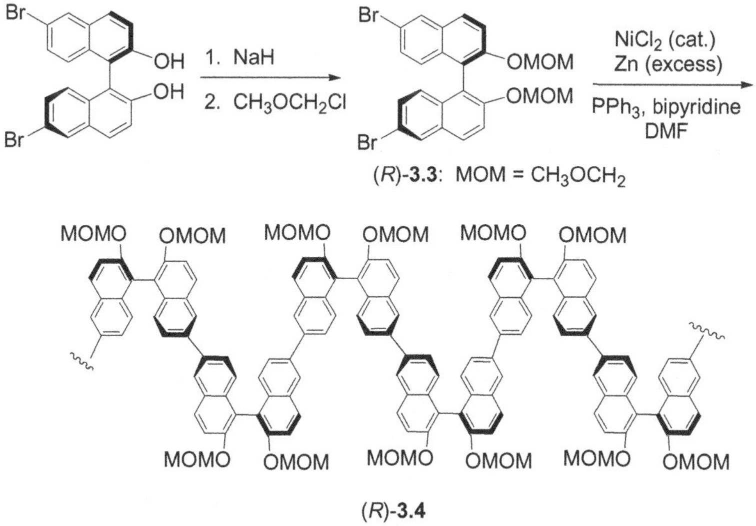

Scheme 3.5. Preparation of polybinaphthyl (R)-**3.4**.

Table 3.1. Molecular weights of the polymers prepared from the polymerization of (R)-**3.3** using various amounts of NiCl$_2$ as the catalyst.[a]

Amount of NiCl$_2$ (mol%)	M_w	M_n	PDI
2	2400	1900	1.2
5	4400	2700	1.6
10	6000	3800	1.6
50	5400	3500	1.5

[a]Determined by GPC relative polystyrene standards.

(R)-**3.4** is obtained with a molecular weight of $M_w = 6000$ and $M_n = 3800$ (PDI = 1.6). Polymer (R)-**3.4** is soluble in common organic solvents such as THF, methylene chloride and chloroform, and its specific optical rotation is $[\alpha]_D = -301.9$ ($c = 1.0$, THF). The molecular weight of (R)-**3.4** is significantly lower than that of (R)-**3.2** ($M_w = 15,000$ and $[\alpha]_D = -215$), but its optical rotation is much higher. This indicates that the methoxy methyl protecting groups have significant effect on the stereo and electronic structure of the polymer.

Monomer (R)-**3.5** that contains two acyl protecting groups is prepared from the reaction of (R)-6,6'-Br$_2$BINOL with acetic anhydride in the presence of Et$_3$N (Scheme 3.6). Polymerization of (R)-**3.5** with 10 mol%

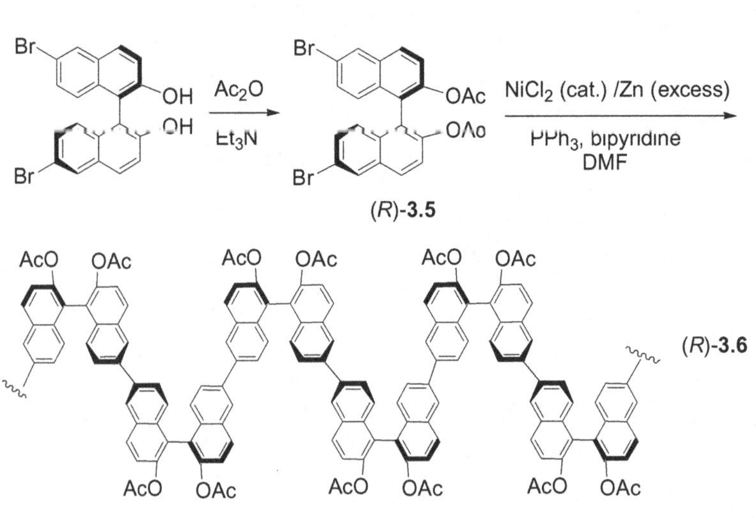

Scheme 3.6. Preparation of polybinaphthyl (R)-**3.6**.

of NiCl$_2$ and excess zinc leads to the formation of polymer (R)-**3.6** with $M_w = 6400$ and $M_n = 3600$ (PDI = 1.8) determined by GPC relative to polystyrene standards. This polymer is soluble in methylene chloride, chloroform and THF. The specific optical rotation of (R)-**3.6** is $[\alpha]_D = -353$ ($c = 0.5$, THF), which is greater than those of (R)-**3.4** and (R)-**3.2**. Polymerization of the racemic monomer leads to the formation of polymer rac-**3.6** with $M_w = 7400$ and $M_n = 3200$ (PDI = 2.3) as determined by GPC relative to polystyrene standards.

A laser light scattering study of polymer (R)-**3.6** and rac-**3.6** shows that the molecular weights of these polymers deviate only 10–20% from the GPC data. This demonstrates that in spite of the very different structures, the polystyrene standards used in the GPC analysis can provide good estimates for the molecular weights of these polybinaphthyls. This is similar to that observed for the polybinaphthyl rac-**2.38** described in Section 2.7.1 of Chapter 2, but different from that observed for the polydendrimer (R)-**2.44** described in Section 2.7.2 of Chapter 2.

3.2.2. Hydrolysis of the Polybinaphthyls to Generate Poly(BINOL)s[5]

Various conditions are applied in order to remove the protecting groups of polymers (R)-**3.2**, (R)-**3.4** and (R)-**3.6** to generate poly(BINOL)s. The alkyl groups of polymer (R)-**3.2** are removed by treatment with excess BBr$_3$ followed by reaction with water. The MOM groups of polymer (R)-**3.4** are removed by using hydrochloric acid and the acetate groups of polymer (R)-**3.6** are removed by using aqueous KOH. Because of the convenience of preparation and deprotection, the route from the acetate-protected polymer (R)-**3.6** to the poly(BINOL) is the method of choice.

Deprotection of polymer (R)-**3.6** is conducted by heating a mixture of the THF solution of this polymer and aqueous KOH at reflux, which gives the optically active poly(BINOL) (R)-**3.7** (Scheme 3.7). This polymer is insoluble in organic solvents but soluble in basic water solution and DMSO. It is purified by precipitating its aqueous KOH solution with HCl. Its specific optical rotation $[\alpha]_D$ is -139.8 ($c = 0.5, 0.5$ M aqueous KOH), which is much smaller than that of (R)-**3.6**. This polymer gives relatively well-defined NMR signals in basic D$_2$O solution, indicating a well-defined structure. Its UV spectrum in 0.5 M aqueous KOH solution displays $\lambda_{\max} = 274$ and 340 nm. This UV spectrum of (R)-**3.7** shows a significant red shift of maximum absorptions from the longest wavelength

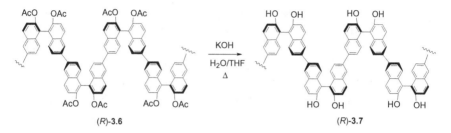

Scheme 3.7. Preparation of poly(BINOL) (R)-**3.7**.

absorption of (R)-**3.2** ($\lambda_{\max} = 316$ nm) due to the negative charges on the oxygen atoms of (R)-**3.7** in the strong basic solution.

3.2.3. Synthesis of Poly(BINOL)s Containing Alkyl-Substituted Phenylene Linkers[6]

The poly(BINOL) (R)-**3.7** described above is not soluble in common organic solvents such as chloroform, methylene chloride, benzene and THF. In order to increase the solubility of the major-groove poly(BINOL), linkers with flexible alkyl groups are incorporated. The Suzuki coupling of (R)-**3.3** with the aryl diboronic acid **3.8** in the presence of Pd(PPh$_3$)$_4$ and K$_2$CO$_3$ followed by hydrolysis with HCl (aq) gives the poly(BINOL) (R)-**3.9** (Scheme 3.8). Introduction of the hexyloxy groups in the linker makes this polymer soluble in common organic solvents such as THF, chloroform and methylene chloride. GPC analysis shows its molecular weight as $M_w = 18,500$ and $M_n = 9000$ (PDI = 2.1) relative to polystyrene standards. Its specific optical rotation $[\alpha]_D$ is -398.6 ($c = 1.00$, CH$_2$Cl$_2$). This polymer has a well-defined structure as shown by its very well-resolved ^{1}H NMR spectrum. The enantiomeric polymer (S)-**3.9** is made from monomer (S)-**3.3**.

A pyridine-containing monomer (R)-**3.10** is prepared from the reaction of (R)-6,6'-Br$_2$BINOL with 2-picolyl chloride hydrochloride salt in the presence of K$_2$CO$_3$ in refluxing acetone (Scheme 3.9). Polymerization of (R)-**3.10** with a long chain alkoxy-substituted aryl dibromide **3.11** gives polymer (R)-**3.12**. This polymer contains pyridine groups that can be also used to bind Lewis acidic metals. Polymer (R)-**3.12** is soluble in common organic solvents. GPC analysis gives its molecular weight as $M_w = 9900$ and $M_n = 8100$ (PDI = 1.2). Its specific optical rotation

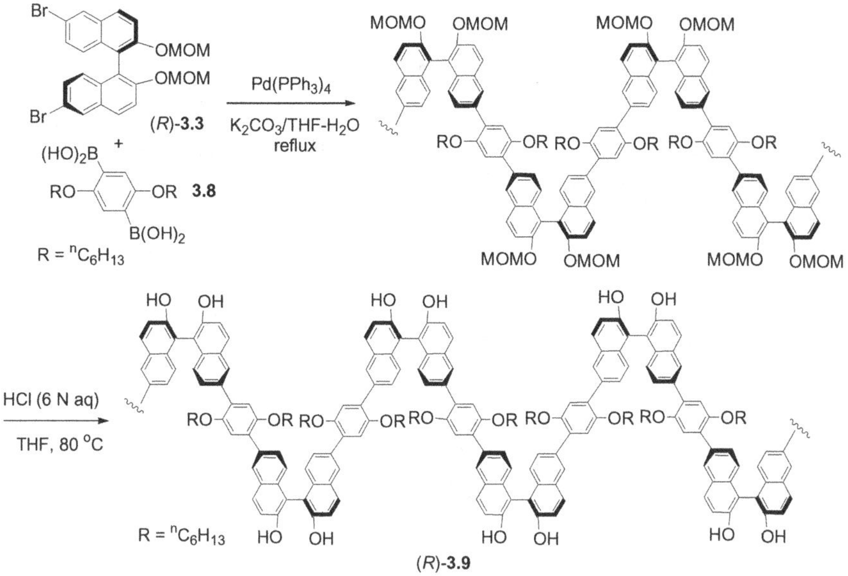

Scheme 3.8. Preparation of the major-groove poly(BINOL) (R)-**3.9**.

$[\alpha]_D$ is -73.1 ($c = 0.20$, CH_2Cl_2), which is much smaller than those of (R)-**3.9**. When the linker molecule **3.8** of a shorter alkyl substituent ($R = {^n}C_6H_{13}$) is used, the resulting polymer has very poor solubility. The polymer obtained from the direct polymerization of (R)-**3.10** in the presence of $NiCl_2$ and Zn is also mostly insoluble. Since the molecular weight of (R)-**3.9** is much greater than that of (R)-**3.12**, the picolyl groups in (R)-**3.12** might have generated a more rigid polymer structure than that of (R)-**3.9** and its precursor polymer, rendering its hexyl-substituted analog insoluble.

3.3. Application of the Major-Groove Poly(BINOL)s to Catalyze the Mukaiyama Aldol Reaction[5]

The Mukaiyama aldol reaction of benzaldehyde with 1-phenyl-1-(trimethylsilyloxy)ethylene (PTSE) in the presence of a Lewis acid catalyst can produce the β-silyloxy ketone product **3.13** (Scheme 3.10). The poly(BINOL)s in combination with a Lewis acid metal complex are used to catalyze this reaction. Treatment of (R)-**3.7** with Et_2AlCl and $AlMe_3$

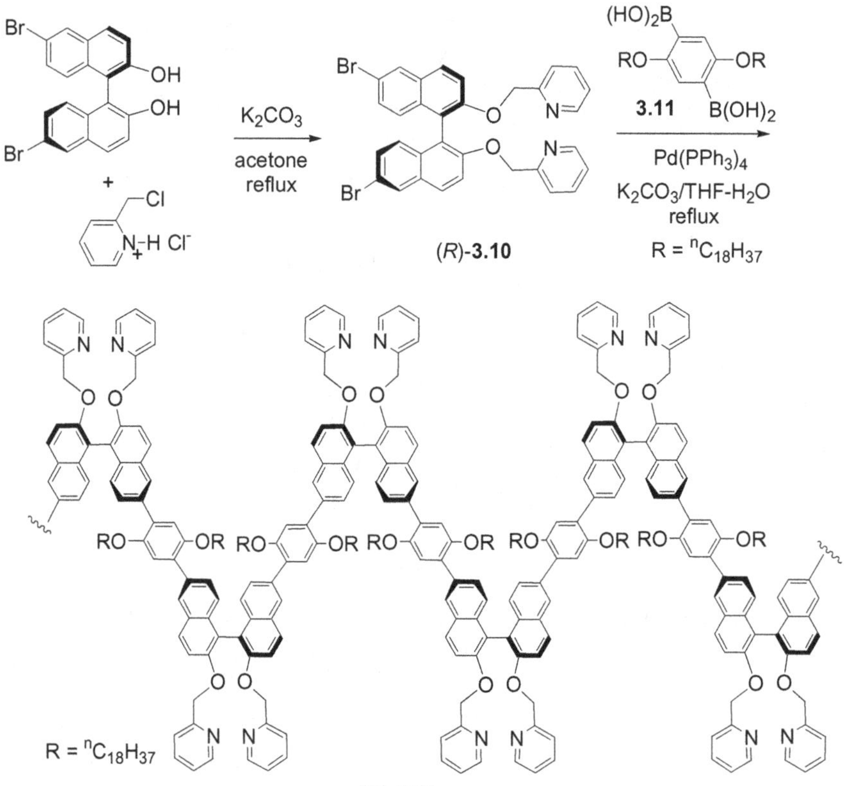

(*R*)-**3.12**

Scheme 3.9. Synthesis of pyridine-containing chiral polybinaphthyl (*R*)-**3.12**.

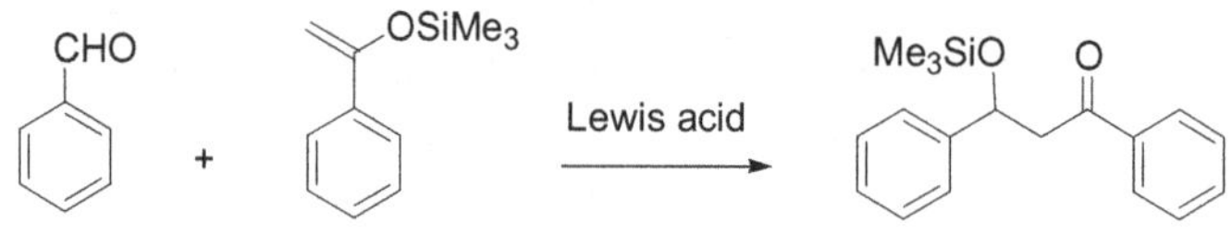

PTSE **3.13**

Scheme 3.10. The Mukaiyama aldol condensation of benzaldehyde with PTSE.

could generate the polymeric aluminum complexes (*R*)-**3.14** and (*R*)-**3.15**, respectively (Scheme 3.11). The aluminum complexes of (*R*)-**3.7** are insoluble in organic solvents and they are used as heterogeneous Lewis acid catalysts for the Mukaiyama aldol reaction shown in Scheme 3.10.

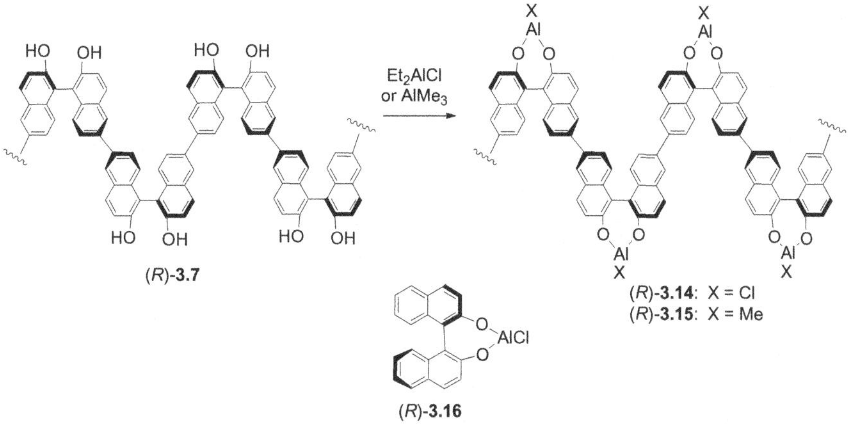

Scheme 3.11. Preparation of poly(BINOL)–Al(III) catalysts (*R*)-**3.14** and (*R*)-**3.15**.

After 3.5 h in methylene chloride at −78°C, (*R*)-**3.14** (16 mol%) gives 100% conversion for the reaction, but (*R*)-**3.15** (16 mol%) gives only 30% conversion. The catalytic activity of (*R*)-**3.14** is also much higher than that of (*R*)-**3.16**, the complex prepared from (*R*)-BINOL and Et$_2$AlCl. In the presence of (*R*)-**3.16** (16 mol%) at −78°C, there is only ∼5% conversion for the same reaction. Thus from BINOL to the poly(BINOL), there is a dramatic enhancement in catalytic activity.

The polymeric Ti(IV) complex (*R*)-**3.17** prepared from the reaction of the poly(BINOL) ligand (*R*)-**3.7** with Ti(O^{i}Pr)$_4$ in the presence of 4 Å molecular sieves is also used to catalyze the Mukaiyama aldol reaction of benzaldehyde with PTSE (Scheme 3.12). However, no reaction is observed even after the reaction mixture is heated under reflux in methylene chloride for one day. This is in contrast with the use of the BINOL complex (*R*)-**3.18** prepared from the reaction of (*R*)-BINOL with Ti(O^{i}Pr)$_4$, which can catalyze this reaction at room temperature.

As shown above, the poly(BINOL) ligand deactivates the titanium catalyst but greatly activates the aluminum complex. Previously, it has been shown that when the BINOL titanium complexes such as (*R*)-**3.18** are used as the Lewis acid catalysts, the catalytically active species are probably their dimeric complexes. The rigid polymeric nature of (*R*)-**3.17** isolates its Ti(IV) centers on the polymer chain. Thus, (*R*)-**3.17** cannot form the catalytically active dimeric BINOL–Ti(IV) structure and exhibits

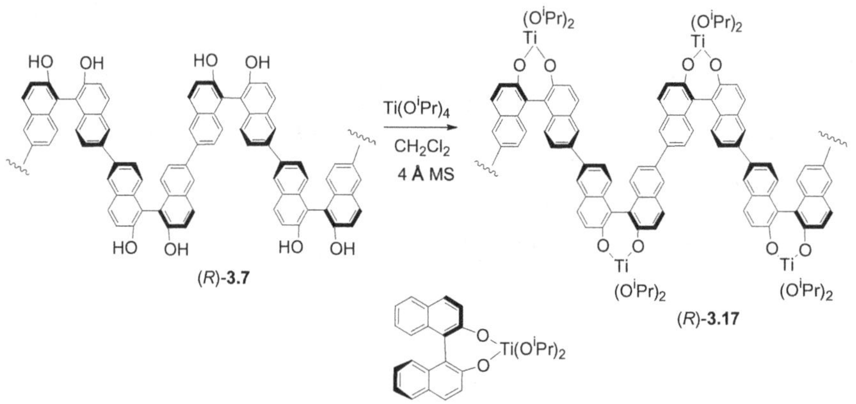

Scheme 3.12. Synthesis of a poly(BINOL) titanium complex.

no catalytic activity for this reaction. In the case of the BINOL–Al(III)-based catalysts such as (R)-**3.16**, the catalytically active species should be monomeric utilizing the empty p orbital of the Al(III) center. In solution, (R)-**3.16** can form oligomers through the intermolecular bridging Al–O–Al bonds. These Al–O–Al bridging bonds make the empty p orbital of the Al(III) centers unavailable for catalysis, leading to the low catalytic activity of (R)-**3.16**. In the rigid poly(BINOL)–Al(III) complex (R)-**3.14**, because all the catalytically active Al(III) centers are isolated from each other and cannot form the bridging Al–O–Al bonds, the empty p orbital of the Al(III) centers should be available for catalysis. This should lead to the greatly enhanced catalytic activity of the poly(BINOL) (R)-**3.14** over the BINOL complex (R)-**3.16**.

The soluble poly(BINOL) (R)-**3.9** is applied to the Mukaiyama aldol reaction of benzaldehyde with PTSE. In methylene chloride solution, when (R)-**3.9** is treated with Et$_2$AlCl, the resulting poly(BINOL)–Al(III) complex (R)-**3.19** is found to be insoluble and precipitates out of the solution. The insolubility is probably due to the partial formation of the interchain Al–O–Al bridging bonds from the originally soluble polymer. This heterogeneous Al(III) complex catalyzes the Mukaiyama aldol reaction with activity similar to that of the poly(BINOL) (R)-**3.14**. That is, most of the catalytic sites in this polymer are still isolated from each other, which exhibit much better activity over the BINOL complex (R)-**3.16**. No enantioselectivity is observed for the Mukaiyama aldol reaction in the presence of the BINOL and poly(BINOL)–Al(III) complexes.

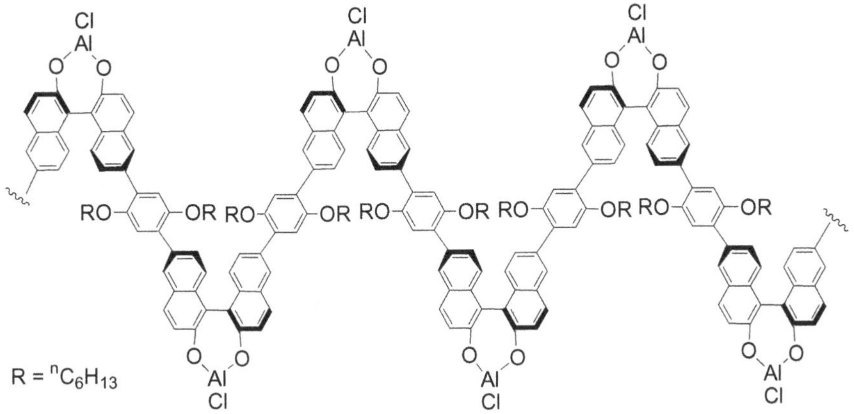

(R)-**3.19**, R = $^nC_6H_{13}$

3.4. Application of the Major-Groove Poly(BINOL)s to Catalyze the Hetero-Diels–Alder Reaction[7]

In collaboration with Jørgensen, we have studied the use of the poly(BINOL)s (R)-**3.7** and (S)-**3.9** combined with Me$_3$Al or Me$_2$AlCl to catalyze the hetero-Diels–Alder (HDA) reaction of 2,3-dimethyl-1,3-butadiene with ethyl glyoxylate (Scheme 3.13). Table 3.2 summarizes the reaction conditions explored. Besides the expected HDA reaction product, an ene reaction product is also observed. It is found that the soluble poly(BINOL) (S)-**3.9** shows better enantioselectivity than the insoluble poly(BINOL) (R)-**3.7**. The use of AlMe$_3$ is better than Me$_2$AlCl. The optimized conditions are given in entry 4 in which (S)-**3.9** (10 mol%)

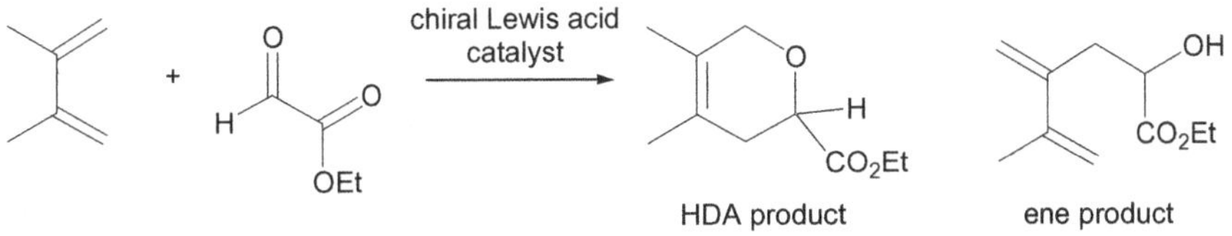

Scheme 3.13. The HAD reaction of 2,3-dimethyl-1,3-butadiene with ethyl glyoxylate.

Table 3.2. The HDA reaction in the presence of 10 mol% of the poly(BINOL)–Al(III) complexes.

Entry	Polymer	X (Me$_2$AlX)	Solvent	Temp/°C	Total yield/%	HDA product ee/%[a]	Ene product ee/%	Product ratio[b]
1	(R)-**3.7**	Me	CH$_2$Cl$_2$	rt	49	27 (S)	7	2:1
2	(R)-**3.7**	Me	CH$_2$Cl$_2$	−40	45	46 (S)	7	3:1
3	(S)-**3.9**	Me	CH$_2$Cl$_2$	−40	10[c]	84 (R)	nd[d]	nd[d]
4	(S)-**3.9**	Me	Et$_2$O	rt	69	89 (R)	46	5:1
5[e]	(S)-**3.9**	Me	Et$_2$O	rt	80	88 (R)	38	5:1
6	(S)-**3.9**	Me	Et$_2$O	−40 to −10	10	95 (R)	42	nd[d]
7	(S)-**3.9**	Cl	Et$_2$O	rt	10	42 (R)	4	2:1
8	(S)-**3.9**	Cl	Et$_2$O	−78 to rt	<10	76 (R)	18	3:1
9	(S)-**3.9**[f]	Me	Et$_2$O	rt	63	85 (R)	39	5:1

[a] Absolute configuration in brackets determined by comparison of optical rotation to reported values.
[b] Ratio of HDA/ene product.
[c] Isolated yield of HDA adduct.
[d] Not determined.
[e] 20 mol% catalyst.
[f] Recovered polymer (second run).

in combination with AlMe$_3$ catalyzes the reaction in diethyl ether at room temperature to give the HDA product with 89% ee and an overall yield of 69%. Increasing the amount of (S)-**3.9** to 20 mol% increases the yield but does not improve the selectivity (entry 5). Although lower temperature leads to higher ee, the yield is diminished (entry 6). In the reactions catalyzed by (S)-**3.9**, the chiral polymer ligand can be easily recovered with the addition of methanol at the end of the reaction. After washing the precipitate with 1 N HCl and methanol, the polymer is dissolved in toluene and filtered. Removal of the solvent under vacuum gives the recovered poly(BINOL) (S)-**3.9**. Entry 9 demonstrates that the recovered polymer exhibits similar catalytic properties as the original polymer.

Previously, BINOL in combination with AlMe$_3$ was used to catalyze the same HDA reaction shown in Scheme 3.13. At −78°C, the BINOL–Al(III) complex leads to the formation of the HDA product with 97% ee and the HDA/ene product ratio of 7:1. Thus, the catalytic properties of the poly(BINOL) (S)-**3.9** are similar to those of BINOL, and the polymer provides the added advantage of easy recovery and reuse.

3.5. Using the Ti(IV) Complex of the Major-Groove Poly(BINOL) to Catalyze the Diethylzinc Addition to Aldehydes[8]

The soluble poly(BINOL) (S)-**3.9** in combination with Ti(O^{i}Pr)$_4$ is used to catalyze the reaction of ZnEt$_2$ with benzaldehyde and 1-naphthaldehyde (Scheme 3.14). The results are summarized in Table 3.3. This polymer shows high enantioselectivity for the ZnEt$_2$ additions in toluene at 0°C. The reactions use 20 mol% of (S)-**3.9**, 1.4 equivalent of Ti(O^{i}Pr)$_4$ and three equivalent of ZnEt$_2$. The catalytic properties of (S)-**3.9** are similar to those of BINOL in the ZnEt$_2$ addition to aldehydes in the presence of Ti(O^{i}Pr)$_4$. This indicates that when BINOL is used, the catalytically active species

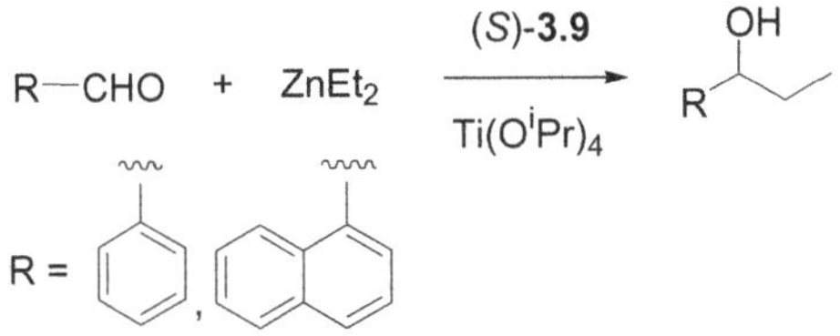

Scheme 3.14. Asymmetric diethylzinc addition to aldehydes.

Table 3.3. The asymmetric diethylzinc addition to aldehydes catalyzed by (S)-**3.9** and Ti(O^{i}Pr)$_4$.

Aldehyde	Solvent	Conversion[a] (%)	ee (%)	Configuration[b]
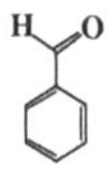	CH$_2$Cl$_2$	84	86[c]	S
benzaldehyde	Toluene	100	86[c]	S
	THF	19	82[c]	S
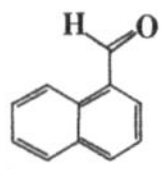 1-naphthaldehyde	Toluene	94	92[d]	S

[a] Determined by GC.
[b] Determined by comparing the GC and HPLC data with the known compounds.
[c] Determined by chiral GC (β-Dex capillary column).
[d] Determined by HPLC-Chiracel OD column.

for the ZnEt$_2$ addition to aldehydes may be the monomeric BINOL–Ti(IV) complex rather than its dimer or oligomers.

Both the insoluble poly(BINOL) (R)-**3.7** and the soluble poly(BINOL) (S)-**3.9** are used to catalyze the ZnEt$_2$ addition to benzaldehyde without Ti(O^{i}Pr)$_4$, but low enantioselectivities are observed (see Table 3.4).

3.6. Synthesis of the Minor-Groove Poly(BINOL)s[6]

Besides the major-groove poly(BINOL)s (R)-**3.7** and (S)-**3.9**, we have also synthesized the minor-groove poly(BINOL)s. Scheme 3.15 shows the synthesis of the 3,3'-diiodosubstituted binaphthyl monomer (R)-**3.20** from the MOM protection of (R)-BINOL followed by *ortho*-metalation and iodonation. The optical purity of this molecule is over 99.9% as determined by HPLC analysis on a Chiracel OD column. The specific optical rotation $[\alpha]_D$ of (R)-**3.20** is -11.05 ($c = 1.01$, THF). The specific optical rotation $[\alpha]_D$ of the MOM-protected (R)-BINOL is $+97.75$ ($c = 1.00$, THF). The 3,3'-iodonation significantly reduces the optical rotation and changes its sign.

The Suzuki polymerization of (R)-**3.20** with the alkylated aryl diboronic acid **3.8** is conducted (Scheme 3.16). Polymers of two different molecular weights are obtained by using two different catalyst systems. In the presence of Pd(OAc)$_2$, tris-*o*-tolylphosphine and Ba(OH)$_2 \cdot$ 8H$_2$O, the polymerization is carried out in DMF and H$_2$O at reflux under nitrogen over 42 h to give (R)-**3.21a** in 93% yield. GPC shows its molecular weight as $M_w = 5900$ and $M_n = 3900$ (PDI = 1.5). Its specific optical rotation $[\alpha]_D$ is -63.4 ($c = 0.50$, THF). In the presence of Pd(PPh$_3$)$_4$ and aqueous K$_2$CO$_3$, the polymerization is conducted in THF at reflux over 36 h to give (R)-**3.21b** in 97% yield. GPC shows its molecular weight as $M_w = 20{,}200$ and $M_n = 7300$ (PDI = 2.8). Its specific optical rotation $[\alpha]_D$ is -90.1 ($c = 0.51$, THF). Both polymers (R)-**3.21a** and (R)-**3.21b** are yellow solids.

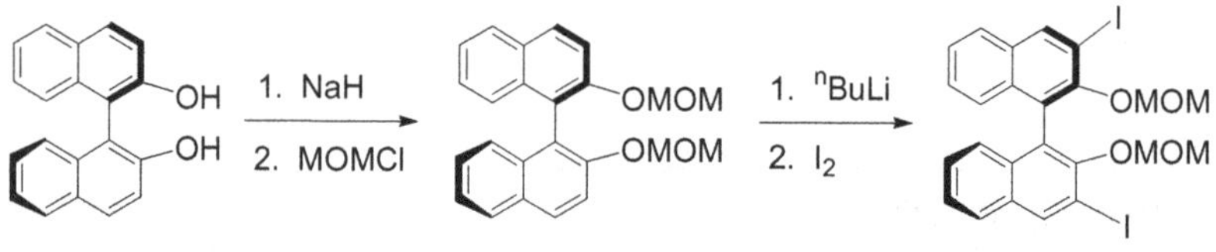

Scheme 3.15. Synthesis of minor-groove binaphthyl monomer (R)-**3.20**.

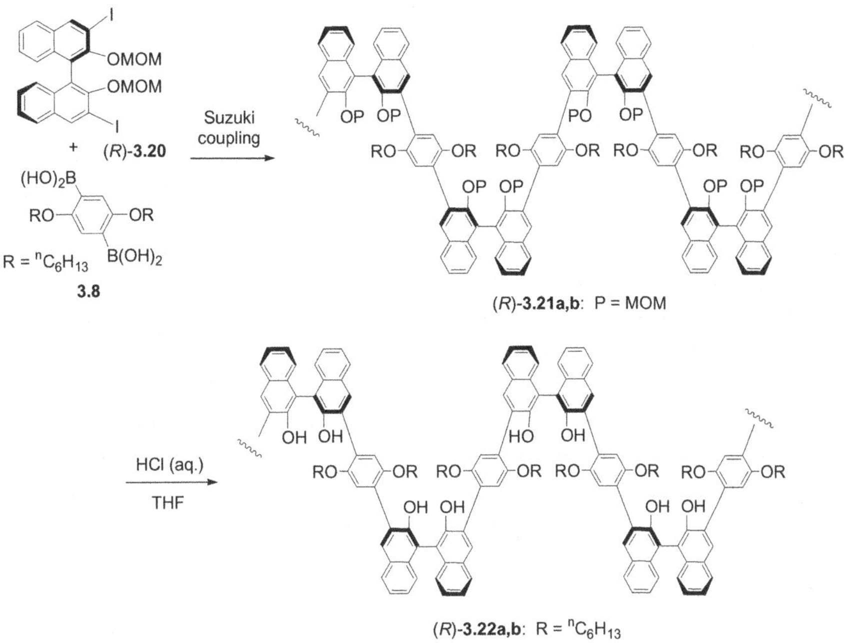

Scheme 3.16. Synthesis of minor-groove poly(BINOL) (R)-**3.22a,b**.

The MOM-protecting groups of polymers (R)-**3.21a,b** are hydrolyzed by heating their THF solutions at reflux in the presence of aqueous HCl, which generates the minor-groove poly(BINOL)s (R)-**3.22a,b** respectively. Both polymers are soluble in common organic solvents such as THF, chloroform, methylene chloride and toluene. GPC shows the molecular weight of (R)-**3.22a** as $M_w = 6700$ and $M_n = 4600$ (PDI = 1.5) and that of (R)-**3.22b** as $M_w = 24{,}300$ and $M_n = 9900$ (PDI = 2.45). The specific optical rotation $[\alpha]_D$ of (R)-**3.22a** is $+11.8$ ($c = 0.50$, THF) and that of (R)-**3.22b** is -16.6 ($c = 0.5$, THF). These polymers have well-defined structures as demonstrated by their very well-resolved ^{1}H NMR signals. Although both (R)-**3.22a** and (R)-**3.22b** contain the same (R)-BINOL units, their optical rotations have the opposite signs. The circular dichroism (CD) spectra of these two polymers are almost identical except their intensity differences. Thus, the optical rotation at 569 nm is not a good representation for the chirality of this polymer. The end groups probably strongly influence the optical rotations of these polymers. The higher molecular weight (R)-**3.22b** has less end group than the lower molecular weight (R)-**3.22a**, which could lead to their very different optical rotations. The optical rotation of the minor-groove polymers have much smaller value than the major-groove polymer (R)-**3.9**. Thus, the substitution at the 3,3′-positions of the BINOL units leads to very different steric and electronic environments in comparison with the substitution at the 6,6′-positions.

There are two distinctive structural differences between the major-groove poly(BINOL) (R)-**3.9** and the minor-groove poly(BINOL) (R)-**3.22**. In the minor-groove poly(BINOL) (R)-**3.22**, two aryl groups are introduced to the 3,3′-positions of the BINOL units, which should increase the steric interaction at the catalytic site when this polymer is used in asymmetric catalysis. In addition, the alkoxy groups on the phenylene linker of (R)-**3.22** not only provide its solubility but can also participate in the coordination when this polymer is treated with a Lewis acid complex. Thus, the Lewis acid complexes of the minor-groove poly(BINOL) (R)-**3.22** should be both sterically and electronically very different from those of the major-groove poly(BINOL) (R)-**3.9**. This should lead to different catalytic properties.

The UV spectrum of the minor-grove poly(BINOL) (R)-**3.22** shows λ_{max} at 244, 260, 322 nm. The longest wavelength absorption of (R)-**3.22** is 14 nm blue shifted from that of the major-groove poly(BINOL) (R)-**3.9**. This indicates that the increased steric interaction in the repeating unit of (R)-**3.22** has disrupted its conjugation. Although both (R)-**3.9** and (R)-**3.22** are derived from (R)-BINOL and contain the same functional groups, their CD signals are inverted. This demonstrates that substitution at the 3,3′-positions and 6,6′-positions might have generated the polymers with the opposite conformation, with the minor-groove one probably favoring a transoid conformation due to steric congestion, and the major-groove one probably favoring a cisoid conformation due to the hydrogen bond within a BINOL unit.

By using the same experimental procedure as the preparation of (R)-**3.22b**, the enantiomeric poly(BINOL) (S)-**3.22** is obtained from the polymerization of (S)-**3.20** with **3.8** followed by hydrolysis. GPC shows its molecular weight as $M_w = 10,000$ and $M_n = 4600$ (PDI = 2.17). Its specific optical rotation $[\alpha]_D$ is -14.2 ($c = 1.01$, CH_2Cl_2). In THF, the optical rotation of (S)-**3.22** is almost zero $\{[\alpha]_D = +0.5$ ($c = 0.51$, THF)$\}$. However, the CD spectrum of this polymer exhibits very strong Cotton effects that are almost exact mirror images of those of (R)-**3.22b**. The near zero optical rotation of (S)-**3.22** is probably due to a significant end group effect. Polymer (S)-**3.22** has a higher molecular weight than (R)-**3.22a**

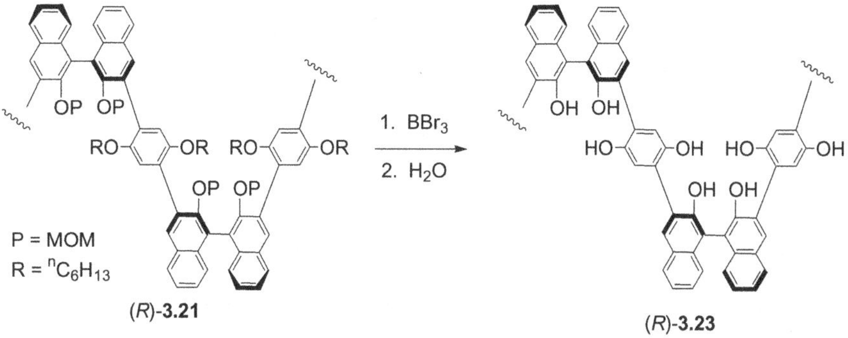

Scheme 3.17. Removal of both the alkyl and MOM groups of (R)-**3.21** to generate the poly(BINOL) (R)-**3.23**.

and lower than (R)-**3.22b**. Thus, the end group effect of (S)-**3.22** should be greater than that of (R)-**3.22a** and smaller than that of (R)-**3.22b**. This is consistent with that the optical rotation value of (S)-**3.22** is in between those of (R)-**3.22a** (+11.8) and (R)-**3.22b** (−16.6).

When polymer (R)-**3.21** is treated with excess BBr$_3$, both the MOM and hexyl groups can be hydrolyzed to generate polymer (R)-**3.23** that contains tetrahydroxyl groups around each BINOL unit (Scheme 3.17). This polymer is insoluble in both organic solvents and basic aqueous solution.

3.7. Application of the Major- and Minor-Groove Poly(BINOL)s to Catalyze the Asymmetric Organozinc Addition to Aldehydes[6,7]

3.7.1. *Asymmetric Diethylzinc Addition to Aldehydes Catalyzed by the Poly(BINOL)s*

Both the major-groove and minor-groove poly(BINOL)s are used to catalyze the asymmetric reaction of benzaldehyde with diethylzinc without the use of Ti(O^{i}Pr)$_4$ (Scheme 3.18). The results are summarized in Table 3.4. All of the reactions are carried out under nitrogen in the presence of ∼5 mol% (based on the BINOL unit) of the polymer and 1–2 equivalents of diethylzinc. Formations of a chiral alcohol product 1-phenyl propanol (PPO) and a side product benzyl alcohol (BnOH) are observed in many cases. As shown in Table 3.4, the major-groove polybinaphthyls (R)-**3.2**, (R)-**3.6** and (R)-**3.12** do not contain free hydroxyl groups and are found to have no or very low catalytic activity for the ZnEt$_2$ addition.

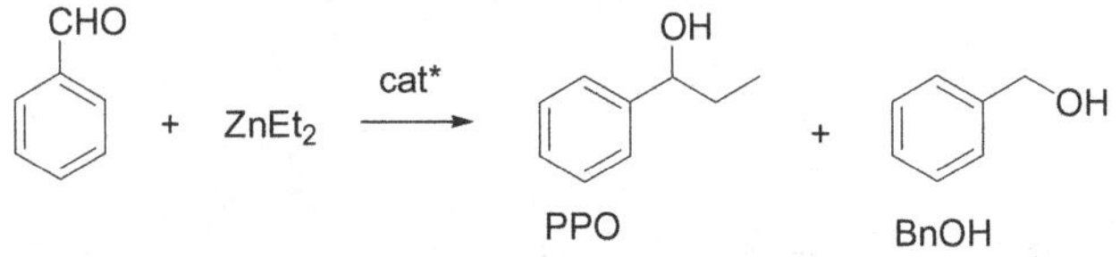

Scheme 3.18. Reaction of benzaldehyde with ZnEt$_2$.

Table 3.4. Reaction of benzaldehyde with diethylzinc in the presence of chiral polybinaphthyls.

Polymer	Solvent	Temp	Time (h)	Conv (%)	PPO:BnOH	ee (%)[a]
(R)-**3.6**	Toluene	rt	24	Trace		
(R)-**3.2**	Toluene	rt	48	54	19:35	8
(R)-**3.7**	CH$_2$Cl$_2$	rt	112	100	53:47	13
(R)-**3.9**	CH$_2$Cl$_2$	rt	20	33	71:29	40
(R)-**3.12**	Toluene	rt	48	No reaction		
(R)-**3.21b**	Toluene	rt	24	85	97.6:2.3	19
(R)-**3.22a**	CH$_2$Cl$_2$	rt	70	100	94:6	67.0
(R)-**3.22a**	Toluene	rt	10	100	99.6:0.4	89.4
(R)-**3.22a**	THF	0°C	10	79	100:0	84.7
(R)-**3.22a**	Hexane: toluene (2:1)	0°C	16	100	100:0	91.7
(R)-**3.22a**	Toluene	0°C	12	100	100:0	92.2

[a] All the ee's were determined by GC with a chiral column (β-Dex capillary column, Supelco Company).

The insoluble major-groove poly(BINOL) (R)-**3.7** is catalytically active but with low chemical selectivity and enantioselectivity. The soluble major-groove poly(BINOL) (R)-**3.9** shows higher chemical selectivity and enantioselectivity. The minor-groove polybinaphthyl (R)-**3.21b**, though containing no free hydroxyl groups, is a much more active as well as more chemical selective than all the major-groove polymers. The minor-groove poly(BINOL) (R)-**3.22a** is found to be an excellent catalyst for the ZnEt$_2$ addition to benzaldehyde. In toluene solution at 0°C, in the presence of 5 mol% of (R)-**3.22a**, the reaction of ZnEt$_2$ with benzaldehyde proceeds with 100% chemical selectivity and 92.2% ee. No BnOH side product is observed. Further lowering of the temperature to −40°C shows no change in ee. These experiments demonstrate that from the major-groove poly(BINOL)s to the minor-groove poly(BINOL) there is a dramatic increase in the catalytic activity, chemical selectivity and enantioselectivity. This is attributed to the differences in both steric and electronic environments of the catalytic sites in these two types of poly(BINOL)s.

The minor-groove poly(BINOL) (R)-**3.22a** is used to catalyze the asymmetric reaction of various aldehydes with diethylzinc and the results are listed in Table 3.5. The ee's are determined relative to the racemic mixture of the alcohol products. These reactions are carried out in toluene solution at 0°C in the presence of 5 mol% (based on the BINOL units) of (R)-**3.22a**. All of the alcohol products except the one from the reaction of 3,3,7-trimethyl-6-octenal have the R configuration as determined by comparing their specific optical rotations with the reported data. The reactions are completed in 12–26 h for aromatic aldehydes and trans-cinnamaldehyde. The addition to aliphatic aldehydes is much slower and takes 70–100 h to complete. As shown in the table, excellent enantioselectivity for the para-substituted benzaldehydes as well as for trans-cinnamaldehyde is achieved. But much lower ee's are observed for the reaction of the ortho-substituted aromatic aldehydes. The ee's observed for the addition to the aliphatic aldehydes are also very significant since few polymeric catalysts are good for the reaction of aliphatic aldehydes.

In these reactions, (R)-**3.22a** can be easily recovered from the reaction mixture by precipitation with methanol. The recycled polymer shows a similar enantioselectivity. When the higher molecular weight poly(BINOL) (R)-**3.22b** is used, the ee for the reaction of benzaldehyde is 92.7% and for the reaction of p-chlorobenzaldehyde it is 93.8%. Therefore, both the molecular weight of the polymer and the method of polymer preparation have little effect on the asymmetric process. This is an excellent property for practical application.

The enantiomeric poly(BINOL) (S)-**3.22** is also used to catalyze the reaction of benzaldehyde with diethylzinc. In the presence of (S)-**3.22**, PPO is obtained with 93.0% ee, similar to that obtained from (R)-**3.22**. As expected, the configuration of the resulting chiral alcohol is opposite to that obtained with (R)-**3.22**. Thus, the R poly(BINOL) produces the R chiral alcohol, and the S poly(BINOL) generates the S chiral alcohol.

Another poly(BINOL) (R)-**3.24** containing further increased steric hindrance at the BINOL unit is prepared from the polymerization of (R)-**3.20** with 2,5-diisopropoxy-1,4-dibromobenzene.[9] In this polymer, the more bulky isopropyl groups are used in place of the hexyl groups in (R)-**3.22**. GPC shows the molecular weight of the poly(BINOL) (R)-**3.24** as M_w = 15,100 and M_n = 8600 (PDI = 1.8). The specific optical rotation $[\alpha]_D$ is −25.3 (c = 0.78, CH_2Cl_2). In toluene at 0°C, (R)-**3.24** catalyzes the diethylzinc addition to benzaldehyde with 89% ee, and to

Table 3.5. The asymmetric reaction of aldehydes with diethylzinc in the presence of the minor-groove poly(BINOL)s (R)-**3.22a,b**.

Polymer	Aldehyde	Isolated yield (%)	ee (%)[a]
(R)-**3.22a**	PhCHO	89	92.2
		91	92.0[b]
	4-H$_3$C-C$_6$H$_4$-CHO	90	92.5[b]
	4-Cl-C$_6$H$_4$-CHO	94	93.4[b]
	4-H$_3$CO-C$_6$H$_4$-CHO	84	88.3[b,c]
	4-Bu^t-C$_6$H$_4$-CHO	63	74.3[b]
	2-F-C$_6$H$_4$-CHO	86	35.1[b]
	2-OCH$_3$-C$_6$H$_4$-CHO	90	59.0[b]
	PhCH=CH-CHO	86	89.7[b,d]
	(CH$_3$)$_2$C=CHCH$_2$CH$_2$CH(CH$_3$)CH$_2$CHO	67	82.8[b,d,e]
	CH$_3$(CH$_2$)$_6$CHO	89	73.5[d]
	C$_6$H$_{11}$-CHO	70	83.3[b,d]
	CH$_3$(CH$_2$)$_3$CH(CH$_3$)CHO	65	74.3[b,d]
(R)-**3.22b**	PhCHO	90	92.7
	4-Cl-C$_6$H$_4$-CHO	95	93.8
(S)-**3.22**	PhCHO	94	93.0

[a] All the ee's were determined by GC with a chiral column (β-Dex capillary column, Supelco Company) except those specifically indicated.
[b] The recycled polymer was used.
[c] The ee was measured by HPLC-chiracel OD column.
[d] The ee was measured by GC analysis (β-Dex capillary column) of the corresponding acetate derivative.
[e] The absolute configuration of the product was not determined. $[\alpha]_D = -13.25$ ($c = 1.95$, THF).

cyclohexanecarboxaldehyde with 79% ee. These ee's are slightly lower than those obtained by using poly(BINOL) (R)-**3.22**.

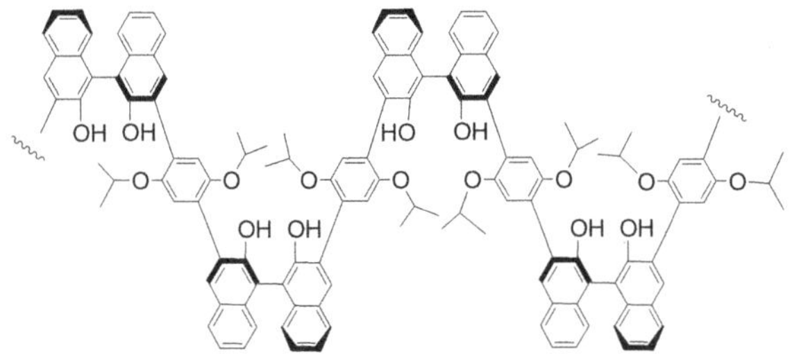

(R)-**3.24**

3.7.2. Study of the Reactions of the Minor-Groove Poly(BINOL) and a Few Monomeric BINOL Derivatives with Diethylzinc

In order to understand the asymmetric reaction catalyzed by the minor-groove poly(BINOL) (R)-**3.22**, the reaction of (R)-**3.22** and a few monomeric BINOL derivatives with diethylzinc are studied. When a toluene solution of (R)-**3.22** is treated with one equivalent of ZnEt$_2$ at room temperature, it produces a viscous solution. In Scheme 3.19, a possible

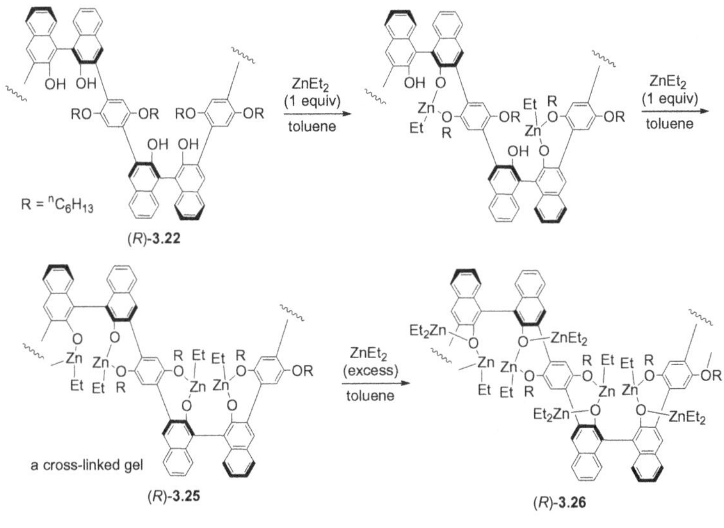

Scheme 3.19. Reaction of poly(BINOL) (R)-**3.22** with diethylzinc.

structure of the resulting polymeric zinc complex is shown in which only one of the ethyl groups of ZnEt$_2$ is protonated off. Other study has shown that when one equivalent of ZnEt$_2$ is treated with a BINOL derivative at room temperature, only one of the ethyl groups can be protonated off. Addition of a second equivalent of ZnEt$_2$ converts the solution to an unmovable gel. A polymeric zinc complex (R)-**3.25** is probably produced, which may form a cross-linked network through the interchain bridging O–Zn–O bonds. When 6–16 equivalents of ZnEt$_2$ are added, the unmovable gel is converted back to a viscous solution. The excess ZnEt$_2$ probably saturates the coordination ability of the oxygen atoms in the polymer chain to form a polymeric zinc complex such as (R)-**3.26**. This inhibits the interchain O–Zn–O bridging bonds, leading to the collapse of the cross-linked polymer network.

The reaction of a monomeric BINOL derivative, (R)-**3.27**, with ZnEt$_2$ is studied (Scheme 3.20). On the basis of ^{1}H and ^{13}C NMR analyses and a cryoscopic molecular weight determination, the resulting compound (R)-**3.28** is determined to be a dimeric zinc complex. In the presence of excess amount of ZnEt$_2$, no observable change in the aromatic NMR signals of (R)-**3.28** is observed except additional peaks, which indicate a reversible ZnEt$_2$ coordination. Thus, this dimeric structure is more stable than the interchain cross-link of (R)-**3.25** that can be destroyed with excess ZnEt$_2$. This is probably due to the increased steric hindrance around the Zn centers in (R)-**3.25**, which reduces the stability of the interchain O–Zn–O bonds.

When BINOL is treated with one equivalent of ZnEt$_2$ in CDCl$_3$ at room temperature, it produces a white precipitate. Addition of a second equivalent of ZnEt$_2$ converts the mixture to a clear solution. Two complexes are generated as shown by ^{1}H and ^{13}C NMR analyses. They are probably oligomeric zinc complexes with intermolecular O–Zn–O bridging bonds. The exact structures of these complexes have not been determined.

Without Ti(O^{i}Pr)$_4$, BINOL is found to be a poor catalyst for the reaction of ZnEt$_2$ with benzaldehyde. In the presence of 5 mol% of (R)-BINOL, benzaldehyde reacts with ZnEt$_2$ in toluene at room temperature

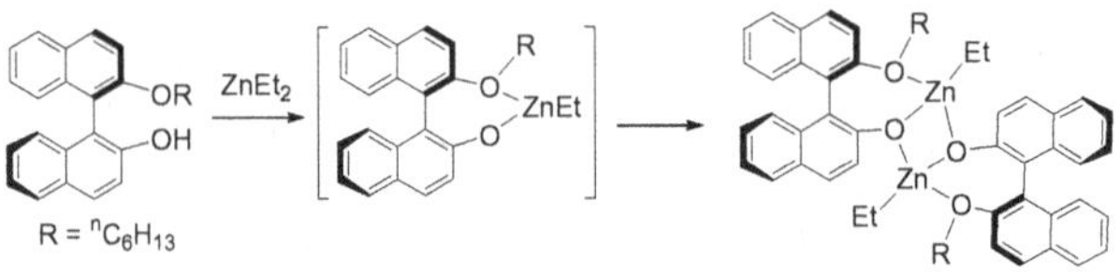

Scheme 3.20. Reaction of a monoalkylated BINOL (R)-**3.27** with ZnEt$_2$.

with only 37% conversion over 24 h. It produces the ethyl addition product PPO (32.9% *ee*) and the reduction product BnOH with a ratio of 95.9:4.1. In methylene chloride solution at room temperature, there is 84% conversion of benzaldehyde over 5 days in the presence of 10 mol% of (*R*)-BINOL. It produces PPO and BnOH with a ratio of 61:39 and the *ee* of PPO is 15%.

Compound (*R*)-**3.27** is a more active catalyst than BINOL. In the presence of 5 mol% of (*R*)-**3.27**, benzaldehyde reacts with ZnEt$_2$ in toluene at room temperature with 72% conversion over 24 h. It produces PPO and BnOH with a ratio of 90.6:9.4 and the *ee* of PPO is 57.0%.

As discussed earlier, the O–Zn–O bridging bonds in (*R*)-**3.28**, the dimeric Zn complex generated from the reaction of (*R*)-**3.27** with ZnEt$_2$, is much harder to dissociate in the presence of excess ZnEt$_2$ than those in the cross-linked poly(BINOL)–Zn complex (*R*)-**3.25**. Treatment of polymer (*R*)-**3.25** with excess ZnEt$_2$ can break its interchain O–Zn–O bridging bonds. In (*R*)-**3.25**, its monoethyl Zn centers are coordinatively saturated just like those in (*R*)-**3.28**. After dissociation of the interchain O–Zn–O bridging bonds, the monoethyl Zn centers in the resulting polymer (*R*)-**3.26** are coordinatively unsaturated. This could contribute to the much higher catalytic activity of the poly(BINOL) over the mono BINOL compound (*R*)-**3.27**. The increased steric hindrance in the poly(BINOL) contributes to its high enantioselectivity.

3.7.3. *Synthesis of the Monomeric Model Compound of the Minor-Groove Poly(BINOL) to Catalyze the Dialkylzinc Addition to Aldehydes*[10b,c]

From the Suzuki coupling of (*R*)-**3.20** with the monoboronic acid compound **3.29** followed by acidic hydrolysis, ligand (*R*)-**3.30** is obtained as the monomeric model compound of the minor-groove poly(BINOL) (*R*)-**3.22** (Scheme 3.21). The specific optical rotation $[\alpha]_D$ of this compound is 95.0 ($c = 0.962$, THF). The optical purity of this compound is over 99% as determined by HPLC Chiracel OD column.

This monomeric BINOL compound is found to be a highly enantioselective catalyst for the ZnEt$_2$ addition to aldehydes. In the presence of 5 mol% of (*R*)-**3.30**, ZnEt$_2$ (two equivalents) reacts with benzaldehyde at 0°C in toluene to give the ethyl addition product (*R*)-PPO with over 99% *ee* and 95% isolated yield. The same conditions are applied to the reaction of many aldehydes and the results are summarized in Table 3.6. As shown

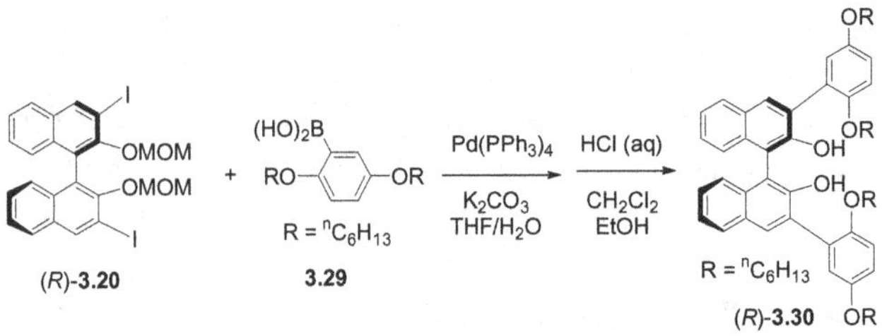

Scheme 3.21. Synthesis of the monobinaphthyl model compound (R)-**3.30**.

in the table, (R)-**3.30** catalyzes the reaction of ZnEt$_2$ with a broad range of aldehydes, including o-, m- or p-substituted benzaldehydes, linear or branched aliphatic aldehydes, α,β-unsaturated aldehydes, with very high enantioselectivity. It is the most generally enantioselective catalyst for the ZnEt$_2$ addition. This compound also catalyses the ZnMe$_2$ addition to aldehydes with high enantioselectivity, though the reactions are much slower than the ZnEt$_2$ additions.

When (R)-**3.30** is treated with two equivalents of ZnEt$_2$ in toluene-d_8, the resulting zinc complex gives a complicated ^{1}H NMR spectrum, indicating a mixture of structural isomers. Cryoscopic analysis shows that the complex is probably a monomeric binaphthyl biszinc complex such as (R)-**3.31**. The intermolecular coordination of (R)-**3.31** through the zinc centers should be very weak. This complex has a number of possible structural isomers because the chirality of the biaryl units at the 3,3'-positions can produce three possible diastereomers and each of the zinc centers in (R)-**3.31** can also be linked with different oxygen atoms. No reaction is observed when benzaldehyde is added to (R)-**3.31**. That is, the Et group on the monoethylzinc center in (R)-**3.31** is not nucleophilic enough for the addition reaction. The ethyl addition product can be obtained only when there is excess ZnEt$_2$.

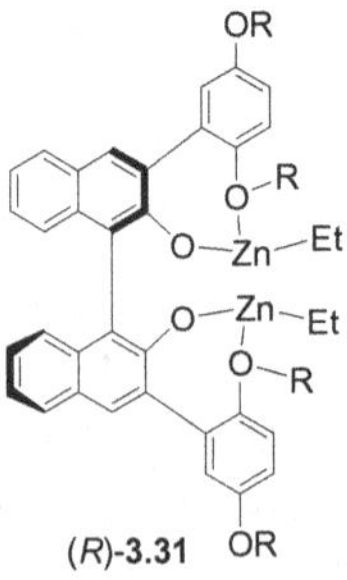

Table 3.6. Reaction of aldehydes with dialkylzincs catalyzed by (R)-**3.30**.

Aldehyde	Dialkylzinc	Time (h)	Isolated yield (%)	ee (%)	Configuration
Ph–CHO	ZnEt$_2$	4	95	99[a]	R
H$_3$C–C$_6$H$_4$–CHO	ZnEt$_2$	4	91	98[b]	R
H$_3$CO–C$_6$H$_4$–CHO	ZnEt$_2$	6	92	97[c]	R
Cl–C$_6$H$_4$–CHO	ZnEt$_2$	4	96	>99[a]	R
3-Cl–C$_6$H$_4$–CHO	ZnEt$_2$	4	97	98[d]	R
3-H$_3$CO–C$_6$H$_4$–CHO	ZnEt$_2$	6	95	99[a]	R
2-F–C$_6$H$_4$–CHO	ZnEt$_2$	4	93	94[b]	R
2-OCH$_3$–C$_6$H$_4$–CHO	ZnEt$_2$	8	90	94[b]	R
1-naphthyl–CHO	ZnEt$_2$	6	92	>99[a]	R
2-naphthyl–CHO	ZnEt$_2$	5	94	99[a]	R
2-furyl–CHO	ZnEt$_2$	6	90	91[d]	R
nC$_5$H$_{11}$–CHO	ZnEt$_2$	40	89	98[d]	R
nC$_6$H$_{12}$–CHO	ZnEt$_2$	24	86	98[d]	R
nC$_8$H$_{17}$–CHO	ZnEt$_2$	45	91	98[d]	R
cyclohexyl–CHO	ZnEt$_2$	40	90	98[d]	R

(*Continued*)

Table 3.6. (*Continued*)

Aldehyde	Dialkylzinc	Time (h)	Isolated yield (%)	ee (%)	Configuration
iPr-CHO (isobutyraldehyde)	$ZnEt_2$	30	73	98[e]	R
PhCH=CH-CHO (cinnamaldehyde)	$ZnEt_2$	24	91	92[a]	R
PhCH=C(Me)-CHO	$ZnEt_2$	27	86	98[d]	R
Crotonaldehyde	$ZnEt_2$	18	66	91[b,f]	R
(CH$_3$)$_2$C=CH-CHO	$ZnEt_2$	40	62	93[b,f,g]	R
CH$_2$=C(iPr)-CHO	$ZnEt_2$	18	64	97[b]	R
alkenyl-CHO	$ZnEt_2$	18	90	98[d]	R
PhC≡C-CHO	$ZnEt_2$	15	90	93[a,h]	R
PhCHO	$ZnMe_2$[i]	96	90	90[a]	R
2-naphthyl-CHO	$ZnMe_2$[i]	96	86	92[a]	R
nC_7H_{15}–CHO	$ZnMe_2$[i]	165	62	88[d]	R

[a] Determined by HPLC-Chiracel OD column.
[b] Determined by chiral GC (β-Dex capillary column).
[c] Determined by HPLC-Chiracel AD column.
[d] Determined by analyzing the acetate derivative of the product on the GC-β-Dex capillary column.
[e] Determined by analyzing the benzoate derivative of the product on the GC-β-Dex capillary column.
[f] Et_2O was used as the solvent.
[g] The reaction was carried out at −40°C and 0.3 equivalent catalyst was used.
[h] 0.2 Equivalent catalyst was used. The reaction was carried out in THF at −10°C and the aldehyde was distilled before use.
[i] Two equivalents of $ZnMe_2$ were used.

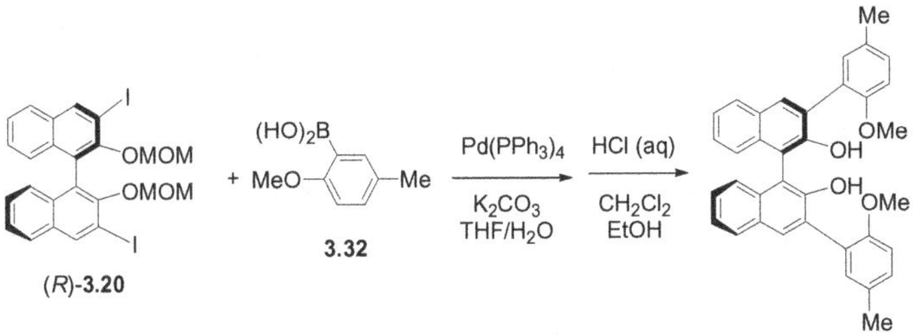

Scheme 3.22. Synthesis of monobinaphthyl compound (*R*)-**3.33**.

It is found that the enantiomeric purity of the product from the reaction of ZnEt$_2$ with benzaldehyde is linearly related with the enantiomeric purity of the catalyst (*R*)-**3.30**. In addition, the enantiomeric purity of the product is also independent of the concentration of (*R*)-**3.30**. This demonstrates that the catalytically active species in the ZnEt$_2$ addition catalyzed by (*R*)-**3.30** is probably monomeric and the dimers or oligomers are not involved in the catalytic process.

The Suzuki coupling of (*R*)-**3.20** with **3.32** followed by hydrolysis gives another BINOL derivative (*R*)-**3.33** (Scheme 3.22). The specific optical rotation $[\alpha]_D$ of this compound is 139.7 ($c = 1.01$, CH$_2$Cl$_2$). This compound contains two 3,3′-anisyl groups that are less electronically rich than those in (*R*)-**3.30**. It catalyzes the reaction of ZnEt$_2$ with benzaldehyde with 98% *ee* at room temperature. That is, (*R*)-**3.33** is as enantioselective as (*R*)-**3.30**.

3.7.4. *Converting the Highly Enantioselective Mono(BINOL) Catalyst to a Highly Enantioselective Poly(BINOL) Catalyst for the Asymmetric Organozinc Additions*[10]

As described above, the monomeric model compound (*R*)-**3.30** is found to be a much more generally enantioselective catalyst than the minor-groove poly(BINOL) (*R*)-**3.22**. In the reaction catalyzed by (*R*)-**3.30**, evidences have shown that the catalytically active species is probably monomeric. The difference in the catalytic properties of the mono(BINOL) ligand (*R*)-**3.30** and the poly(BINOL) (*R*)-**3.22** may be due to the structural difference of their zinc complexes. In the poly(BINOL)–Zn complex (*R*)-**3.25**, each of the two alkoxy groups in one of the phenylene linkers coordinates to a zinc center separately on the adjacent BINOL units. One phenylene linker in (R)-**3.25** generates two chiral biaryl units because it connects two naphthalene rings. The conformation of these two chiral biaryl units is interdependent. Thus, the adjacent binaphthyl units in (R)-**3.25** interfere with each other both electronically and sterically. This makes the catalytic sites in (R)-**3.25** structurally different from those in the mono(BINOL) complex (R)-**3.31**, leading to their differences in the catalytic $ZnEt_2$ addition to aldehydes.

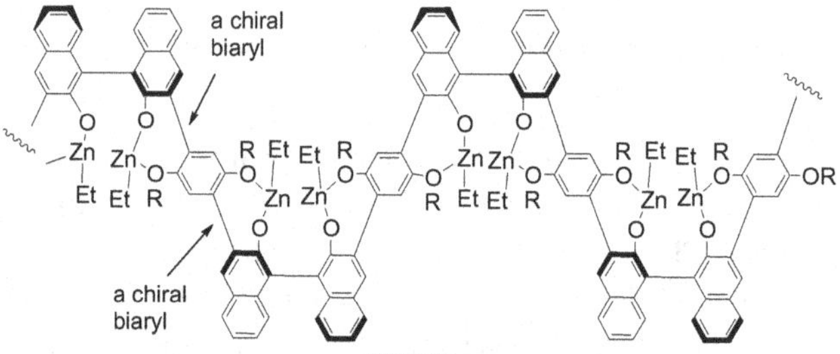

(R)-**3.25**

It is therefore hypothesized that if the interference between the catalytic sites in (R)-**3.25** could be removed, it could generate a poly(BINOL) catalyst as generally enantioselective as the mono(BINOL) ligand (R)-**3.30**. Scheme 3.23 shows the synthesis of a long linker minor-groove

(R)-**3.35**

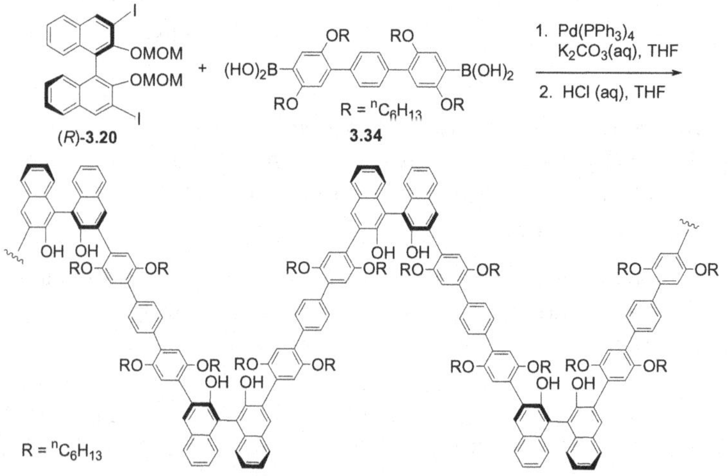

Scheme 3.23. Synthesis of the long linker minor-groove poly(BINOL) (R)-**3.35**.

poly(BINOL) (*R*)-**3.35** in which the adjacent units of the monomeric ligand (*R*)-**3.30** are separated by a rigid phenylene linker. In comparison with the poly(BINOL) (*R*)-**3.22**, there should be minimum interference between the BINOL sites in (*R*)-**3.35**. The Suzuki coupling of (*R*)-**3.20** with the triphenylene diboronic acid **3.34** is conducted with a slight excess of **3.34**. When polymerization is completed, the polymer chain end is capped with the addition of *t*-butylbromobenzene. The resulting polymer is hydrolyzed in the presence of HCl to give the poly(BINOL) (*R*)-**3.35** in 90% yield over the two steps. The ^{1}H and ^{13}C NMR of spectra of this polymer are very well resolved and are consistent with the expected sterically regular structure. Its molecular weight is determined by GPC analysis as $M_w = 25{,}800$ and $M_n = 14{,}300$ (PDI = 1.8). This polymer has very good solubility in common organic solvents such as chloroform, THF, toluene and methylene chloride. Its specific optical rotation $[\alpha]_D$ is -92.9 ($c = 1.01$, CH_2Cl_2).

When this long linker poly(BINOL) is used to catalyze the $ZnEt_2$ addition to aldehydes, it shows much more general and high enantioselectivity than the short linker poly(BINOL) (*R*)-**3.22**. The results for the reactions catalyzed by (*R*)-**3.35** are summarized in Table 3.7. Most of the reactions use 5 mol% of (*R*)-**3.35** (based on the BINOL unit) and two equivalents of $ZnEt_2$, and they are conducted in toluene solution at 0°C. High enantioselectivity is observed for diverse aldehydes including *o*-, *m*- or *p*-substituted benzaldehydes, linear or branched aliphatic aldehydes, and α,β-unsaturated aldehydes. This polymer also shows high enantioselectivity for the $ZnMe_2$ addition to aldehydes. The catalytic properties of (*R*)-**3.35** are very close to those of the monomeric model compound (*R*)-**3.30**. The advantage of (*R*)-**3.35** is its easy recovery and reuse. The recovered (*R*)-**3.35** shows the same catalytic properties as those of the original polymer. This polymer represents the most generally enantioselective polymeric catalyst for the asymmetric alkylzinc addition to aldehydes.

When the long linker poly(BINOL) (*R*)-**3.35** is treated with two equivalents of $ZnEt_2$ in toluene, the original homogeneous solution is converted to a gel. This is similar to that observed for the reaction of the short linker poly(BINOL) (*R*)-**3.22** with $ZnEt_2$. Polymer (*R*)-**3.36** is a structure proposed for the zinc complex generated from the reaction of (*R*)-**3.35** with two equivalents of $ZnEt_2$. The coordinately unsaturated zinc centers in (*R*)-**3.36** can form interchain O–Zn–O bridging bonds to generate the cross-linked gel network. This interchain interaction is also weak, which is similar to that observed for the poly(BINOL) (*R*)-**3.22**. Addition of

Table 3.7. Reaction of aldehydes with dialkylzincs in the presence of the poly(BINOL) (R)-**3.35**.

Aldehyde	Dialkylzinc	Time (h)	Isolated yield (%)	ee (%)	Configuration
Ph–CHO	ZnEt$_2$	5	92	98[a]	R
H$_3$C–C$_6$H$_4$–CHO	ZnEt$_2$	5	90	98[b]	R
Cl–C$_6$H$_4$–CHO	ZnEt$_2$	5	94	98[b]	R
H$_3$CO–C$_6$H$_4$–CHO	ZnEt$_2$	3.5	89	97[a]	R
2-F–C$_6$H$_4$–CHO	ZnEt$_2$	4	88	91[b]	R
2-OCH$_3$–C$_6$H$_4$–CHO	ZnEt$_2$	4	90	93[b]	R
3-H$_3$CO–C$_6$H$_4$–CHO	ZnEt$_2$	4	93	98[a]	R
2-naphthyl–CHO	ZnEt$_2$	12	95	96[a]	R
1-naphthyl–CHO	ZnEt$_2$	5	93	98[a]	R
nC$_5$H$_{11}$–CHO	ZnEt$_2$	20	71	98[c]	R
nC$_7$H$_{15}$–CHO	ZnEt$_2$	20	85	97[c]	–
nC$_8$H$_{17}$–CHO	ZnEt$_2$	24	88	97[c]	R
cyclohexyl–CHO	ZnEt$_2$	24	81	98[d]	R
iPr–CHO	ZnEt$_2$	24	65	98[e]	R
PhCH=CH–CHO	ZnEt$_2$	5	93	92[a,f]	R

(*Continued*)

Table 3.7. (Continued)

Aldehyde	Dialkylzinc	Time (h)	Isolated yield (%)	ee (%)	Configuration
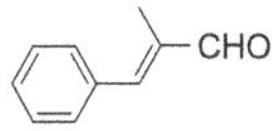 PhCH=CH–CHO	ZnEt$_2$	6	92	97[a]	–
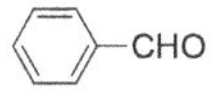 Ph–CHO	ZnMe$_2$[g]	48	92	93[a]	R
nC$_7$H$_{15}$–CHO	ZnMe$_2$[g]	72	78	89[c]	R

[a] Determined by HPLC-Chiracel OD column.
[b] Determined by chiral GC (β-Dex capillary column).
[c] Determined by analyzing the acetate derivative of the product on the GC-β-Dex capillary column.
[d] Determined by analyzing the propionate derivative of the product on the GC-β-Dex capillary column.
[e] Determined by analyzing the Mosher's ester of the product on the GC-β-Dex capillary column.
[f] The solvent was a 1:1 mixture of toluene:diethyl ether.
[g] Three equivalents of ZnMe$_2$ and 0.2 equivalents of polymer were used.

excess ZnEt$_2$ breaks this cross-linked network and converts the gel to a clear solution.

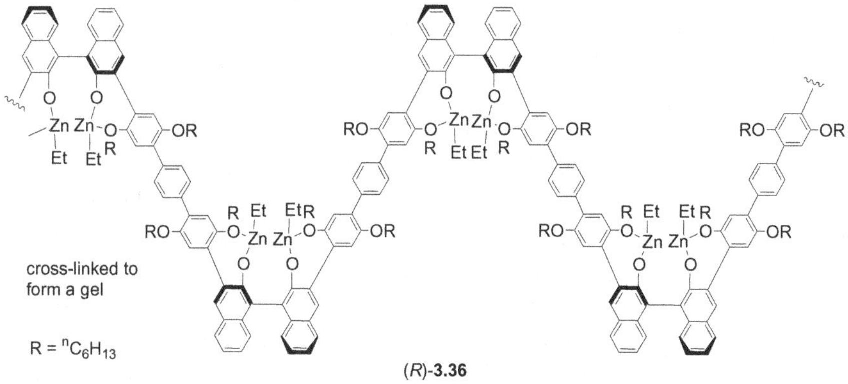

(R)-**3.36**

cross-linked to form a gel

R = ^{n}C$_6$H$_{13}$

The same high and general enantioselectivity of the long linker poly(BINOL) (R)-**3.35** and the mono(BINOL) ligand (R)-**3.30** demonstrates that the rigid and sterically regular polymer structure can be used to preserve the catalytic properties of a monomer catalyst as long as the catalytically active species of the monomer catalyst are not its intermolecular aggregates. This study provides a new strategy to convert a monomeric catalyst to a polymer-supported catalyst.

We have used polymer (R)-**3.35** to catalyze the diphenylzinc addition to propionaldehyde and p-anisaldehyde and the results are summarized in Table 3.8. In these reactions, high enantioselectivity is achieved with the use of 0.2–0.4 equivalents (based on the BINOL unit) of the polymer. In the presence of 0.2 equivalent of the polymer, diphenylzinc reacts with propionaldehyde at 0°C with 85% ee. For the reaction with p-methoxybenzaldehyde, a slow mixing of the reactant with the catalyst is required (Table 3.8, entries 3 and 4). In these experiments, two solutions are prepared, one containing (R)-**3.35**, diethylzinc and diphenylzinc in toluene and another containing the aldehyde in toluene. Both the solutions are simultaneously added into a flask containing toluene at −30°C over 20 h via a syringe pump. After the addition, the reaction mixture is stirred further at −30°C. Up to 92% ee is observed when the double slow addition technique is applied. In these reactions, the polymer is pretreated with excess diethylzinc. Since the diphenylzinc addition is much faster than the diethylzinc addition, no product from the reaction of diethylzinc with the aldehyde is observed even though the amount of diethylzinc is 2–3 times more than diphenylzinc during the reaction. The excess diethylzinc is added in order to dissolve the gel formed from the reaction of (R)-**3.35** with two equivalents [versus the BINOL unit of (R)-**3.35**] of diethylzinc.

The mono(BINOL) ligand (R)-**3.30** is also found to be a highly enantioselective catalyst for the diphenylzinc addition to aldehydes (see Section 4.3.1 in Chapter 4).[11] It shows higher catalytic activity than the poly(BINOL) (R)-**3.35**. A greater amount of the poly(BINOL) (R)-**3.35** (20 or 40 mol%) than the mono(BINOL) (R)-**3.30** (10 or 20 mol%) is required in order to achieve high enantioselectivity. The following explanation is proposed to account for the differences between the catalytic properties of the polymer and the monomer in the asymmetric diphenylzinc addition. Since the catalytic sites in (R)-**3.35** are almost identical to the structure of the monomeric ligand (R)-**3.30**, in principle, the polymer should have the same stereoselectivity as that of the monomer. This is true if there is no competition from the uncatalyzed background reaction as demonstrated for the dialkylzinc addition catalyzed by polymer (R)-**3.35** and monomer (R)-**3.30**. However, in the diphenylzinc addition, the rate difference between the catalyzed and the uncatalyzed reactions is small. Because of the mobility differences between the polymer and the monomer, there may be less collision between the substrate and the catalytic sites of the polymer,

Table 3.8. Asymmetric diphenylzinc addition to aldehydes catalyzed by the poly(BINOL) (R)-**3.35**.

Entry	Aldehyde	ZnPh$_2$ (equiv.)	(R)-**3.35** or (R)-**3.35** + ZnEt$_2$ (mol%)	Aldehyde (mM)	Solvent	Temp (°C)	Time (h)	Isolated yield (%)	ee^a (%)	Configuration^b
1	⌒CHO	2	20	75	Toluene	0	66	82	85	S
2	H$_3$CO–C$_6$H$_4$–CHO	1	20 + 40 ZnEt$_2$	50	Toluene	−30	20	65	76	R
3^c		1	20 + 200 ZnEt$_2$	5	Toluene	−30	40	59	82	R
4^c		1	40 + 320 ZnEt$_2$	5	Toluene	−30	56	72	92	R

^{a}Determined by HPLC–Chiracel OD column.
^{b}Determined by comparing the optical rotation with the literature data.
^{c}Slow mixing of the catalyst, reagent and substrate *via* a syringe pump.

which will make the uncatalyzed reaction more competitive. Therefore, higher concentration of the polymer and lower concentration of the substrate are needed to suppress the competing uncatalyzed diphenylzinc addition in order to maintain the high enantioselectivity of the monomeric catalyst.

3.8. Asymmetric Reduction of Prochiral Ketones Catalyzed by the Chiral BINOL Monomer and Polymer Catalysts[10b]

The zinc complexes of the poly(BINOL)s and the monomeric model compound are used to catalyze the asymmetric reduction of prochiral ketones with catecholborane to generate the optically active secondary alcohols. Scheme 3.24 shows the reduction of acetophenone with catecholborane. In the presence of 5 mol% of the short linker minor-groove poly(BINOL) (S)-**3.22** and 10 mol% of $ZnEt_2$, acetophenone is reduced with 54% ee and 77% yield at room temperature in toluene. At −30°C, the ee is 67%. Other solvents such as THF, diethyl ether and methylene chloride give poorer results.

The monomeric model compound (R)-**3.30** is tested for this reaction. This compound (5 mol%) in combination with $ZnEt_2$ (10 mol%) catalyzes the reduction of acetophenone to (S)-1-phenylethanol at −30°C in toluene with 87% yield and 81% ee. When the long linker minor-groove poly(BINOL) (R)-**3.35** is used to catalyze this reaction under the same conditions, it gives almost the same result as (R)-**3.30**. This demonstrates that the catalytic properties of the monomeric ligand (R)-**3.30** is preserved in the rigid and sterically regular polymer. The lower enantioselectivity of the poly(BINOL) (S)-**3.22** indicates a negative interference between the adjacent BINOL units in this polymer, which is similar to that observed in the asymmetric diethylzinc addition to aldehydes. The poly(BINOL) (R)-**3.35** is used to catalyze the reduction of other prochiral ketones. As the results summarized in Table 3.9 show, moderate to good enantioseletivities

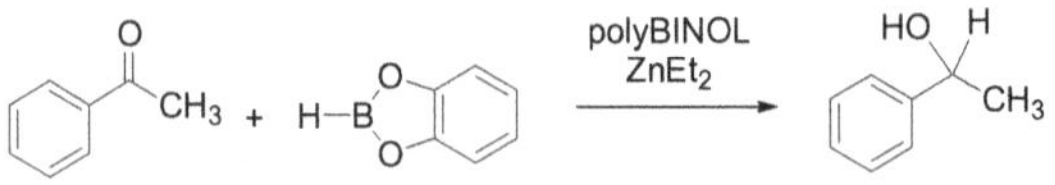

Scheme 3.24. Reduction of acetophenone with caltecholborane in the presence of a chiral poly(BINOL)–zinc complex.

Table 3.9. The asymmetric reduction of ketones in the presence of polymer (S)-**3.22**, monomer (R)-**3.30** and polymer (R)-**3.35**.

Entry[a]	Ketone	Ligand	Solvent	Yield (%)	ee (%)	Product configuration
1[b]	PhC(O)Me	(S)-**3.22**	Toluene	77	54[c]	R
2	PhC(O)Me	(S)-**3.22**	Toluene	85	67[c]	R
3	PhC(O)Me	(S)-**3.22**	CH_2Cl_2	90	62[c]	R
4	PhC(O)Me	(S)-**3.22**	Et_2O	88	59[c]	R
5	PhC(O)Me	(S)-**3.22**	THF	51	56[c]	R
6	PhC(O)Me	(R)-**3.30**	Toluene	87	**81**[c]	S
7[d]	PhC(O)Me	(R)-**3.30**	Toluene	88	77[c]	S
8[e]	PhC(O)Me	(R)-**3.30**	Toluene	86	14[c]	S
9[f]	PhC(O)Me	(R)-**3.30**	Toluene	90	8.5[c]	S
10	PhC(O)Me	(R)-**3.35**	Toluene	89	**80**[c]	S
11	MeO-C$_6$H$_4$-C(O)Me	(R)-**3.35**	Toluene	78	70[g]	S
12	Br-C$_6$H$_4$-C(O)Me	(R)-**3.35**	Toluene	86	79[g]	S
13	2-Br-C$_6$H$_4$-C(O)Me	(R)-**3.35**	Toluene	88	74[g]	S
14	2-naphthyl-C(O)Me	(R)-**3.35**	Toluene	92	76[g]	S

(Continued)

Table 3.9. (*Continued*)

Entry[a]	Ketone	Ligand	Solvent	Yield (%)	ee (%)	Product configuration
15	(cinnamyl phenyl ketone)	(*R*)-**3.35**	Toluene	89	78[g]	*S*
16	(chloroacetophenone)	(*R*)-**3.35**	Toluene	90	14[g]	*R*
17	(isobutyrophenone)	(*R*)-**3.35**	Toluene	91	4.3[g]	*R*
18	(2-methyl-3-heptanone)	(*R*)-**3.35**	Toluene	79	36[g]	*S*

[a] The reactions were carried out at −30 °C in the presence of 5 mol% chiral ligand and 10 mol% ZnEt$_2$ using catecholborane as the reducing agent unless otherwise specified.
[b] The reaction was performed at room temperature.
[c] Determined by GC (β-Dex capillary column).
[d] About 5 mol% ZnMe$_2$ was used in place of ZnEt$_2$.
[e] BH$_3 \cdot$ SMe$_2$ was the reducing agent.
[f] 0.2 equivalent of ZnEt$_2$ was used.
[g] Determined by HPLC-Chiracel OD column.

are observed for the reaction of methyl aryl or methyl vinyl ketones. For nonmethyl ketones, the enantioselectivities are low. Polymer (*R*)-**3.35** can be easily recovered from the reaction mixture and the three-time recovered polymer still gives 78% ee and 80% yield for the acetophenone reduction. The ^{1}H NMR spectrum and optical rotation of the recovered polymer are also very close to those of the original one. Thus there is little structural or functional change for the recycled polymer.

The isopropyl-substituted poly(BINOL) (*R*)-**3.24** is used in combination with Ti(O^{i}Pr)$_4$ to catalyze the reduction of acetophenone with catachobrane.[9] In the presence of 11 mol% of (*R*)-**3.24** and 10 mol% of Ti(O^{i}Pr)$_4$ in CH$_2$Cl$_2$ at −30°C, acetophenone is reduced with 94% conversion and 48% ee over 18 h. The reduction of other ketones such as 1′-acetonaphthone, 4′-chloroacetophenone and 2-chloroacetophenone gives lower than 15% ee. Using other reducing agent such as BH$_3 \cdot$ THF, 9-BBN and BH$_3 \cdot$ SMe$_2$ also gives lower ee. No enantioselectivity is observed when TiCl$_2$(O^{i}Pr)$_2$ is used in place of Ti(O^{i}Pr)$_4$.

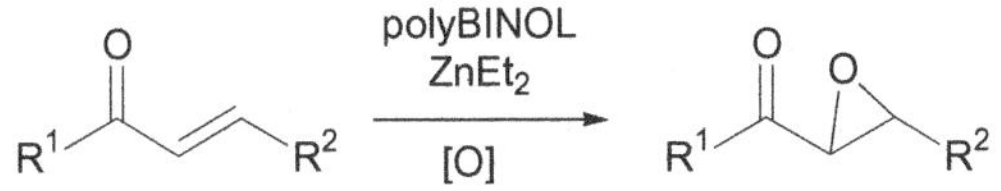

Scheme 3.25. Asymmetric epoxidation of α,β-unsaturated ketones.

3.9. Asymmetric Epoxidation of α,β-Unsaturated Ketones Catalyzed by the Minor- and Major-Groove Poly(BINOL)s[12]

The zinc complexes of the minor- and major-groove poly(BINOL)s are used to catalyze the asymmetric epoxidation of α,β-unsaturated ketones to generate chiral α,β-epoxy ketones (Scheme 3.25).

3.9.1. *Asymmetric Epoxidation of α,β-Unsaturated Ketones in the Presence of a Stoichiometric Amount of the Major-Groove Poly(BINOL)s, Diethylzinc and Oxygen*

In the presence of oxygen as the oxidant, the major-groove poly(BINOL) (R)-**3.9** in combination with ZnEt$_2$ is used to carry out the epoxidation of chalcone, an α,β-unsaturated ketone with $R^1 = R^2 =$ Ph. As shown in entry 1 of Table 3.10, up to 71% ee is observed for the epoxidation of chalcone using one equivalent of the polymer and 0.95 equivalent of ZnEt$_2$ in methylene chloride at 0°C. Even though a stoichiometric amount of the poly(BINOL) ligand is used, the yield is only 41%. The absolute configuration of the product is 2S and 3R. Although the yield can be improved with the use of 1.9 equivalent of ZnEt$_2$, the ee is significantly reduced (50% ee, entry 2). The use of 0.95 equivalent of ZnEt$_2$ in combination with one equivalent of the poly(BINOL) probably generates a catalyst that is different from the use of 1.9 equivalent of ZnEt$_2$. Other reaction conditions and other substrates give poorer results. For example, THF inhibits the epoxidation probably by coordination to the zinc centers in the poly(BINOL)–zinc complex (entry 6).

Scheme 3.26 shows a possible mechanism for the asymmetric epoxidation promoted by the poly(BINOL) (R)-**3.9** in combination with ZnEt$_2$ and oxygen. Reaction of (R)-**3.9** with one equivalent of ZnEt$_2$ can generate the intermediate **3.37**. In the presence of oxygen, the Zn–Et bond can be oxidized to the zinc peroxide **3.38**, which can conduct the epoxidation

Table 3.10. Epoxidation of α,β-unsaturated ketones in the presence of the major-groove poly(BINOL) (R)-**3.9**, diethylzinc and oxygen.

Entry	Ketone	Yield (%)	ee (%)	Solvent	Temp (°C)	Mole ratio [(R)-**3.9**/ZnEt$_2$/ketone]
1	Ph-CO-CH=CH-Ph	41	71	CH$_2$Cl$_2$	0	1:0.95:0.90
2	Ph-CO-CH=CH-Ph	99	50	CH$_2$Cl$_2$	0	1:1.9:0.90
3	Ph-CO-CH=CH-Ph	64	49	CH$_2$Cl$_2$	−30	1:1.9:0.90
4	Ph-CO-CH=CH-Ph	11	58	CH$_2$Cl$_2$	r.t.	1:1.9:0.90
5	Ph-CO-CH=CH-Ph	34	37	Toluene	0	1:0.95:0.90
6	Ph-CO-CH=CH-Ph	0	0	THF	0	1:1.9:0.90
7	Ph-CO-CH=CH-C$_6$H$_4$-Me	34	54	CH$_2$Cl$_2$	−15	1:1.9:0.90
8	Ph-CO-CH=CH-(2-naphthyl)	75	54	CH$_2$Cl$_2$	0	1:1.9:0.90
9	(2-naphthyl)-CO-CH=CH-Ph	91	47	CH$_2$Cl$_2$	0	1:1.9:0.90
10	Ph-CO-CH=CH-CH(CH$_3$)$_2$	18	25	Toluene	0	1:0.95:0.90

of chalcone *via* a Michael addition mechanism to give the epoxy ketone product. The resulting zinc alkoxide cannot be oxidized by oxygen to regenerate the zinc peroxide complex **3.38**. Thus, this reaction is not catalytic.

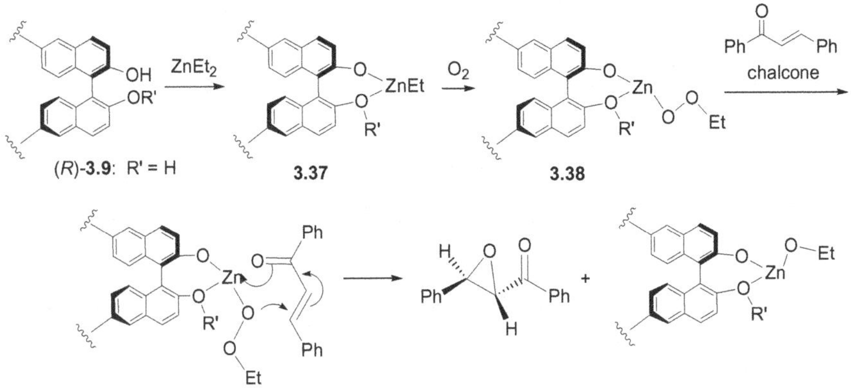

Scheme 3.26. A proposed mechanism for the poly(BINOL)-mediated epoxidation of chalcone.

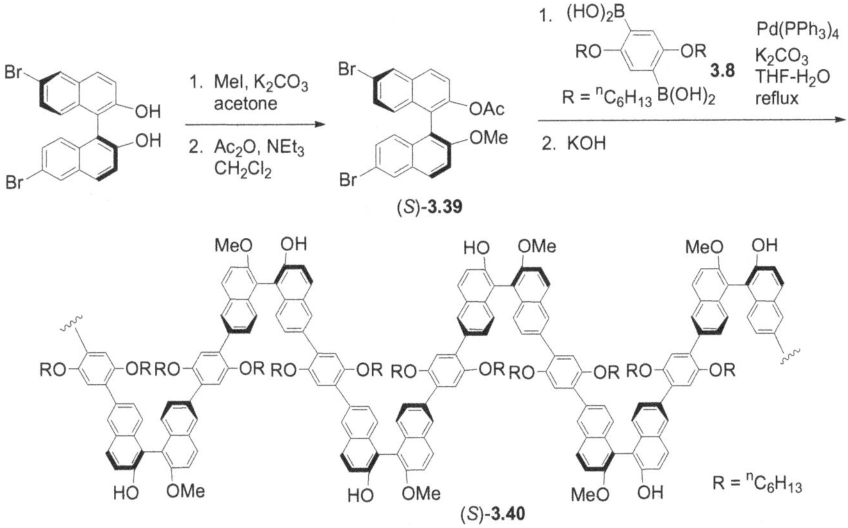

Scheme 3.27. Synthesis of a monomethylated major-groove poly(BINOL) (*S*)-**3.40**.

Scheme 3.27 shows the synthesis of a structurally modified major-groove poly(BINOL). Monomer (*S*)-**3.39** containing two different protecting groups is polymerized with the diboronic acid **3.8**. After removal of the acetyl group under basic condition, the major-groove poly(BINOL)

Table 3.11. The asymmetric epoxidation of α,β-unsaturated ketones in the presence of polymer (S)-**3.40**, diethylzinc and oxygen.

Entry	Ketone	Yield (%)	ee (%)	Solvent	Temp (°C)	Mole ratio [(S)-**3.40**/ZnEt$_2$/ketone]
1	Ph−CO−CH=CH−Ph	>90	35	CH$_2$Cl$_2$	0	1:0.95:0.9
2	Ph−CO−CH=CH−Ph	>90	38	CH$_2$Cl$_2$	0	1:0.95:0.9
3	Ph−CO−CH=CH−(2-naphthyl)	95	33	CH$_2$Cl$_2$	0	1:0.95:0.9

(S)-**3.40** is obtained. In (S)-**3.40**, each of its BINOL units is monomethylated. GPC gives the molecular weight of this polymer as $M_w = 36{,}000$ and $M_n = 12{,}800$ (PDI = 2.8). Its specific optical rotation $[\alpha]_D$ is -301.1 ($c = 0.5$, CH$_2$Cl$_2$). This polymer shows better solubility than (R)-**3.9** in organic solvents.

Polymer (S)-**3.40** is used in combination with ZnEt$_2$ and oxygen for the asymmetric epoxidation of α,β-unsaturated ketones. As the results in Table 3.11 show, polymer (S)-**3.40** gives greatly improved yields over the use of (R)-**3.9** without using excess ZnEt$_2$. The diastereoselectivity for these reactions remains to be high (>99% de), but the enantioselectivity is low.

3.9.2. Development of a Catalytic Asymmetric Epoxidation of α,β-Unsaturated Ketones Using Poly(BINOL)s, Diethylzinc and t-Butyl Hydroperoxide

The use of oxygen as the oxidant in the above epoxidation requires a stoichiometric amount of the chiral ligand–Zn(II) complex. When we replace oxygen with ^{t}BuOOH as the oxidant, we discover that only a catalytic amount of the chiral ligand–Zn(II) complex is needed for the asymmetric epoxidation of α,β-unsaturated ketones. For example, chalcone can be oxidized to the corresponding epoxy ketone when 5 mol% of polymer (R)-**3.9** and 10 mol% of ZnEt$_2$ are used. The reaction is conducted in

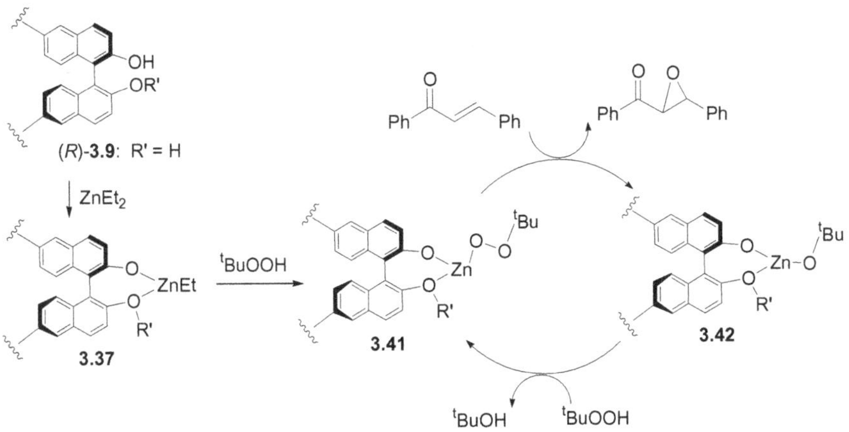

Scheme 3.28. A proposed mechanism for the poly(BINOL)-*catalyzed* asymmetric epoxidation of chalcone.

methylene chloride at 0°C, which gives the product with 95% isolated yield, >99% de and 28% ee.

Scheme 3.28 shows a proposed mechanism for the reaction catalyzed by (*R*)-**3.9**. The zinc complex **3.37** generated from the reaction of (*R*)-**3.9** with ZnEt$_2$ can react with ^{t}BuOOH to form the zinc peroxide intermediate **3.41**. Reaction of **3.41** with chalcone *via* a Michael addition mechanism will give the epoxy ketone product and the zinc *t*-butyloxide complex **3.42**. Complex **3.42** can react with ^{t}BuOOH to regenerate the zinc peroxide **3.41** and allow the epoxidation to be conducted catalytically.

Two monomeric BINOL ligands (*R*)-**3.30** and (*R*)-**3.33** are tested for the catalytic asymmetric epoxidation of α,β-unsaturated ketones. For the epoxidation of a β-propyl substituted enone, 1-phenyl-hex-2-en-1-one, (*R*)-**3.30** gives 75% yield and 30% ee, and (*R*)-**3.33** gives 83% yield and 21% ee.

In spite of the low enantioselectivity of the monomeric ligands, when the short linker minor-groove poly(BINOL) (*R*)-**3.22** is used for the catalytic asymmetric epoxidation of the β-propyl-substituted enone, it shows greatly improved enantioselectivity. This polymer gives the epoxy ketone product with 91% yield and 76% ee. The results for the epoxidation of various α,β-unsaturated ketones catalyzed by (*R*)-**3.22** in combination with ZnEt$_2$ are summarized in Table 3.12. In the presence of 20 mol% of (*R*)-**3.22**, up to 81% ee has been achieved for the asymmetric epoxidation of the α,β-unsaturated ketones.

Table 3.12. Asymmetric epoxidation of α,β-unsaturated ketones catalyzed by the minor-groove poly(BINOL) (R)-**3.22**.[a]

Entry	Ketone	Solvent	Temp	Time (h)	Isolated yield (%)	ee[b] (%)	$[\alpha]_D$ (c, CH_2Cl_2)
1	Ph-CO-CH=CH-Et	Et$_2$O	rt	5	92	76	1.7 (1.0)
2	Ph-CO-CH=CH-Et	^{t}BuOMe	rt	4.5	89	71	
3	Ph-CO-CH=CH-Et	^{n}BuOMe	rt	3.5	87	62	
4	Ph-CO-CH=CH-Et	Anisole	rt	6	98	48	
5	Ph-CO-CH=CH-Et	THF	rt	6	97	0	
6[c]	Ph-CO-CH=CH-Et	Et$_2$O	rt	4.5	81	73	
7	Ph-CO-CH=CH-iPr	Et$_2$O	rt	3.5	93	78	26.6 (0.9)
8[d]	Ph-CO-CH=CH-iPr	Et$_2$O	rt	3.5	94	81	27.1 (0.9)
9	Ph-CO-CH=CH-tBu	Et$_2$O	rt	8	67	64	14.0 (0.36)[e]
10	Ph-CO-CH=CH-Ph	Toluene	0°C	3	89	39	
11	Ph-CO-CH=CH-Ph	Et$_2$O	0°C	5	95	74	
12	Ph-CO-CH=CH-Ph	Et$_2$O	rt	3	92	73	−151.1 (1.0)

(*Continued*)

Table 3.12. (Continued)

Entry	Ketone	Solvent	Temp	Time (h)	Isolated yield (%)	ee^b (%)	$[\alpha]_D$ (c, CH$_2$Cl$_2$)
13	Ph-CO-CH=CH-C$_6$H$_4$-Cl	Et$_2$O	rt	5.5	81	79	−176.8 (0.7)
14	Ph-CO-CH=CH-C$_6$H$_4$-Me	Et$_2$O	rt	8	93	70	−171.0 (1.1)

[a] All reactions used 20 mol% of (R)-**3.22** (based on the repeating unit), 36 mol% of ZnEt$_2$, 1.2 equivalent of ^{t}BuOOH unless indicated otherwise.
[b] Determined by HPLC with a Daicel-OD column.
[c] 1.2 Equivalent of (R)-**3.22**, 1.8 equivalent of ZnEt$_2$ and 1.0 equivalent of ^{t}BuOOH were used.
[d] About 40 mol% of the polymer and 72 mol% of ZnEt$_2$ were used.
[e] Measured in CHCl$_3$.

When the long linker minor-groove poly(BINOL) (R)-**3.35** is used for the asymmetric epoxidation, under the same conditions as the use of the short linker poly(BINOL) (R)-**3.22** in entry 1 of Table 3.12, it gives 37% ee and 99% yield for the epoxidation of β-propyl-substituted enone, trans-1-phenyl-hex-2-en-1-one, and 55% ee and 88% yield for the β-isopropyl-substituted enone, 4-methyl-1-phenyl-pent-2-en-1-one. That is, the catalytic properties of the long linker polymer (R)-**3.35** are very similar to those of the monomeric ligand (R)-**3.30**, and the structure and function of (R)-**3.30** are mostly preserved in (R)-**3.35**.

The significantly enhanced enantioselectivity of polymer (R)-**3.22** over polymer (R)-**3.35** and monomer (R)-**3.30** is attributed to a collaborative effect between the adjacent catalytic sites in the zinc complex of (R)-**3.22**. The interaction between the adjacent catalytic sites in the zinc complex (R)-**3.25** generated from the reaction of (R)-**3.22** with ZnEt$_2$ decreases the enantioselectivity of (R)-**3.22** in the asymmetric ZnEt$_2$ addition to aldehydes as described in Section 3.7 and in the asymmetric reduction of ketones as described in Section 3.8. However, this interaction apparently greatly increases the enantioselectivity of (R)-**3.22** in the epoxidation of α,β-unsaturated ketones. In the zinc complex of the long linker polymer (R)-**3.35**, there should be minimum collaborative effect between the catalytic sites because of its long and rigid linkers.

3.9.3. Asymmetric Epoxidation of α,β-Unsaturated Ketones by Using the Yb and La Complexes of Poly(BINOL)s[13]

The Yb(III) and La(III) complexes of the poly(BINOL)s are studied in collaboration with Sasai. Scheme 3.29 shows the use of these complexes to catalyze the asymmetric epoxidation of α,β-unsaturated ketones. The results for the epoxidation of *trans*-4-phenyl-but-3-en-2-one [*trans*-PhCH=CHC(CO)CH$_3$] in the presence of poly(BINOL)s (R)-**3.7**, (R)-**3.9**, (R)-**3.22** and (R)-**3.35** in combination with Yb(O^{i}Pr)$_3$ are summarized in Table 3.13. Among these chiral polymers, it is found that the major-groove poly(BINOL) (R)-**3.9** is the most enantioselective one for this

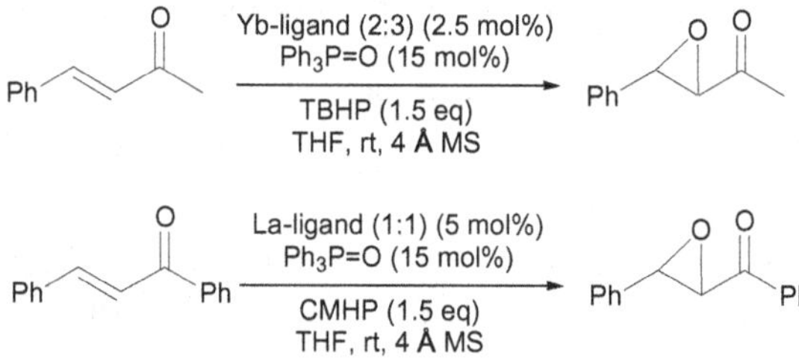

Scheme 3.29. Asymmetric epoxidation of α,β-unsaturated ketones catalyzed by the poly(BINOL)–Yb(III) and –La(III) complexes.

Table 3.13. Asymmetric epoxidation of 4-phenyl-but-3-en-2-one with TBHP catalyzed by Yb–ligand complex.[a]

Entry	Ligand	Additive (mol%)		Time (h)	Yield (%)[b]	ee (%)[c]
		H$_2$O	Ph$_3$P=O			
1	(R)-**3.7**	—	—	18	84	34
2	(R)-**3.9**	—	—	22	91	49
3	(R)-**3.22**	—	—	20	47	39
4	(R)-**3.35**	5	—	18	51	41
5	(R)-**3.9**	—	—	24	73	50
6	(R)-**3.9**	—	15	24	86	71
7	(R)-**3.9**	5	15	18	59	63
8	(R)-**3.9**	—	15	18	76	72

[a] The metal to ligand ratio is 2:3 and the catalyst is generated at 40°C.
[b] Isolated yields after column chromatography.
[c] By HPLC analysis.

Table 3.14. Asymmetric epoxidation of chalcone with CMHP catalyzed by the lanthanum complex of (R)-**3.9**.

Entry	La:(R)-**3.9**	Temp^a (°C)	Time (h)	Ph$_3$P=O (mol%)	Time^b (h)	Yield (%)	ee^c (%)
1	2:3	rt	12	—	4	95	35
2	2:3	40	12	—	4	97	37
3	1:1	40	1	15	15	98	62
4	1:1	40	1	30	14	97	52
5	1:1	40	1	50	14	97	53
6	1:1	40	1	15 + 15^d	1.5	99	73

^{a}Temperature at which the catalyst is generated.
^{b}Time required to generate the active catalyst.
^{c}Reaction time.
^{d}15 mol% of the additive is added before the catalyst generation step and additional 15 mol% is added before the addition of the substrate.

reaction. As shown in entry 6, in the presence of 2.5 mol% of polymer (R)-**3.9**, the epoxide product is obtained in 86% yield and 71% ee. In this reaction, a 2:3 ratio of the ligand to Yb(O^{i}Pr)$_3$ is heated at 40°C first. The resulting complex (2.5 mol%) is used in combination with ^{t}BuOOH (TBHP), Ph$_3$P=O and 4 Å molecular sieves to catalyze the epoxidation at room temperature. Increasing the amount of Ph$_3$P=O to 75 mol% only slightly increases the ee (entry 9).

Table 3.14 gives the results for the use of the major-groove poly(BINOL) (R)-**3.9** in combination with La(O^{i}Pr)$_3$ to catalyze the asymmetric epoxidation of chalcone in the presence of cumene hydroperoxide (CMHP). An shown in entry 6, up to 73% ee and 99% yield are achieved in 1.5 h. This reaction also uses 30% Ph$_3$P=O additive and 4 Å molecular sieves.

3.10. Asymmetric Diels–Alder Reaction Catalyzed by Poly(BINOL)–B(III) Complexes[14]

A BINOL-based boron complex (R)-**3.43** was found to be an excellent catalyst for the Diels–Alder reaction by Yamamoto.[15] As shown in Scheme 3.30, (R)-**3.43** catalyzes the reaction of cyclopentadiene methacrolein with very high diastereoselectivity and enantioselectivity.

In Scheme 3.17, we have shown that the poly(BINOL) (R)-**3.23** containing a ligand environment analogous to that of (R)-**3.43** can be obtained from the treatment of polymer (R)-**3.21** with excess BBr$_3$ followed

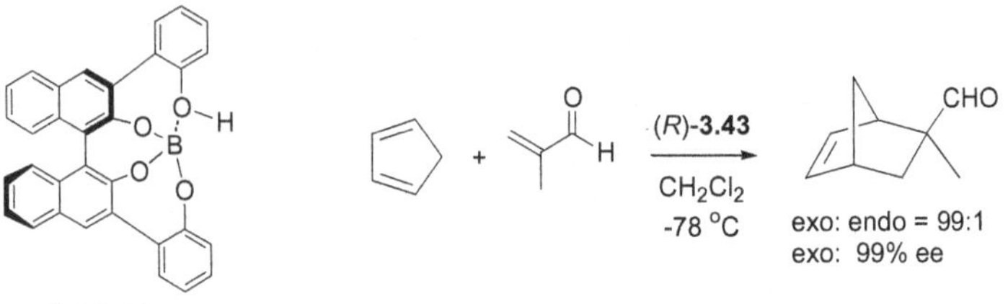

Scheme 3.30. Asymmetric Diels–Alder reaction catalyzed by (R)-**3.43**.

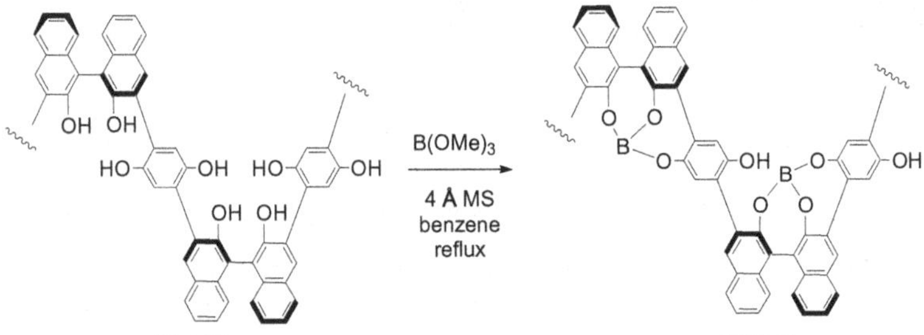

Scheme 3.31. Formation of the polymeric boron complex (R)-**3.44**.

by hydrolysis. Polymer (R)-**3.23** is insoluble in organic solvents. When it is treated with B(OMe)$_3$ in the presence of 4 Å molecular sieves in refluxing benzene, the corresponding insoluble polymeric boron complex (R)-**3.44** is probably produced (Scheme 3.31). Polymer (R)-**3.44** is used to catalyze the Diels–Alder reaction of cyclopentadiene and methacrolein. In methylene chloride at −35°C, it gives the desired cycloaddition product with an exo:endo ratio of 91:9. However, the ee for the exo product is only 4%. Lowering the reaction temperature to −78°C gives an exo:endo ratio of 94:6 and the ee of the exo product is increased to 22%.

In order to improve the enantioselectivity of the poly(BINOL) for the Diels–Alder reaction, a soluble analog of (R)-**3.23** with minimum interference between the adjacent catalytic sites is prepared. Monomer (R)-**3.45** is prepared from the MOM-protected BINOL (Scheme 3.32). Its specific optical rotation $[\alpha]_D$ is 54.3 ($c = 0.52$, THF). The Suzuki coupling of (R)-**3.45** and the terphenylene dibromide **3.46** followed by hydrolysis gives the poly(BINOL) (R)-**3.47** in 88% yield. This polymer is soluble

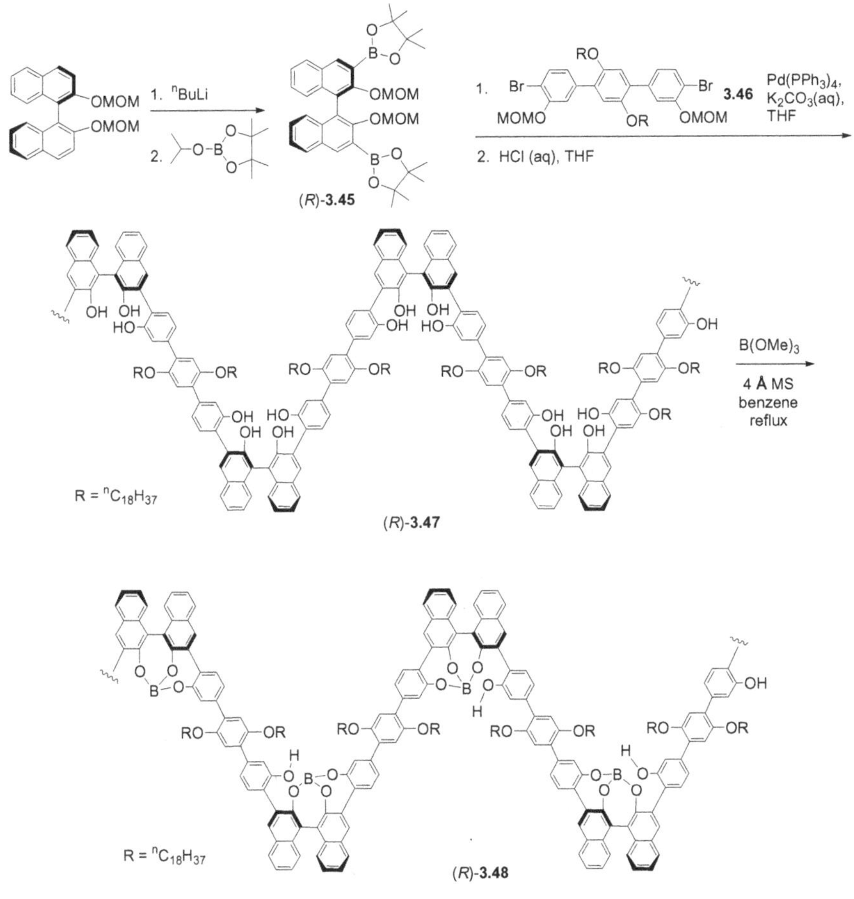

Scheme 3.32. Synthesis of poly(BINOL)–B(III) complex (R)-**3.48**.

in common organic solvents. Its molecular weight is determined by GPC analysis as $M_w = 40,800$ and $M_n = 23,500$ (PDI = 1.74).

Treatment of the poly(BINOL) (R)-**3.47** with B(OMe)$_3$ in the presence of 4 Å molecular sieves in refluxing benzene generates (R)-**3.48**. The catalytic sites in (R)-**3.48** are isolated from each other with minimum interference between the adjacent BINOL units. This polymer is soluble in CH$_2$Cl$_2$/THF (100:1) till −40°C. It turns to unmovable gel at lower temperature. When a cosolvent of CH$_2$Cl$_2$/THF (24:1) is used, (R)-**3.48** forms a suspension at −78°C. In the presence of 10 mol% of (R)-**3.48**, the reaction of cyclopentadiene with methacrolein in CH$_2$Cl$_2$/THF (100/1) at

−35°C gives the cycloaddition product with an exo/endo ratio of 87:13 and 8% ee (exo). At −78°C in CH_2Cl_2/THF (24:1), both the diastereoselectivity and enantioselectivity are greatly enhanced. It gives the product with an exo/endo ratio of 99:1 and 88% ee (exo). The low solubility of (R)-**3.48** at −78°C should have contributed to its lower enantioselectivity than the monomeric catalyst (R)-**3.43**.

3.11. 1,3-Dipolar Cycloaddition Catalyzed by the Minor- and Major-Groove Poly(BINOL)–Al(III) Complexes[16]

The asymmetric 1,3-dipolar cycloaddition of nitrones and vinyl ethers in the presence of the poly(BINOL)-based catalysts is studied in collaboration with Jorgesen (Scheme 3.33). In the reaction, the poly(BINOL) (R)-**3.22** is first treated with $AlMe_3$, which probably generates the polymeric Al(III) complex (R)-**3.49** (Scheme 3.34). In the presence of 20 mol% of (R)-**3.49**, the reaction of a nitrone and a vinyl ether proceeds at room temperature to give an isoxazolidine product in high yield. Very high diastereoselectivity is observed which favors the exo isomer with high enantioselectivity (Table 3.15). After the reaction, the polymer is precipitated out with methanol, which allows the isolation of the product of high purity even without further purification. This significantly simplifies the purification of the product. The recovered polymer ligand can be reused for the catalysis. After four consecutive uses, this polymer can still give the exo product (R^1 = Ph and R^2 = Et) with 88% yield and 92% ee.

The monomeric model compound (R)-**3.30** in combination with $AlMe_3$ probably forms the Al(III) complex (R)-**3.50**. This complex is examined for the reaction of nitrones with vinyl ethers. As the results summarized in Table 3.16 show, this monomeric Al(III) complex exhibits very high stereoselectivity that is similar to those of the poly(BINOL)–Al(III) complex (R)-**3.49**. It demonstrates that the catalytically active species of (R)-**3.50** is probably monomeric and its intermolecular aggregates do not

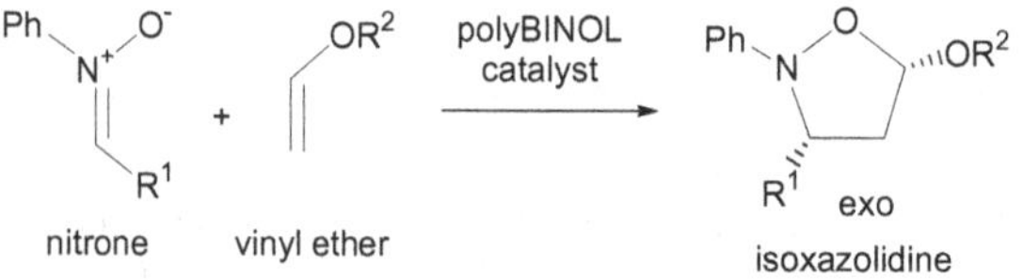

Scheme 3.33. An asymmetric 1,3-dipolar cycloaddition.

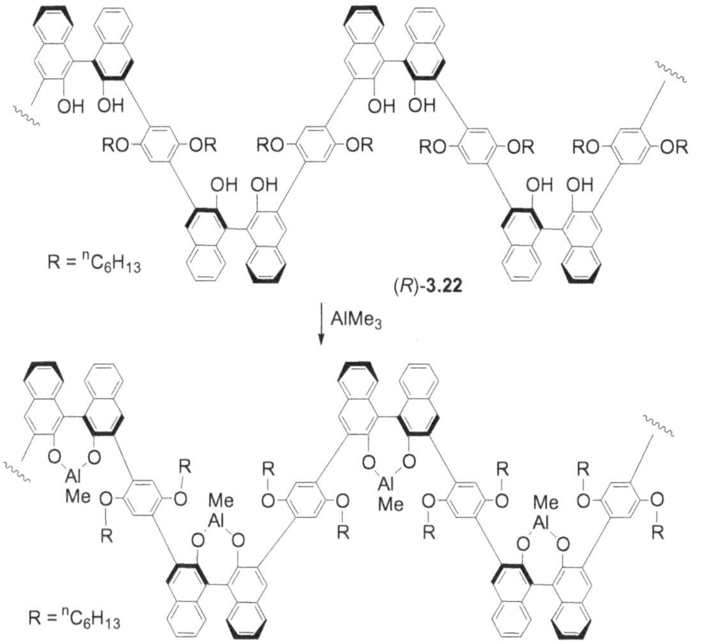

Scheme 3.34. Preparation of poly(BINOL)–Al(III) complex (R)-**3.49**.

play a role in the catalysis. The adjacent catalytic sites in the poly(BINOL)–Al(III) complex (R)-**3.50** also do not interfere with each other, allowing the catalytic properties of (R)-**3.50** to be mostly preserved in the polymer. This behavior is quite different from the Zn(II) complex (R)-**3.25** derived from (R)-**3.22** in which there is significant interference between the adjacent catalytic sites.

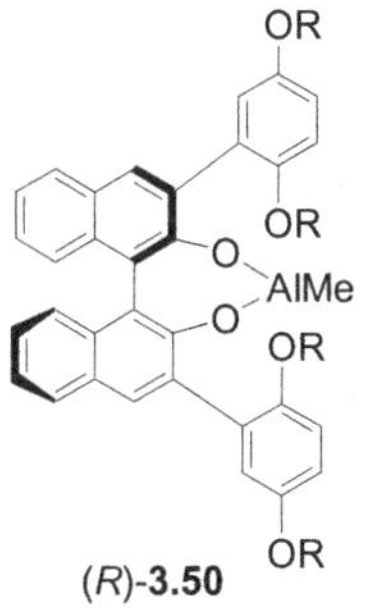

Table 3.15. 1,3-Dipolar cycloadditions of nitrones and vinyl ethers catalyzed by poly(BINOL) (R)-**3.49** (20 mol%).

Entry	Nitrone R^1	Vinyl Ether R^2	Yield[a]	Exo:endo[b]	Exo ee[c]
1	Ph	Et	97	>98:<2	99
2[d]	Ph	Et	80	>98:<2	94
3	Ph	^{t}Bu	86	>98:<2	95
4	Ph	Bn	72	>98:<2	93
5	p-Tolyl	^{t}Bu	77	>98:<2	94

[a] Isolated yield.
[b] Determined by ^{1}H NMR.
[c] Determined by HPLC using a Chiracel OD column.
[d] 10 mol%.

Table 3.16. 1,3-Dipolar cycloadditions of nitrones and vinyl ethers catalyzed by the monomeric Al(III) complex (R)-**3.50** (10 mol%).

Entry	Nitrone R^1	Vinyl ether R^2	Yield[a]	Exo:endo[b]	Exo ee[c]
1	Ph	Et	93	>98:<2	99
2[d]	Ph	Et	83	>98:<2	99
3[d,e]	Ph	Et	76	>98:<2	99
4	p-Tolyl	Et	76	>98:<2	94
5	p-ClPh	Et	85	>98:<2	95
6	Ph	^{t}Bu	89	>98:<2	95
7	p-Tolyl	^{t}Bu	85	98:2	93
8	p-ClPh	^{t}Bu	82	>98:<2	93

[a] Isolated yield.
[b] Determined by ^{1}H NMR.
[c] Determined by HPLC using a Chiracel OD column.
[d] 0 °C.
[e] Solvent = toluene.

The major-groove poly(BINOL) (R)-**3.9** is also treated with AlMe$_3$ to form the polymeric Al(III) complex (R)-**3.51**. Polymer (R)-**3.51** (20 mol%) is used to catalyze the reaction of a nitrone (R^1 = Ph) and a vinyl ether (R^2 = Et). It gives a greatly reduced diastereoselectivity (exo:endo = 75:25) and enantioselectivity (23% ee). That is, the 3,3'-aryl substituents in the BINOL units of polymer (R)-**3.22** are important for the observed high stereoselectivity. In spite of the low stereoselectivity of the poly(BINOL) (R)-**3.51**, it is still significantly better than the use of BINOL–AlMe$_3$, which gives only 5% ee. This indicates that in the BINOL–AlMe$_3$- catalyzed reaction, the catalytically active species should involve its intermolecular aggregates rather than the monomeric complex. In polymer (R)-**3.51**, its catalytic sites should be the isolated BINOL–AlMe units that cannot form significant intermolecular aggregate.

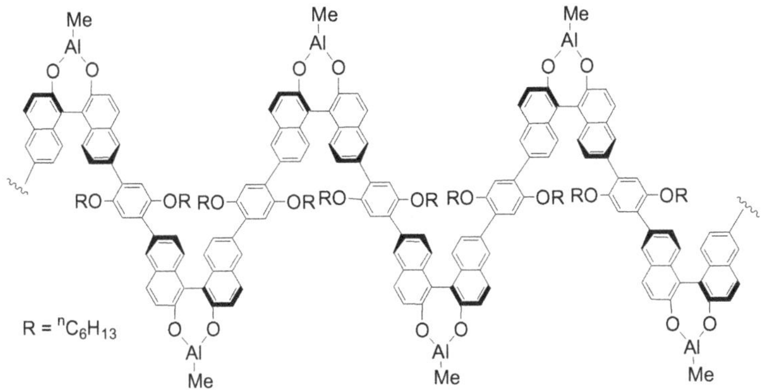

(R)-**3.51**

3.12. Asymmetric Michael Addition Catalyzed by the Poly(BINOL)s[17]

Shibasaki demonstrated that the heterobimetallic BINOL complex **3.52** catalyzed the reaction of malonates with enones with high

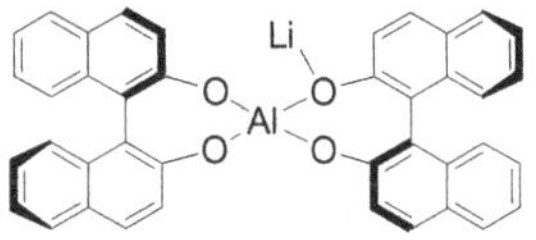

3.52

enantioselectivity.[18] In collaboration with Sasai, we have studied the use of the Al–Li complexes of the poly(BINOL)s for the asymmetric Michael reaction of 2-cyclohexenone with dibenzyl malonate (Scheme 3.35). The insoluble major-groove poly(BINOL) (R)-**3.7** is treated with LiAlH$_4$ (LAH), which generates an insoluble poly(BINOL)–Al–Li complex. When

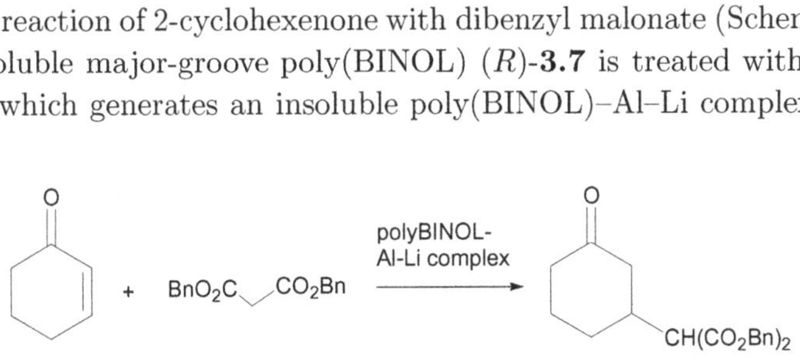

Scheme 3.35. The Michael addition catalyzed by the poly(BINOL) complexes.

Table 3.17. Asymmetric Michael reaction catalyzed by poly(BINOL) complexes.

Entry	Catalyst	Time (h)	Yield (%)	ee (%)
1	(R)-**3.7**+LAH	73	32	55
2	(R)-**3.7**+LAH+^{n}BuLi[a]	26	85	58
3	(R)-**3.9**+LAH	49	69	85
4	(R)-**3.9**+LAH+^{n}BuLi[a]	48	78	93
5	(R)-**3.22**+LAH	49	26	56
6	(R)-**3.22**+LAH+^{n}BuLi[a]	52	66	11

[a]The second generation catalysts: 0.9 molar equivalent of ^{n}BuLi is added to ∼10 mol% of the poly(BINOL)–LAH complex.

the soluble major-groove poly(BINOL) (R)-**3.9** is treated with LAH, its resulting poly(BINOL)–Al–Li complex becomes insoluble. When the minor-groove poly(BINOL) (R)-**3.22** is treated with LAH, the resulting Al–Li complex is still soluble. These minor-groove poly(BINOL)–Al–Li complexes are used to catalyze the reaction shown in Scheme 3.35 and the results are summarized in Table 3.17. Among the three polymers, (R)-**3.9** shows the highest enantioselectivity (85% ee, entry 3). When the Al–Li complex of the poly(BINOL) (R)-**3.9** is treated with one equivalent of ^{n}BuLi, the resulting insoluble polymeric complex exhibits further enhanced catalytic activity and enantioselectivity (93% ee, entry 4).

Since poly(BINOL) (R)-**3.7** is insoluble, when it is treated with LAH, the BINOL units in the polymer cannot associate with each other to form the dimeric complex such as **3.52**. In the case of polymer (R)-**3.22**, the steric hindrance around the BINOL units also prevents their association to generate the dimeric units such as **3.52** when this polymer is treated with LAH. Thus, the polymeric Al–Li complexes of (R)-**3.22** remain to be soluble. However, when the soluble poly(BINOL) (R)-**3.9** is treated with LAH, it could generate the dimeric complex such as **3.52** through inter polymer interaction, leading to the formation of a cross-linked insoluble polymer network. That is, the structure of the highly enantioselective catalyst **3.52** can be generated from the poly(BINOL) (R)-**3.9** but cannot be generated from (R)-**3.7** and (R)-**3.22**. This explains the high enantioselectivity of the complex (R)-**3.9**+LAH and that of (R)-**3.9**+LAH+^{n}BuLi.

The polymer ligand (R)-**3.9** can be recovered after treatment with 1 N HCl and precipitation with methanol. In addition, the cross-linked polymeric Al–Li complex of (R)-**3.9**+LAH+^{n}BuLi can also be recovered by

Table 3.18. Reuse of Al–Li complex from (R)-**3.9**+LAH+^{n}BuLi (10 mol%).

Run	Yield (%)	ee (%)	Recovery of catalyst
1	78	93	Quant.
2	66	89	98%
3	68	88	97%
4	56	76	95%
5	51	73	95%
6[a]	83	85	93%

[a] 5 mol% of ^{n}BuLi is added to the (R)-**3.9**+LAH+^{n}BuLi complex recovered from run 5.

simple filtration after reaction. The recovered ligand–Al–Li complex can be directly reused as the catalyst. This is different from the use of most of the polymer-supported Al(III) catalysts where it is the polymeric ligand being recovered and not the metal complex because of the moisture sensitivity of the Lewis acidic Al(III) complexes. Thus, the cross-linked polymer network not only allows easy recovery but also provides stabilization of the catalysts. Table 3.18 shows the catalytic properties of the recovered catalyst after multiple runs. Most of the catalytic activity of the polymeric catalyst can be restored with the addition of ^{n}BuLi after five runs.

3.13. Synthesis and Study of Poly(BINAP)[19]

BINOL provides the hard oxygen ligands and allows the binding of various hard metal centers to generate highly enantioselective Lewis acid catalysts for many carbon–carbon bond formation reactions. BINAP represents another class of 1,1'-binaphthyl-based compounds. It provides the soft

(R)-BINAP

phosphine ligands and allows the binding of various soft late transition metals to generate highly enantioselective catalysts for many reactions, most notably the asymmetric hydrogenation of alkenes.[20] Besides the poly(BINOL) catalysts described in the previous sections, we have also synthesized a poly(BINAP) catalyst for asymmetric hydrogenation.

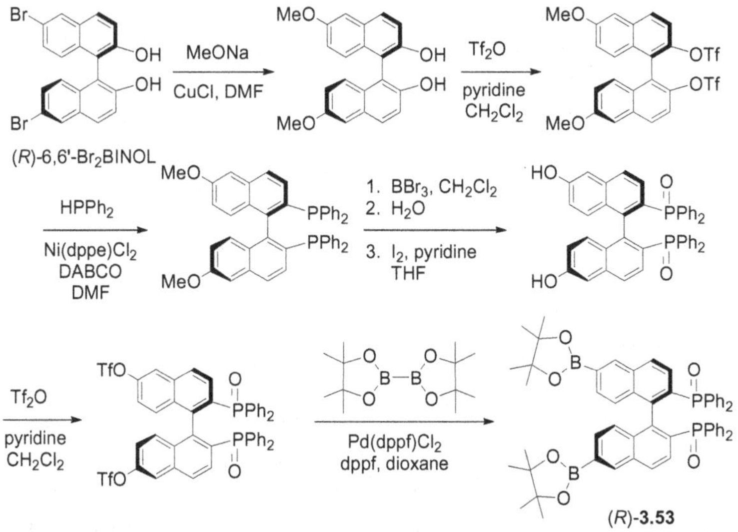

Scheme 3.36. Preparation of binaphthyl phosphinoxy monomer (R)-**3.53**.

Scheme 3.36 shows the multi-step synthesis of the 6,6′-diborate of the binaphthyl diphosphinoxy monomer (R)-**3.53** from (R)-6,6′-Br$_2$BINOL. Scheme 3.37 shows the Suzuki coupling polymerization of (R)-**3.53** with a phenylenebromide generating a phosphinoxy polymer. Reduction of this polymer with HSiCl$_3$ generating the poly(BINAP) (R)-**3.54**. This polymer is a yellow solid. It is soluble in chloroform, methylene chloride, THF and toluene. GPC shows its molecular weight as $M_w = 5800$ and $M_n = 4300$ (PDI = 1.35). Its specific optical rotation $[\alpha]_D$ is 36.6 ($c = 0.12$, THF). This polymer gives a well-resolved ^{1}H NMR spectrum, indicating a well-defined structure. Its ^{31}P NMR gives a predominate singlet at −14.88, which is identical to that of BINAP.

Treatment of the polyBINAP (R)-**3.54** with Rh(COD)$_2$BF$_4$ (COD = 1,5-cyclooctadiene) in a 1.1:1 ratio in THF gives a polymeric Rh complex (R)-**3.55**. This polymeric Rh complex is used to catalyze the asymmetric hydrogenation of dehydroamino esters or acids (Scheme 3.38). The results are summarized in Table 3.19. In the presence of 2 mol% of (R)-**3.55** under 30 psi H$_2$, the hydrogenation proceeds with quantitative conversion and up to 75% ee (entry 1). The catalytic properties of (R)-**3.55** are almost the same as the catalyst derived from BINAP (entry 2). This demonstrates that the structure and function of BINAP are retained in the rigid and sterically

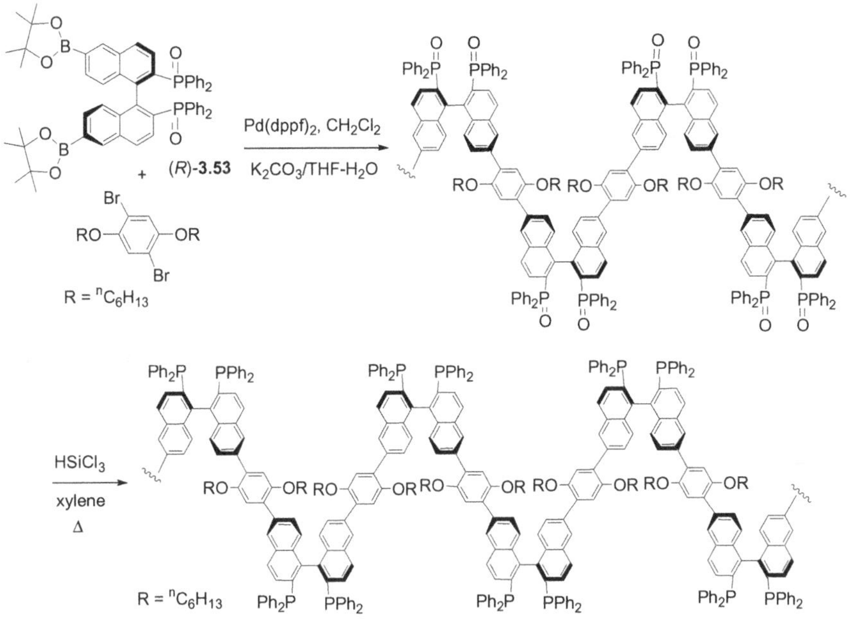

(R)-3.54

Scheme 3.37. Synthesis of a rigid and optically active poly(BINAP).

regular polymer. The polymeric catalyst can be easily recovered by simple filtration after reaction and the recovered catalyst shows the same catalytic properties as the original one (entry 3).

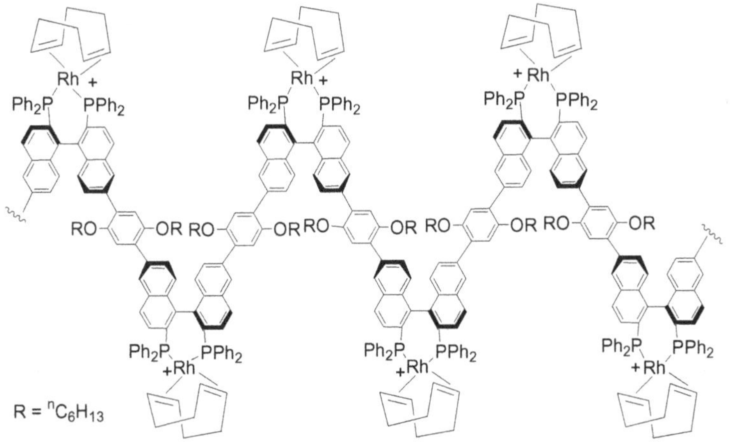

(R)-3.55

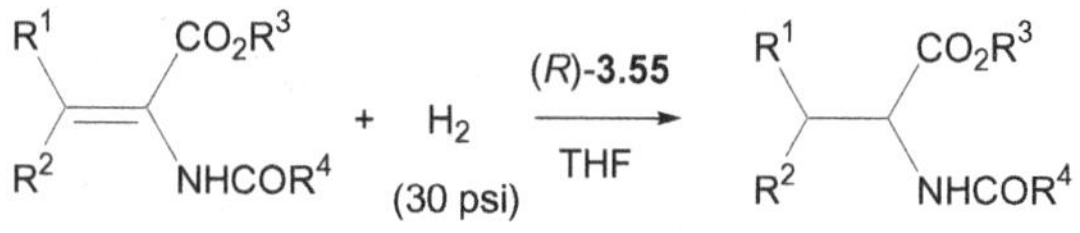

R^1=Ph, H; R^2=H, Ph
R^3=Me, H; R^3=Me, Ph

Scheme 3.38. Asymmetric hydrogenation of dehydroamino acid derivatives by the poly(BINAP)–Rh catalyst.

Table 3.19. Hydrogenation of dehydroamino acid using the poly(BINAP) (R)-**3.54** as the chiral ligand.[a]

Entry	Substrate	Chiral ligand	Conv. (%)[b]	ee (%)[c]	Conf.[d]
1	(Z)-methyl α-(benzamido) cinnamate	(R)-**3.54**	>99	75	S
2	(Z)-methyl α-(benzamido) cinnamate	(R)-BINAP	>99	76	S
3	(Z)-methyl α-(benzamido) cinnamate	(R)-**3.54**[e]	>99	75	S
4	methyl α-(acetamido) acrylate	(R)-**3.54**	>99	32	ND
5	(Z)-α-benzamido-cinnamic acid	(R)-**3.54**	>99	59[f]	S
6	(Z)-α-benzamido-cinnamic acid[g]	(R)-**3.54**	>99	57[f]	S
7	(E)-methyl α-(benzamido) cinnamate	(R)-**3.54**	>99	40	ND

[a] All the reactions were carried out at room temperature under 30 psi of H_2 in the presence of 2 mol% of Rh(COD)$_2$BF$_4$ and 2.2 mol% of (R)-**3.54** in THF.
[b] Determined by ^{1}H NMR.
[c] Determined by HPLC chiral OD column.
[d] The absolute configuration was determined by comparing the sign of the optical rotations with the literature data.
[e] Recovered catalyst (2 mol%) was used.
[f] Determined by HPLC after converted to the methyl ester.
[g] 0.5 Equivalents of Et$_3$N was added.

The poly(BINAP) (R)-**3.54** is also treated with [RuCl$_2$(C$_6$H$_6$)]$_2$ in DMF followed by addition of (R,R)-1,2-diphenylethylenediamine to form an air stable polymeric Ru complex (R)-**3.56**. The ^{31}P NMR spectrum of (R)-**3.56** gives a predominate signal at 46.3 ppm, which is very close to

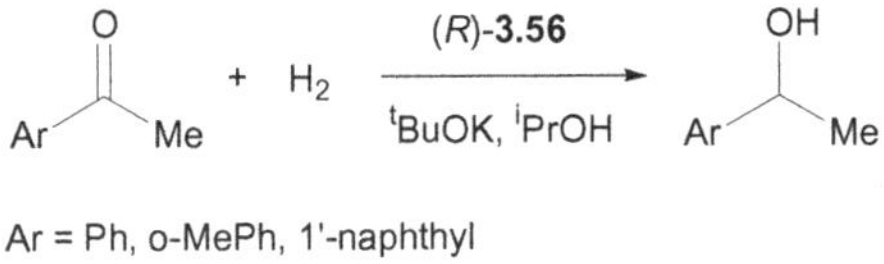

$$\text{Ar}\overset{\text{O}}{\underset{}{\|}}\text{Me} + \text{H}_2 \xrightarrow[{}^t\text{BuOK, }{}^i\text{PrOH}]{(R)\text{-}\mathbf{3.56}} \text{Ar}\overset{\text{OH}}{\underset{}{|}}\text{Me}$$

Ar = Ph, o-MePh, 1'-naphthyl

Scheme 3.39. Asymmetric hydrogenation of ketones by the poly(BINAP)–Ru catalyst (R)-**3.56**.[a]

the corresponding monomeric BINAP complex (47.4 ppm). This polymeric complex is used to catalyze the asymmetric hydrogenation of ketones (Scheme 3.39). The hydrogenation is conducted in isopropanol in the presence of 20 mol% of ^{t}BuOK using 0.5 mol% of (R)-**3.56**. As the results summarized in Table 3.20 show, high enantioselectivity has been observed for the hydrogenation of several aryl methyl ketones. The enantioselectivity is found to be independent of the hydrogen pressure (60–200 psi) but higher pressure leads to faster reaction. The reaction can be worked up in air and the catalyst (R)-**3.56** is recovered by precipitation with methanol. The catalytic properties of the recovered (R)-**3.56** are similar to those of the original catalyst (entry 8). The catalytic properties of this poly(BINAP) catalyst are also similar to those of the complex prepared from the monomeric BINAP. That is, the rigid and sterically regular poly(BINAP) has preserved the structure and function of the monomeric Ru catalyst and it provides the advantage of easy recovery and reuse.

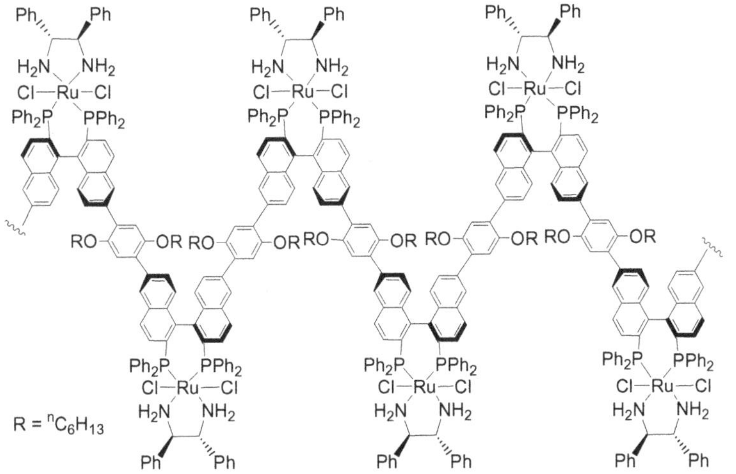

R = ^{n}C$_6$H$_{13}$

(R)-**3.56**

Table 3.20. Asymmetric hydrogenation of ketones catalyzed by the poly(BINAP)–Ru complex.[a]

Entry	Substrates	H_2 (psi)	Time (h)	Conversion (%)[b]	ee (%)[b]
1	acetophenone	60	72	100	80
2	2′-methyl 1-acetophenone	60	100	78	91
3	2′-methyl 1-acetophenone[c]	200	24	86	90
4	2′-methyl 1-acetophenone[c]	200	44	>99	90
5	1′-acetonaphthone	60	96	13	nd.
6	1′-acetonaphthone	200	66.5	>99	92
7	1′-acetonaphthone[c]	200	22	33.5	87
8	1′-acetonaphthone[c]	200	66	>99	89

[a]The reactions are carried out in 2-propanol at room temperature by using a suspension of 0.5 mol% of (R)-**3.56** and 20 mol% of potassium t-butoxide.
[b]Determined by GC with a chiral β-Dexc 120 column.
[c]Recovered (R)-**3.56** catalyst is used.

3.14. Synthesis and Study of a BINOL–BINAP Copolymer[21]

Because of the distinctively different coordination ability of BINOL and BINAP, a polymer containing both the BINOL and BINAP units can allow the incorporation of different metal centers to build multifunctional catalysts that are capable of catalyzing different reactions. Thus, the copolymerization of the BINOL monomer (R)-**3.45**, the BINAP monomer (R)-**3.53** and the terphenylene diiodide linker molecule **3.57** in a 1:1:2 ratio is conducted (Scheme 3.40). The MOM-protecting groups are then hydrolyzed and replaced with the more bulky benzoyl groups in order to increase the stability of the chiral configuration of the binaphthyl units during the subsequent high temperature reduction step. The resulting polymer (R)-**3.58** is reduced with $HSiCl_3$ in xylene. The benzoyl groups are then removed by reduction with LAH. This gives the BINOL–BINAP copolymer (R)-**3.59** as a yellow solid and soluble in chloroform, THF, methylene chloride, toluene and DMF. In this polymer, the BINOL and BINAP units are expected to be randomly distributed. GPC shows its molecular weight as $M_w = 11,600$ and $M_n = 7500$ (PDI = 1.55). Its specific optical rotation $[\alpha]_D$ is -75 ($c = 0.14$, CH_2Cl_2). This polymer gives a well-resolved ^1H NMR spectrum, indicating a well-defined structure. Its ^{31}P NMR gives a predominate singlet at -14.77, consistent with the BINAP units. In (R)-**3.59**, the BINOL side contains the minor-groove binaphthyl units and the BINAP side contains the major-groove binaphthyls.

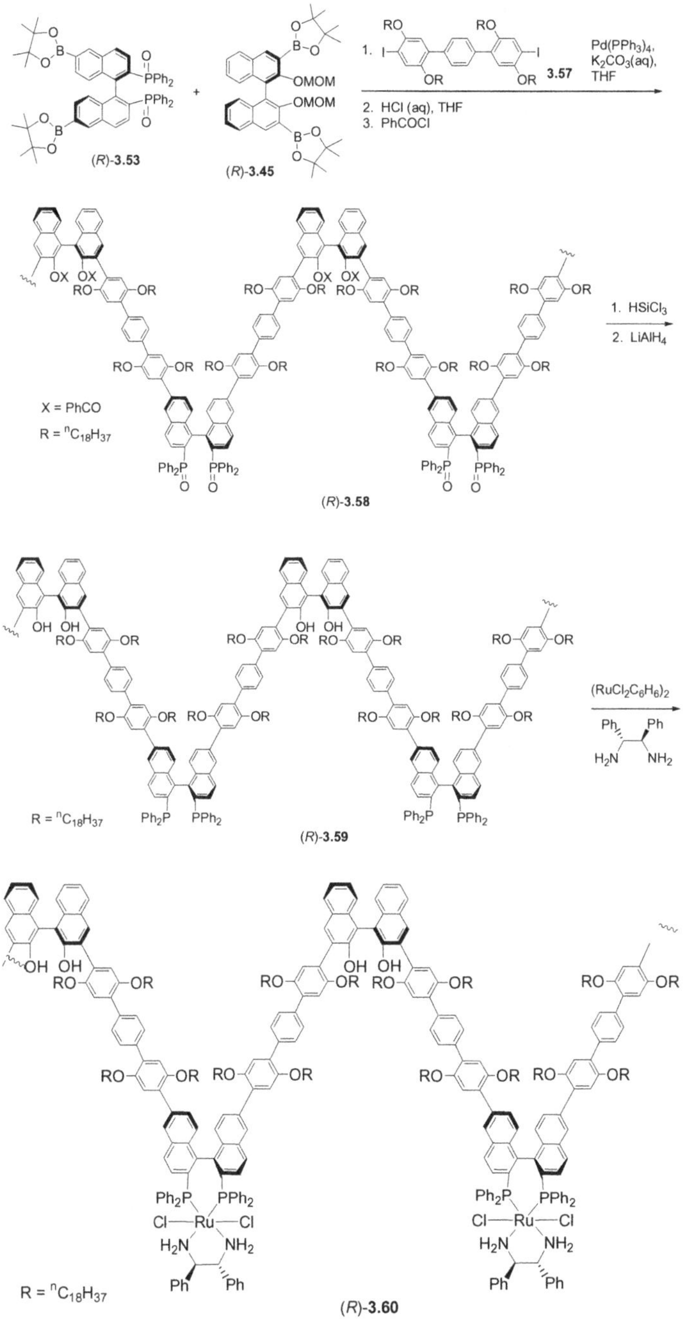

Scheme 3.40. Preparation of BINOL–BINAP copolymer.

When this polymer is treated with 0.5 equivalent of $(RuCl_2C_6H_6)_2$ and (R, R)-1,2-diphenylethylenediamine, the BINAP units of this polymer are converted to the corresponding Ru complexes to generate polymer (R)-**3.60**. Its ^{31}P NMR spectrum gives a predominate singlet at δ 46.33, which is identical to that of the corresponding monomeric BINAP–Ru complex. The remaining BINOL units in (R)-**3.60** can allow an *in situ* introduction of Lewis acidic metal centers upon treatment within metal complexes such as $ZnEt_2$.

Polymer (R)-**3.60** (4 mol% based on the repeating unit) is used to catalyze the tandem catalytic asymmetric $ZnEt_2$ addition and asymmetric hydrogenation of *p*-acetylbenzaldehyde (Scheme 3.41). In the first step, $ZnEt_2$ adds to the aldehyde to form an R alcohol. In the second step, the hydrogenation generates an S alcohol. Table 3.21 summarizes the reaction

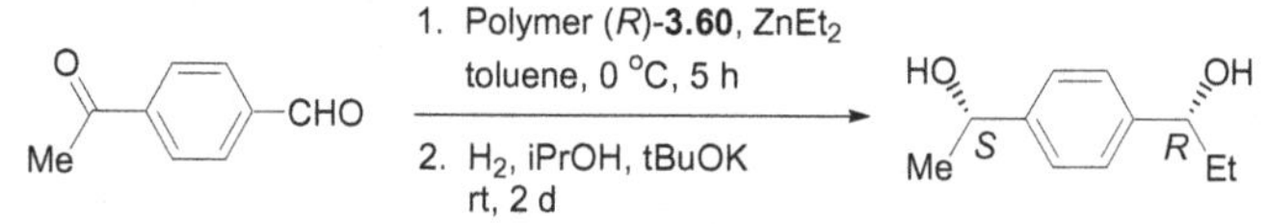

Scheme 3.41. A tandem asymmetric reaction involving diethylzinc addition and hydrogenation.

Table 3.21. The tandem asymmetric reactions of acetyl benzaldehydes in the presence of the multifunctional chiral polymer catalyst.

Entry	Substrate	Solvent		Conversion (%)[a]	ee (%)[b,c]	de (%)[b,d]
		$ZnEt_2$ addition	H_2 addition			
1	*p*-acetyl benzaldehyde	Toluene	^{i}PrOH	>99	92	86
2	*p*-acetyl benzaldehyde	Toluene	Toluene +^{i}PrOH[e]	>99	94	87
3[f]	*p*-acetyl benzaldehyde	Toluene	^{i}PrOH	>99	93	78
4	*m*-acetyl benzaldehyde	Toluene	^{i}PrOH	>99	94	75

[a]Determined by ^{1}H NMR.
[b]Determined by GC-chiral β-Dex 120 column after converted to the diacetate.
[c]For the diethylzinc addition.
[d]For the hydrogenation.
[e]Toluene:^{i}PrOH = 1:1.
[f]The recovered catalyst was used.

results. As shown in entry 1, the ee for the ZnEt$_2$ addition is 92% and the de for the hydrogenation is 86%. The stereoselectivities of the corresponding monomeric BINOL and BINAP catalysts have been mostly preserved in this copolymer. The polymeric catalyst can be easily recovered by precipitation with methanol and the recovered catalyst shows catalytic properties similar to those of the original polymer (entry 3). The tandem asymmetric ZnEt$_2$ addition and hydrogenation of m-acetylbenzaldehyde is also catalyzed by polymer (R)-**3.60**. It gives 94% ee for the ZnEt$_2$ addition and 75% de for the hydrogenation (entry 4).

A typical experimental procedure for the tandem asymmetric reactions is given here. Under nitrogen, ZnEt$_2$ (0.049 g, 0.40 mmol) was added to a toluene (2 mL) solution of polymer (R)-**3.60** (0.020 g, 0.0080 mmol, based on the polymer repeating unit) in a 10 mL flask. The mixture was stirred at room temperature for 15 min, and then p-acetylbenzaldehyde (0.030 g, 0.20 mmol) was added. The reaction mixture was stirred at 0°C for 5 h and was quenched by addition of ^{i}PrOH (0.5 mL). After removal of the solvent under vacuum, ^{t}BuOK (0.010 g, 0.090 mmol) and ^{i}PrOH (1.5 mL) were added. The mixture was then transferred into a 50 mL stainless steel pressure reactor and was degassed by two freeze-pump-thaw cycles. Hydrogen pressure (150 psi) was applied to the reactor, which was stirred at room temperature for 2 days. After the release of the hydrogen pressure, CH$_2$Cl$_2$ (10 mL) was added and the organic solution was washed with 1 N HCl (5 mL × 2) and brine (5 mL × 3). After concentration under vacuum, the polymer complex (R)-**3.60** was recovered from the organic solution by precipitation with MeOH. Removal of the solvents from the organic solution followed by purification with column chromatography on silica gel (Hexane:EtOAc = 3:1) gave the chiral diol product in quantitative yield. The ee (92%) and de (86%) of the product were determined by GC-chiral column after converted to the corresponding diacetate.

The BINOL–BINAP copolymer is also used to catalyze the individual asymmetric reactions. At a BINAP unit to substrate ratio of 1:4900, polymer (R)-**3.60** catalyzes the asymmetric hydrogenation of acetophenone to (S)-1-phenylethanol with >99% conversion and 84% ee. The reaction is conducted in toluene/^{i}PrOH (1:1) under H$_2$ (175 psi) in the presence of ^{t}BuOK at room temperature. At a BINOL unit to substrate ratio of 1:50, polymer (R)-**3.60** catalyzes the ZnEt$_2$ addition to p-acetylbenzaldehyde to give (R)-1-p-acetylphenyl propanol with 95% ee and complete conversion at 0°C in 5.5 h.

References

1. (a) Blossey EC, Ford WT, in *Comprehensive Polymer Science, The Synthesis, Characterization, Reactions and Applications of Polymers*, Allen G, Bevington JC (eds.), Pergamon Press, New York, Vol. 6, p. 81, 1989. (b) Pittman CU Jr, in *Comprehensive Organometallic Chemistry*, Wilkinson G, Stone FGA, Abel EW (eds.), Pergamon Press, Oxford, Vol. 8, p. 553, 1983. (c) *Chem Rev* **102**(10), 2002.
2. (a) Dumont W, Poulin JC, Dang TP, Kagan HB, *J Am Chem Soc* **95**:8295, 1973. (b) Han H, Janda KD, *Tetrahedron Lett* **38**:1527–1530, 1997.
3. Pu L, *Tetrahedron Asymmetry* **9**:1457–1477, 1998.
4. (a) Juliá S, Masana J, Vega JC, *Angew Chem Int Ed Engl* **19**:929, 1980. (b) Canali L, Karjalainen JK, Sherrington DC, Hormi O, *J Chem Soc Chem Commun* 123–124, 1997.
5. (a) Hu Q-S, Zheng X-F, Pu L, *J Org Chem* **61**:5200–5201, 1996. (b) Hu Q-S, Vitharana D, Zheng X-F, Wu C, Kwan, CMS, Pu L, *J Org Chem* **61**:8370–8377, 1996.
6. (a) Hu Q-S, Huang W-S, Vitharana D, Zheng X-F, Pu L, *J Am Chem Soc* **119**:12454–12464, 1997. (b) Huang W-S, Hu Q-S, Zheng X-F, Anderson J, Pu L, *J Am Chem Soc* **119**:4313–4314, 1997.
7. Johannsen M, Jørgensen KA, Zheng X-F, Hu Q-S, Pu L, *J Org Chem* **64**:299–301, 1999.
8. Yu H-B, Zheng X-F, Hu Q-S, Pu L, *Polymer Preprints* **40**(1):546–547, 1999.
9. Lin Z-M, Hu Q-S, Pu L, *Polymer Preprints* **39**(1):235, 1998.
10. (a) Hu Q-S, Huang W-S, Pu L, *J Org Chem* **63**:2798–2799, 1998. (b) Huang W-S, Hu Q-S, Pu L, *J Org Chem* **64**:7940–7956, 1999. (c) Huang W-S, Hu Q-S, Pu L, *J Org Chem* **63**:1364–1365, 1998.
11. Huang W-S, Pu L, *J Org Chem* **64**:4222–4223, 1999.
12. Yu H-B, Zheng X-F, Lin Z-M, Hu Q-S, Huang W-S, Pu L, *J Org Chem* **64**:8149–8155, 1999.
13. Jayaprakasha D, Kobayashia Y, Araia T, Hu Q-S, Zheng X-F, Pu L, Sasai H, *J Mol Catalysis* **196**:145–140, 2003.
14. Hu Q-S, Pu L, *Polymer Preprints* **41**(1):16–17, 2000.
15. (a) Ishihara K, Yamamoto H, *Eur J Org Chem* 527–538, 1999. (b) Ishihara K, Kurihara H, Matsumoto M, Yamamoto H, *J Am Chem Soc* **120**:6920, 1998. (c) Ishihara K, Yamamoto H, *J Am Chem Soc* **116**:1561, 1994.
16. Simonsen KB, Jørgensen KA, Hu Q-S, Pu L, *J Chem Soc Chem Commun* 811–812, 1999.
17. Arai T, Hu Q-S, Zheng X-F, Pu L, Sasai H, *Org Lett* **2**:4261–4263, 2000.
18. Arai T, Sasai H, Aoe K, Okamura K, Date T, Shibasaki M, *Angew Chem Int Ed Engl* **35**:104–106, 1996.
19. Yu H-B, Hu Q-S, Pu L, *Tetrahedron Lett* **41**:1681–1685, 2000.
20. (a) Noyori R, Takaya H, *Acc Chem Res* **23**:345, 1990. (b) Takaya H, Ohta T, Noyori R, in *Catalytic Asymmetric Synthesis*, Ojima I (ed.), VCH, New York, p. 1, 1993.
21. Yu H-B, Hu Q-S, Pu L, *J Am Chem Soc* **122**:6500–6501, 2000.

Chapter 4

Asymmetric Catalysis by BINOL and Its Non-polymeric Derivatives

4.1. Introduction

In Chapter 3, the development of the rigid and sterically regular polybinaphthyls as chiral catalysts for various asymmetric organic reactions is described. We have also studied the application of BINOL and its functionalized non-polymeric derivatives in asymmetric catalysis. The functionalized BINOLs described in this chapter contain substituents at the 3- or 3,3'-positions. Introduction of substituents to these positions can significantly modify both the steric and electronic environments around the chiral BINOL core, which has led to diverse catalytic properties in many asymmetric reactions. Besides our work in this area, many other researchers have conducted extensive studies on the use of BINOL derivatives for asymmetric catalysis. References 1–6 provide a few reviews on this subject. In this chapter, our work on using BINOL and its functional derivatives in combination with Lewis acidic metal complexes to catalyze the asymmetric reactions of prochiral aldehydes with alkynylzincs, arylzincs, alkylzincs, TMSCN and conjugated dienes are described. Highly enantioselective catalytic processes have been developed.

4.2. Asymmetric Alkyne Additions to Aldehydes

Chiral propargylic alcohols are very useful in the synthesis of diverse organic compounds. Two general methods are used for the asymmetric synthesis of chiral propargylic alcohols. One involves the asymmetric reduction of ynones[7–11] and the other is the nucleophilic addition of metal acetylides to aldehydes. The addition of metal acetylides to aldehydes should be a more efficient method because it simultaneously produces a carbon–carbon bond and a chiral center.

A few asymmetric alkyne additions have been developed. Corey reported an oxazaborolidine-catalyzed reaction of alkynylboranes with aldehydes which showed high enantioselectivity.[12] However, this method requires multiple steps to prepare the alkynylboranes. Carreira reported the use of N-methyl ephedrine in combination with $Zn(OTf)_2$ and Et_3N to

N-methyl ephedrine

catalyze the alkyne addition to α-substituted aliphatic aldehydes with high enantioselectivity.[13] This method is found to be practically useful in the preparation of certain types of propargylic alcohols. However, the catalytic process in the presence of N-methyl ephedrine, $Zn(OTf)_2$ and Et_3N is found to be not good for the reactions of alkynes with linear aliphatic, aromatic and α,β-unsaturated aldehydes. We have developed an array of catalysts based on BINOL and its derivatives to catalyze the highly enantioselective alkyne additions to a variety of aldehydes containing aromatic, aliphatic and vinyl substituents. These results as described in the following sections have significantly expanded the pathways to propargylic alcohols.

4.2.1. Catalysis by $BINOL$–$ZnEt_2$–$Ti(O^iPr)_4$[14,15]

We have used BINOL in combination with $ZnEt_2$ and $Ti(O^iPr)_4$ to catalyze the reaction of phenylacetylene with benzaldehyde for the asymmetric synthesis of propargylic alcohols. A two-step procedure is tested for this reaction: (1) phenylacetylene is treated with $ZnEt_2$ with or without the presence of BINOL and $Ti(O^iPr)_4$; (2) benzaldehyde [or with (S)-BINOL and $Ti(O^iPr)_4$] is then added. The first step probably produces an alkynylzinc intermediate whose addition to benzaldehyde is then subjected to the BINOL-based asymmetric catalysis (Scheme 4.1). Besides the desired propargylic alcohol product 1,3-diphenyl-prop-2-yn-1-ol (DPP), ethyl addition can also occur to generate 1-phenylpropanol (PPO) as a side product.

A variety of reaction conditions are examined and are summarized in Table 4.1. In all of the experiments, 1.1 mmol phenylacetylene is used to

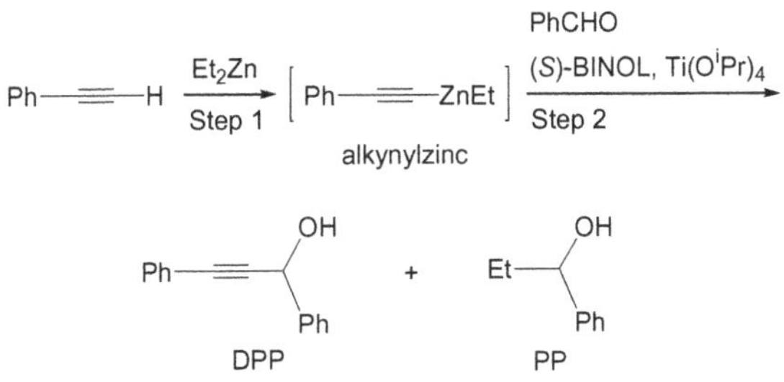

Scheme 4.1. The asymmetric reaction of phenylacetylene with benzaldehyde in the presence of (S)-BINOL, ZnEt$_2$ and Ti(O^{i}Pr)$_4$.

react with 1.0 mmol ZnEt$_2$ and 0.5 mmol benzaldehyde under nitrogen in dry solvents. Heating phenylacetylene and ZnEt$_2$ at reflux in toluene in the first step is necessary in order to avoid the formation of the ethyl addition side product and produce the propargylic alcohol as the major product. The high temperature in the first step allows the complete conversion of phenylacetylene and ZnEt$_2$ to the alkynylzinc intermediate. At lower temperatures in various solvents with or without BINOL and Ti(O^{i}Pr)$_4$ the ethyl addition product is observed as the major product (entries 1–7). Entry 5 indicates that when CH$_2$Cl$_2$ is used as the solvent, although the ee of DPP is high (85%), the main product is from the ethyl addition. However, entry 8 shows that the refluxing in toluene in the first step gives high yield of DPP but a lower ee (80%). Through a systematic variation of the reaction conditions, entry 19 is found to give the optimized result. It uses a dual solvent system in which toluene is used in the first step and methylene chloride is used in the second step. This reaction uses 20 mol% BINOL and 50 mol% Ti(O^{i}Pr)$_4$ and gives the desired proparglic alcohol as the major product with 96% ee.

In these experiments, HPLC-Chiralcel OD column is used to determine the enantiomeric purity of the propargylic alcohol and its configuration is R. Using BINOL in combination with ZnMe$_2$ and Ti(O^{i}Pr)$_4$ is also reported by Chan to show high enantioselectivity for the reaction of phenylacetylene with a few aromatic aldehydes.[16]

Using the optimized procedure of entry 19 in Table 4.1, we have conducted the asymmetric reaction of terminal alkynes with various aromatic aldehydes. The results are summarized in Table 4.2. It shows that in the presence of BINOL, ZnEt$_2$ and Ti(O^{i}Pr)$_4$, high enantioselectivity is achieved for the phenylacetylene addition to benzaldehydes bearing electron-donating and electron-withdrawing substituents at the o-, m- and

Table 4.1. Results for the reaction of phenylacetylene with benzaldehyde (Tol = toluene).

Entry	Condition for step 1	Solvent in step 2	Ti(O^{i}Pr)$_4$ (mol%)	BINOL (mol%)	Temperature for step 2	Major product	ee of DPP (%)
1	Tol (5 mL), BINOL, rt	Tol (5 mL)	25	10	rt	PPO	~85
2	CH$_2$Cl$_2$ (5 mL), BINOL, rt	CH$_2$Cl$_2$ (5 mL)	25	10	rt	PPO	~88
3	THF (5 mL), BINOL, rt	THF (5 mL)	25	10	rt	PPO	~70
4	Ether (5 mL), BINOL, rt	Ether (5 mL)	25	10	rt	PPO	~78
5	CH$_2$Cl$_2$ (5 mL), reflux	CH$_2$Cl$_2$ (5 mL)	25	10	rt	PPO	~85
6	CH$_2$Cl$_2$ (5 mL), BINOL, reflux	CH$_2$Cl$_2$ (5 mL)	25	10	rt	PPO	ND
7	CH$_2$Cl$_2$ (5 mL), BINOL, Ti(O^{i}Pr)$_4$, reflux	CH$_2$Cl$_2$ (5 mL)	25	10	rt	PPO	ND
8	Tol (5 mL), reflux	Tol (5 mL)	25	10	rt	DPP	80
9	Tol (1 mL), reflux	Tol (1 mL)+CH$_2$Cl$_2$ (4 mL)	25	10	rt	DPP	85
10	Tol (1 mL), reflux	Tol (1 mL)+CH$_2$Cl$_2$ (4 mL)	25	10	0°C	DPP	62
11	Tol (1 mL), reflux	Tol (1 mL)+CH$_2$Cl$_2$ (4 mL)	25	10	42°C	DPP	48
12	Tol (1 mL), reflux	Tol (1 mL)+CH$_2$Cl$_2$ (4 mL)	25	10	12°C	DPP	78
13	Tol (1 mL), reflux	Tol (1 mL)+CH$_2$Cl$_2$ (4 mL)	100	10	rt	DPP	87
14	Tol (1 mL), reflux	Tol (1 mL)+CH$_2$Cl$_2$ (4 mL)	10	10	rt	DPP	70
15	Tol (1 mL), reflux	Tol (1 mL)+CH$_2$Cl$_2$ (4 mL)	50	20	rt	DPP	91
16	Tol (2 mL), reflux	Tol (2 mL)+CH$_2$Cl$_2$ (8 mL)	25	10	rt	DPP	92
17	Tol (2 mL), reflux	Tol (2 mL)+CH$_2$Cl$_2$ (18 mL)	25	10	rt	DPP	90
18	Tol (2 mL), reflux	Tol (2 mL)+CH$_2$Cl$_2$ (48 mL)	25	10	rt	DPP	80
19	**Tol (2 mL), reflux**	**Tol (2 mL)+CH$_2$Cl$_2$ (8 mL)**	**50**	**20**	**rt**	**DPP**	**96**

Table 4.2. Results for the terminal alkyne additions to aldehydes in the presence of (S)-BINOL, ZnEt$_2$ and Ti(O^{i}Pr)$_4$.

Entry	Alkyne	Aldehyde	Isolated yield (%)	ee (%)
1	Ph–≡	Ph–CHO	77	96
2	Ph–≡	2-Cl-C$_6$H$_4$-CHO	79	92
3	Ph–≡	4-Cl-C$_6$H$_4$-CHO	81	92
4	Ph–≡	2-Me-C$_6$H$_4$-CHO	81	96
5	Ph–≡	3-Me-C$_6$H$_4$-CHO	77	94
6	Ph–≡	2-OMe-C$_6$H$_4$-CHO	73	93
7	Ph–≡	3-MeO-C$_6$H$_4$-CHO	78	93
8	Ph–≡	4-F-C$_6$H$_4$-CHO	74	96
9	Ph–≡	4-O$_2$N-C$_6$H$_4$-CHO	79[a]	97
10	Ph–≡	2-naphthyl-CHO	77	98
11	Ph–≡	1-naphthyl-CHO	71[a]	92
12	^{i}Pr$_3$Si–≡	Ph–CHO	75	92

[a] Determined by ^{1}H NMR.

p-positions (92–97% ee, entries 1–9). The reactions of other aromatic aldehydes such as 2-naphthaldehyde and 1-naphthaldehyde give 92–98% ee (entries 10 and 11). The triisopropylsilylacetylene addition to benzaldehyde is also highly enantioselective (92% ee, entry 12).

When the above conditions for the highly enantioselective alkyne addition to aromatic aldehydes are applied to the reaction of phenylacetylene with nonyl aldehyde, an aliphatic aldehyde, the propargylic alcohol product is obtained with only 75% ee (Table 4.3, entry 1). In order to expand the scope of the enantioselective alkyne addition, the reaction conditions are further explored. As shown in Table 4.3, when the solvent is changed from CH_2Cl_2 to diethyl ether, the enantioselectivity for the reaction of phenylacetylene with nonyl aldehyde is improved to 83% (entry 4). However, changing the reaction temperature, adding molecular sieves and using $Ti(O^tBu)_4$ in place of $Ti(O^iPr)_4$ all cannot increase ee. Finally, it is found that increasing the amount of all the reagents gives excellent enantioselectivity and high yield (91% ee and 96% yield, entry 14).

The conditions of entry 14 in Table 4.3 are applied to the reaction of phenylacetylene with other aliphatic aldehydes and the results are summarized in Table 4.4. Table 4.4 shows that the reactions of phenylacetylene with linear, α-branched and β-branched aliphatic aldehydes give very high enantioselectivity (91–95% ee, entries 1–9). In addition, excellent enantioselectivity is also observed for α,β-unsaturated aldehydes (96–99% ee, entries 10–13). Further increasing the amount of the reagents further increases ee (entries 14–18).

We have also identified a small non-linear effect between the enantiomeric purity of BINOL and that of the propargylic alcohol product in this asymmetric alkyne addition process. Thus, the aggregates of the BINOL-based metal complexes may have certain effect on this reaction.

4.2.2. Catalysis by BINOL–$ZnEt_2$–$Ti(O^iPr)_4$–HMPA and Functional Alkyne Addition[17–19]

As described in the previous section, the reaction catalyzed by BINOL–$ZnEt_2$–$Ti(O^iPr)_4$ requires the reflux of a terminal alkyne with $ZnEt_2$ in toluene in order to generate the nucleophilic alkynylzinc reagents. The high temperature in this step limits the scope of the alkynes since certain functional terminal alkynes undergo undesirable side reactions with $ZnEt_2$ under this condition. Thus reducing the temperature is necessary for the formation of certain functional alkynylzincs.

It was reported that $ZnEt_2$ can react with terminal alkynes rapidly at room temperature to generate alkynylzincs in highly polar aprotic solvents such as DMSO, DMF and HMPA.[20] We have thus tested the use of these compounds as additives in the asymmetric alkyne addition.

Table 4.3. Attempted reactions of phenylacetylene with nonyl aldehyde (Tol = toluene).[a]

Entry	Conditions for step 1	Conditions for step 2	Phenylacetylene/diethylzinc (mol%)	Ti(O^{i}Pr)$_4$ (mol%)	BINOL (mol%)	ee (%)
1	2 mL Tol, reflux 5 h	8 mL CH$_2$Cl$_2$, rt, 4 h	200/200	50	20	75
2	2 mL Tol, reflux 5 h	4 mL Tol, rt, 4 h	200/200	50	20	75
3	2 mL Tol, reflux 5 h	8 mL THF, rt, 4 h	200/200	50	20	79
4	2 mL Tol, reflux 5 h	8 mL Et$_2$O, rt, 4 h	200/200	50	20	83
5	2 mL Tol, reflux 5 h	8 mL Et$_2$O, 0 °C, 4 h	200/200	50	20	59
6	2 mL Tol, reflux 5 h	8 mL Et$_2$O, reflux, 4 h	200/200	50	20	77
7	2 mL Tol, reflux 5 h	8 mL Et$_2$O, MS, rt, 4 h	200/200	50	20	83
8	2 mL Tol, reflux 5 h	8 mL Et$_2$O, rt, 4 h	200/200	50	20	63[b]
9	2 mL Tol, reflux 5 h	8 mL Et$_2$O, rt, 4 h	200/200	140	20	68
10	2 mL Tol, reflux 1 h	8 mL Et$_2$O, rt, 4 h	200/200	50	20	89[c]
11	2 mL Tol, reflux 0.5 h	8 mL Et$_2$O, rt, 4 h	200/200	50	20	82[d]
12	2 mL Tol, reflux 1 h	8 mL Et$_2$O, rt, 4 h	400/400	50	20	85
13	2 mL Tol, reflux 1 h	8 mL Et$_2$O, rt, 4 h	200/200	100	40	84[e]
14	2 mL Tol, reflux 1 h	8 mL Et$_2$O, rt, 4 h	400/400	100	40	91[f]
15	1 mL Tol, reflux 1 h	8 mL Et$_2$O, rt, 4 h	400/400	100	40	91[g]

[a] Redistilled nonyl aldehyde was used unless otherwise indicated.
[b] Ti(O^{t}Bu)$_4$ was used.
[c] Yield: 54%.
[d] Yield: 9%.
[e] Yield: 35.9%.
[f] Yield: 96%.
[g] Undistilled aldehyde was used. Yield: 66%.

Table 4.4. Results for the addition of phenylacetylene to aliphatic and α,β-unsaturated aldehydes.[a]

Entry	Aldehyde	Isolated yield (%)	ee (%)
1	C_7H_{15}CHO	66	91
2[b]	C_7H_{15}CHO	96	91
3	C_6H_{13}CHO	70	93
4	C_3H_7CHO	91	93
5	iPr-CHO	84	97
6	Et-CHO	60	94
7	Cy-CHO	58	95
8	PhCH₂-CHO	99	93
9[c]	PhCH₂-CHO	93	91
10	CH₃CH=CH-CHO	92	96
11	(CH₃)₂C=CH-CHO	93	96
12	PhCH=CH-CHO	89	97
13	PhCH=C(CH₃)-CHO	96	99
14[b,d]	C_7H_{15}CHO	94	94
15[b,e]	C_7H_{15}CHO	92	96

(*Continued*)

Table 4.4. (*Continued*)

Entry	Aldehyde	Isolated yield (%)	ee (%)
16[b,f]	M₇CHO	92	95
17[f]	M₇CHO	77	96
18[g]	M₇CHO	96	96

[a] Conditions of entry 15 in Table 4.2 were used with undistilled aldehydes unless otherwise indicated.
[b] Redistilled aldehyde was used. An aliquot of 2 mL of toluene was used in the first step.
[c] Aldehyde was added dropwise.
[d] phenylacetylene:ZnEt$_2$:Ti(O^{i}Pr)$_4$:BINOL:aldehyde = 6:6:1.5:0.6:1.
[e] phenylacetylene:ZnEt$_2$:Ti(O^{i}Pr)$_4$:BINOL:aldehyde = 6:6:2:0.8:1.
[f] phenylacetylene:ZnEt$_2$:Ti(O^{i}Pr)$_4$:BINOL:aldehyde = 6:6:2.5:1:1.
[g] The same condition as entry 17 in this table except that redistilled aldehyde was used.

Table 4.5 summarizes the results for the reaction of phenylacetylene with benzaldehyde in the presence of BINOL–Ti(O^{i}Pr)$_4$–ZnEt$_2$ and the additives. The reactions are conducted by mixing (*S*)-BINOL, phenylacetylene, ZnEt$_2$ and an additive in a solvent at room temperature for 1 h and then with Ti(O^{i}Pr)$_4$ for 1 h followed by the addition of benzaldehyde unless otherwise indicated. Without an additive, only a very small amount of the propargylic alcohol product is produced (entry 1). We find that the use of HMPA as the additive is better than the use of DMF and DMSO. As shown in entry 22, using two equivalents of HMPA and 40 mol% of BINOL in methylene chloride at room temperature gives good yield and excellent enantioselectivity (93% *ee* and 76% yield). The configuration of the propargylic alcohol product is *R*.

Table 4.6 summarizes the results obtained when the conditions of entry 22 in Table 4.5 are applied to the reaction of several alkynes with aldehydes. At room temperature, high enantioselectivity is observed for the reactions of a variety of aromatic aldehydes (88–93% *ee*). Particularly interesting is that good results are observed for the reactions of several functional alkynes (entries 14–16).

1,1′-Binaphthyl-Based Chiral Materials

Table 4.5. Reactions of phenylacetylene with benzaldehyde in the presence of (S)-BINOL, ZnEt$_2$, Ti(O^{i}Pr)$_4$ and an additive.

Entry	BINOL (mol%)	Additive	PhCCH+ZnEt$_2$	Ti(O^{i}Pr)$_4$ (mol%)	Solvent	T (°C)	Yield (%)	ee (%)
1	20	None	Two equiv.	50	CH$_2$Cl$_2$ (3 mL)	rt	–[e]	88
2	20	DMF (two equivalents)	Two equiv.	50	CH$_2$Cl$_2$ (3 mL)	rt	51	75
3	20	DMSO (two equiv)	Two equiv.	50	CH$_2$Cl$_2$ (3 mL)	rt	43	71
4	20	DMF (four equiv.)	Two equiv.	50	CH$_2$Cl$_2$ (3 mL)	rt	51	72
5	20	DMSO (four equiv.)	Two equiv.	50	CH$_2$Cl$_2$ (3 mL)	rt	~60	68
6	20	HMPA (two equiv)	Two equiv.	50	CH$_2$Cl$_2$ (3 mL)	rt	88	83
7	20	HMPA (1 equiv.)	Two equiv.	50	CH$_2$Cl$_2$ (3 mL)	rt	88	82
8[a]	20	HMPA (two equiv)	Two equiv.	50	CH$_2$Cl$_2$ (3 mL)	rt	73	69
9	40	HMPA (two equiv)	Two equiv.	50	CH$_2$Cl$_2$ (3 mL)	rt	77	86
10	20	HMPA (two equiv)	Two equiv.	100	CH$_2$Cl$_2$ (3 mL)	rt	73	85
11	20	HMPA[b] (two equiv.)	Two equiv.	50	CH$_2$Cl$_2$ (3 mL)	0	46	93
12	20	HMPA (two equiv)	Two equiv.	50	THF (3 mL)	−40 to −50	49	76
13	20	HMPA (two equiv)	Two equiv.	50	Ether (3 mL)	rt	–	52
14	20	HMPA[b] (two equiv.)	Two equiv.	50	Tol (3 mL)	rt	67	76
15	20	HMPA[b] (two equiv.)	Two equiv.	50	CH$_2$Cl$_2$ (3 mL)	rt	90	86
16	20	HMPA[b] (two equiv.)	Two equiv.	100	CH$_2$Cl$_2$ (3 mL)	rt	76	87
17	20	HMPA[b] (two equiv.)	Two equiv.	150	CH$_2$Cl$_2$ (3 mL)	rt	83	83
18[c]	20	HMPA[b] (two equiv.)	Two equiv.	50	CH$_2$Cl$_2$ (3 mL)	rt	37	82
19[d]	20	HMPA[b] (two equiv.)	Two equiv.	50	CH$_2$Cl$_2$ (3 mL)	rt	90	85
20	20	HMPA[b] (two equiv.)	Two equiv.	50	CH$_2$Cl$_2$ (1 mL)	rt	98	76
21	20	HMPA[b] (two equiv.)	Two equiv.	50	CH$_2$Cl$_2$ (5 mL)	rt	85	86
22	**40**	**HMPA[b] (two equiv.)**	**Four equiv.**	**100**	**CH$_2$Cl$_2$ (3 mL)**	**rt**	**72**	**93**
23	40	HMPA[b] (two equiv.)	Four equiv.	100	CH$_2$Cl$_2$ (6 mL)	rt	75	93
24	40	HMPA[b] (four equiv.)	Four equiv.	100	CH$_2$Cl$_2$ (6 mL)	rt	~80	89

[a] Phenylacetylene, ZnEt$_2$, and HMPA in methylene chloride were stirred for 1 h, and then (S)-BINOL and Ti(O^{i}Pr)$_4$ were added. After an additional hour, benzaldehyde was added.
[b] Redistilled HMPA was used.
[c] (S)-BINOL, HMPA, phenylacetylene, ZnEt$_2$ and Ti(O^{i}Pr)$_4$ in methylene chloride were stirred for 1 h, and then benzaldehyde was added.
[d] ZnMe$_2$ was used in place of ZnEt$_2$.
[e] The main product was that of the ZnEt$_2$ addition.

Table 4.6. Asymmetric reactions of various alkynes with aromatic aldehydes in the presence of (S)-BINOL–ZnEt$_2$–Ti(O^{i}Pr)$_4$–HMPA at room temperature.

Entry	Alkyne	Aldehyde	Isolated yield (%)	ee (%)
1	Ph–≡	Ph–CHO	72	93
2	Ph–≡	2-Me-C$_6$H$_4$–CHO	77	93
3	Ph–≡	3-Me-C$_6$H$_4$–CHO	75	93
4	Ph–≡	4-Me-C$_6$H$_4$–CHO	69	93
5	Ph–≡	3-Cl-C$_6$H$_4$–CHO	57	93
6	Ph–≡	4-Cl-C$_6$H$_4$–CHO	57	93
7	Ph–≡	4-MeO-C$_6$H$_4$–CHO	56	93
8	Ph–≡	4-F-C$_6$H$_4$–CHO	67	93
9	Ph–≡	4-Br-C$_6$H$_4$–CHO	72	93
10	Ph–≡	C$_6$F$_5$–CHO	66	88
11	Ph–≡	1-naphthyl–CHO	86	95
12	Ph–≡	2-furyl–CHO	69	88
13	Ph–≡	PhCH=CH–CHO	56	92
14	(EtO)$_2$CH–C≡CH	Ph–CHO	51	91
15	Cl–CH$_2$CH$_2$CH$_2$–C≡CH	Ph–CHO	53	93
16	AcO–CH$_2$–C≡CH	Ph–CHO	52	88

Among the reaction of functional alkynes, we have conducted a more detailed study on the addition of methyl propiolate to aldehydes because the resulting γ-hydroxy-α,β-acetylenic esters from these reactions have been extensively used in the synthesis of highly functional organic molecules. Previously, the optically active γ-hydroxy-α,β-acetylenic esters were prepared by a three-step reaction sequence, that is, the addition of the lithiated methyl propiolate to aldehydes at very low temperature followed by oxidation and asymmetric reduction.[21,22] No asymmetric methyl propiolate addition to aldehyde was reported before.[23]

We have studied the use of the above BINOL–Ti(O^{i}Pr)$_4$–ZnEt$_2$–HMPA system for the asymmetric methyl propiolate addition. It is found that treatment of methyl propiolate with ZnEt$_2$ in the presence of (R)-BINOL and HMPA at room temperature at a longer time (16 h) before the addition of Ti(O^{i}Pr)$_4$ and benzaldehyde gives the desired product with both good yield and excellent enantioselectivity (Scheme 4.2). Other reaction conditions are the same as those shown in entry 22 in Table 4.5. The absolute configuration of the product is S determined by studying the ^{19}F NMR of its Mosher esters. This is consistent with the products of other alkyne additions to aldehydes when BINOL is used as the catalyst. That is, the ester function of methyl propiolate probably does not change the mechanism of the BINOL-based asymmetric alkyne addition.

The above reaction conditions are applied to the methyl propiolate addition to a variety of aromatic and α,β-unsaturated aldehydes and the results are summarized in Table 4.7. The ee's are determined by using HPLC-Diacel Chiralcel OD column. Excellent enantioselectivities are achieved for the reactions of benzaldehydes containing o-, m- and p-electron-withdrawing or electron-donating groups (85–95% ee, entries 1–12). The reactions of other aromatic aldehydes including 2-furaldehyde

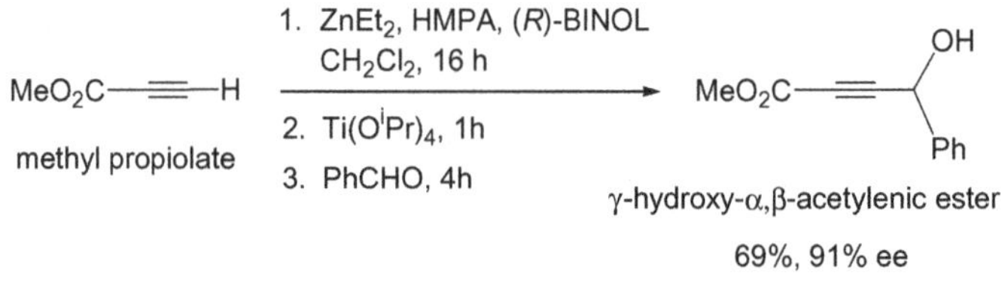

Scheme 4.2. The asymmetric reaction of methyl propiolate with benzaldehyde using the BINOL–Ti(O^{i}Pr)$_4$–ZnEt$_2$–HMPA catalyst system.

Table 4.7. Enantioselective addition of methyl propiolate to aldehydes.[a]

Entry	Aldehyde	Product	Isolated yield (%)	ee (%)[b]
1	PhCHO	PhCH(OH)C≡CCO$_2$Me	69	91
2	2-MeC$_6$H$_4$CHO	2-MeC$_6$H$_4$CH(OH)C≡CCO$_2$Me	96	91
3	3-MeC$_6$H$_4$CHO	3-MeC$_6$H$_4$CH(OH)C≡CCO$_2$Me	91	93
4	4-MeC$_6$H$_4$CHO	4-MeC$_6$H$_4$CH(OH)C≡CCO$_2$Me	81	93
5	2-MeOC$_6$H$_4$CHO	2-MeOC$_6$H$_4$CH(OH)C≡CCO$_2$Me	91	90
6	3-MeOC$_6$H$_4$CHO	3-MeOC$_6$H$_4$CH(OH)C≡CCO$_2$Me	52	90
7	4-MeOC$_6$H$_4$CHO	4-MeOC$_6$H$_4$CH(OH)C≡CCO$_2$Me	82	91
8	2-ClC$_6$H$_4$CHO	2-ClC$_6$H$_4$CH(OH)C≡CCO$_2$Me	94	91
9	3-ClC$_6$H$_4$CHO	3-ClC$_6$H$_4$CH(OH)C≡CCO$_2$Me	90	93
10	4-ClC$_6$H$_4$CHO	4-ClC$_6$H$_4$CH(OH)C≡CCO$_2$Me	84	95
11	4-BrC$_6$H$_4$CHO	4-BrC$_6$H$_4$CH(OH)C≡CCO$_2$Me	82	93
12	4-FC$_6$H$_4$CHO	4-FC$_6$H$_4$CH(OH)C≡CCO$_2$Me	76	85

(*Continued*)

Table 4.7. (*Continued*)

Entry	Aldehyde	Product	Isolated yield (%)	ee (%)[b]
13	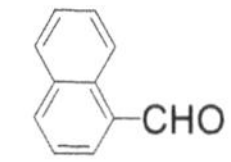 1-naphthyl-CHO	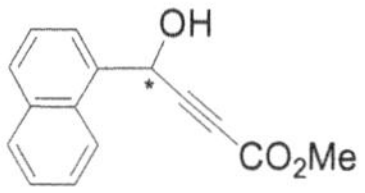 1-naphthyl-CH(OH)-C≡C-CO₂Me	87	95
14	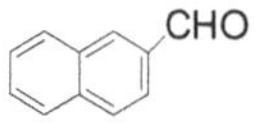 2-naphthyl-CHO	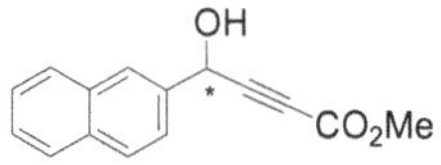 2-naphthyl-CH(OH)-C≡C-CO₂Me	84	93
15	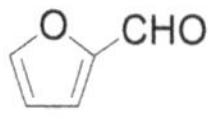 furyl-CHO	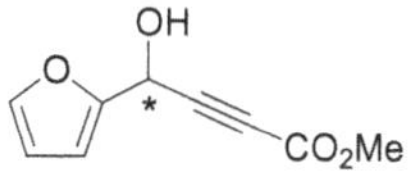 furyl-CH(OH)-C≡C-CO₂Me	55[a]	87
16	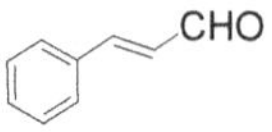 PhCH=CH-CHO	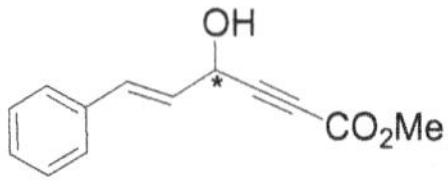 PhCH=CH-CH(OH)-C≡C-CO₂Me	65[b]	91
17	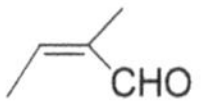 (CH₃)₂C=CH-CHO	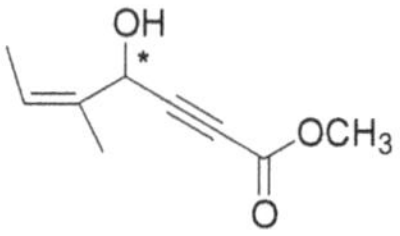 (CH₃)C=CH-CH(OH)-C≡C-C(O)OCH₃	38	90

[a] Reaction conditions. Under argon, (*R*)-BINOL (40 mol%) methylene chloride (3 mL), HMPA (0.50 mmol), methyl propiolate (1.0 mmol) and ZnEt₂ (1.0 mmol) were stirred at room temperature for 16 h. Ti(O^{i}Pr)₄ (0.25 mmol) was added and stirred for 1 h. An aldehyde (0.25 mmol) was then added and stirred for 4 h.
[b] Determined by NMR analysis.

and 1-/2-naphthaldehyde also give high *ee*'s (87–95% *ee*, entries 13–15). High enantioselectivity is observed for the reaction of α,β-unsaturated aldehydes as well. The methyl propiolate addition to cinnamaldehyde gives 91% *ee* (entry 16), and to 2-methyl-2-butenal gives 90% *ee* but lower yield (entry 17).

Under the same conditions as above, the reaction of methyl propiolate with aliphatic aldehydes is conducted in the presence of (*S*)-BINOL and the results are summarized in Table 4.8. The *ee*'s are measured by analyzing the ^{1}H NMR spectrum of the mandelate acetate. Good enantioselectivites are observed for the reactions of linear, α-branched and β-branched aliphatic aldehydes.

The BINOL-based Zn–Ti complex shown below is proposed as one of the possible intermediates for the alkyne addition to aldehydes catalyzed by the (*S*)-BINOL system. The molecular modeling structure is obtained by

Table 4.8. Results for the enantioselective addition of methyl propiolate to aliphatic aldehydes.

Entry	Aldehyde	Product	Isolated yield (%)	ee (%)
1	CH$_3$(CH$_2$)$_6$CHO	HO-CH(C$_6$H$_{13}$)-C≡C-CO$_2$CH$_3$	72	81
2	CH$_3$(CH$_2$)$_3$CHO	HO-CH(C$_3$H$_7$)-C≡C-CO$_2$CH$_3$	76	89
3	Cy-CHO	HO-CH(Cy)-C≡C-CO$_2$CH$_3$	60	81
4	iPr-CH$_2$-CHO	HO-CH(CH$_2$iPr)-C≡C-CO$_2$CH$_3$	73	83

using the PC Spartan Semiemperical PM3 program. In this intermediate, the aldehyde carbonyl oxygen is coordinated to both the Ti(IV) and Zn(II) center. The ethyl group on Zn is down in order to avoid the steric interaction with the 3-hydrogen of (S)-BINOL, and the aldehyde ^{n}Bu group is up in order to avoid the steric interaction with the ethyl group. Thus the subsequent alkyne migration to the si face of the aldehyde will give the observed R product.

4.2.3. Reactivity of γ-Hydroxy-α,β-acetylenic Esters

After the preparation of the chiral γ-hydroxy-α,β-acetylenic esters from the asymmetric alkyne addition, we have studied their conversions to other functional products. We find that a platinum(II) complex, Ziese's dimer, can catalyze a regiospecific hydration of the chiral γ-hydroxy-α,β-acetylenic esters.[19] As shown in Scheme 4.3, in the presence of a catalytic amount of Ziese's dimer and CH_3OH/H_2O (3:1) under reflux, a chiral γ-hydroxy-α,β-acetylenic ester undergoes hydration to generate a tetronic acid product. This compound exists as a mixture of the keto and enol forms in solution as shown by ^1H NMR spectroscopy. Tetronic acids are a class of biologically significant compounds as exemplified by vitamin C (ascorbic acid).

Treatment of the keto–enol mixture of the tetronic acid product with acetic anhydride in the presence of pyridine gives the corresponding acetate product quantitatively (Scheme 4.3). The enantiomeric purity of the acetate is determined by using GC-β-cyclodextrin column. Table 4.9 summarizes the results for the conversion of the chiral γ-hydroxy-α,β-acetylenic esters to the tetronic acetates. These results show that the enantiomeric purity of the chiral γ-hydroxy-α,β-acetylenic esters is mostly maintained in the tetronic acid products. The reaction works for those substrates derived from aliphatic aldehydes, α,β-unsaturated aldehydes and an *ortho*-substituted benzaldehyde. There is significant reduction in enantiomeric purity for the products derived from benzaldehyde and its derivatives containing *para*- or *meta*-substituents.

For the Pt-catalyzed hydration of the chiral γ-hydroxy-α,β-acetylenic esters, a Pt-complex shown below is proposed as a possible intermediate. Because of the electron-withdrawing effect of the ester group and the Lewis

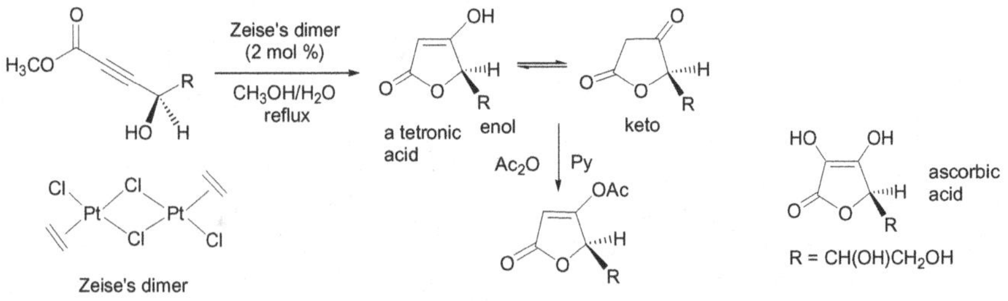

Scheme 4.3. Regiospecific hydration of γ-hydroxy-α,β-acetylenic esters catalyzed by the Zeise's dimer.

Table 4.9. Results for the conversion of the γ-hydroxy-α,β-acetylenic esters to the acetate of the tetronic acids.

Entry	γ–Hydroxy-α,β-acetylenic ester	Acetate of tetronic acid	Yield (%)	ee[a] (%)
1			55	80
2			66	83
3			61	74
4			78	82
5			46	90
6			41	92[b]

[a] Obtained by using GC-β-cyclodextrin column.
[b] Estimated (>95% reliable) since baseline resolution was not achieved.

acidic metal activation, water attacks the β-position of the alkyne leading to the regiospecific hydration. The subsequent lactonization with the attack of the γ-hydroxy group to the ester carbonyl carbon gives the final tetronic acid product.

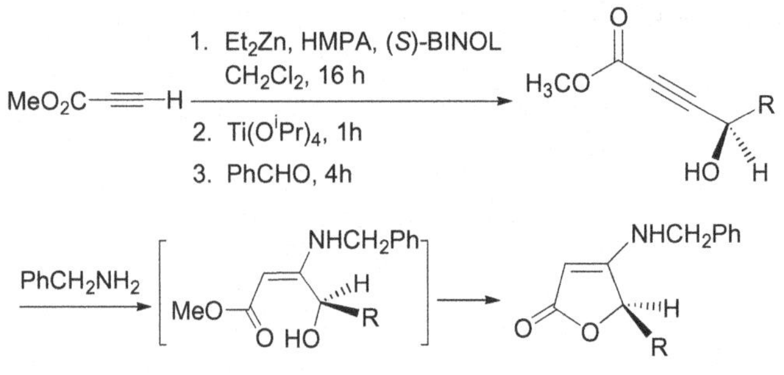

Scheme 4.4. Reaction of γ-hydroxy-α,β-acetylenic esters with benzylamine.

We also find that when the optically active γ-hydroxy-α,β-acetylenic esters prepared from the methyl propiolate addition to aldehydes is treated with benzylamine at room temperature, optically active 4-amino-2(5H)-furanones are obtained (Scheme 4.4).[24] Table 4.10 gives the results obtained in two steps from methyl propiolate. It shows good enantioselectivity for the reaction of methyl propiolate with various aliphatic aldehydes followed by treatment with benzylamine. The reaction of a few aromatic aldehydes also gives good enantioselectivity. However, it is found when the aromatic aldehydes such as p- or m-methylbenzaldehyde and o-methoxylbenzaldehyde are used, racemized products are obtained. It is not clear what factors are contributed to the observed racemization.

In a later study, You and coworkers used N-methyl imidazole to replace HMPA and found that the reaction of methyl propiolate with benzaldehyde in the presence of BINOL, ZnEt$_2$ and Ti(O^{i}Pr)$_4$ can give the corresponding γ-hydroxy-α,β-acetylenic ester in 68% yield and 88% ee.[25] This procedure needs only a catalytic amount of N-methyl imidazole and 10 mol% of BINOL. On the basis of this, we further find that without using BINOL and Ti(O^{i}Pr)$_4$, racemic γ-hydroxy-α,β-acetylenic esters can also be prepared at room temperature by treatment of methyl propiolate with ZnEt$_2$, N-methyl imidazole and then aldehydes (Scheme 4.5).[26] That is, it is not necessary to use BINOL and Ti(O^{i}Pr)$_4$ when preparing the racemic products. Generally, racemic γ-hydroxy-α,β-acetylenic esters are prepared by treatment of a large excess amount of methyl propiolate with ^{n}BuLi at very low temperature followed by reaction with aldehydes. Using ZnEt$_2$ and a catalytic amount of N-methyl imidazole at room temperature provides a very convenient way to the synthetically useful γ-hydroxy-α,β-acetylenic esters from the reaction of methyl propiolate with various aldehydes. These results are listed in Table 4.11.

Table 4.10. Results for the asymmetric methyl propiolate addition to various aldehydes followed by treatment with benzylamine.

Entry	Aldehyde	Product	Yield (%)	ee (%)
1	CH$_3$(CH$_2$)$_6$CHO	(furanone, R = (CH$_2$)$_6$CH$_3$)	75	84
2	CH$_3$(CH$_2$)$_3$CHO	(furanone, R = (CH$_2$)$_3$CH$_3$)	70	87
3	CH$_3$(CH$_2$)$_7$CHO	(furanone, R = (CH$_2$)$_7$CH$_3$)	31	86
4	isobutyl-CHO	(furanone, R = isobutyl)	51	85
5	isopropyl-CHO	(furanone, R = isopropyl)	62	84
6	cyclohexyl-CHO	(furanone, R = cyclohexyl)	69	86
7	sec-butyl-CHO	(furanone, R = sec-butyl)	53	84; 87
8	PhCHO	(furanone, R = Ph)	78	87
9	o-MeC$_6$H$_4$CHO	(furanone, R = o-MeC$_6$H$_4$)	57	90
10	p-MeOC$_6$H$_4$CHO	(furanone, R = p-MeOC$_6$H$_4$)	67	90

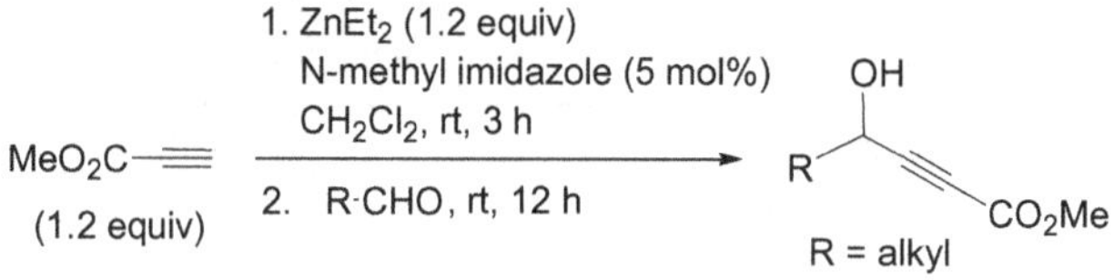

Scheme 4.5. A convenient synthesis of racemic γ-hydroxy-α,β-acetylenic esters.

Table 4.11. Synthesis of γ-hydroxy-α,β-acetylenic esters from the reaction of methyl propiolate with aldehydes in the presence of $ZnEt_2$ and N-methyl imidazole.

Entry	Aldehyde	γ-Hydroxy-α,β-acetylenic ester	Yield (%)
1	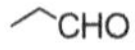 ⁀CHO	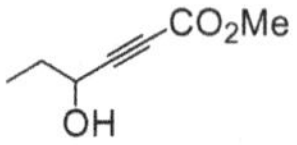	78
2	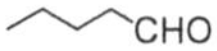 ⁀⁀⁀CHO	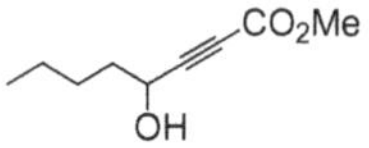	71
3	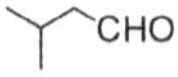 iPr-CHO	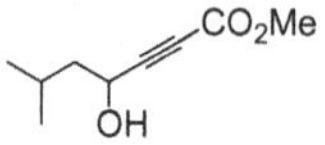	73
4	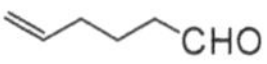 CH₂=CH-(CH₂)₃-CHO	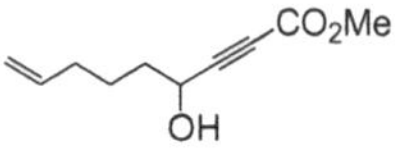	73
5	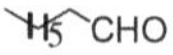 H₅-CHO	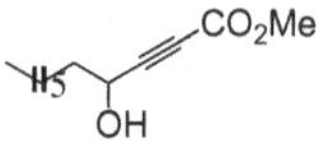	74
6	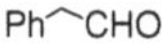 Ph⁀CHO	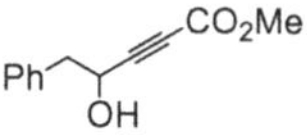	55
7	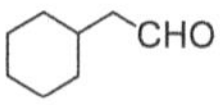 Cy-CH₂-CHO	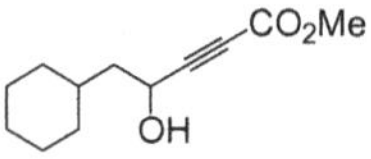	73
8	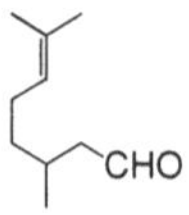	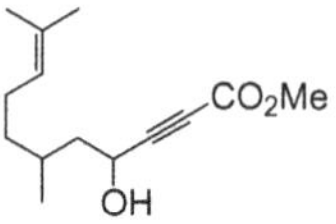	68
9	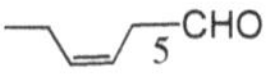 ⁀⁀=⁀(5)-CHO	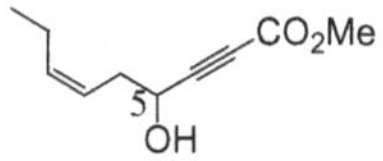	67

(*Continued*)

Table 4.11. (*Continued*)

Entry	Aldehyde	γ-Hydroxy-α,β-acetylenic ester	Yield (%)
10	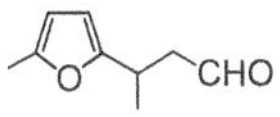 furyl-CH(CH₃)-CHO	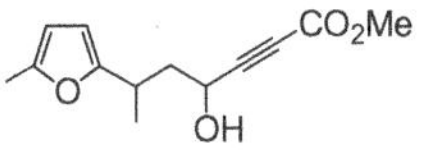 furyl-CH(CH₃)-CH(OH)-C≡C-CO₂Me	63
11	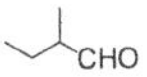 iPr-CH₂-CHO	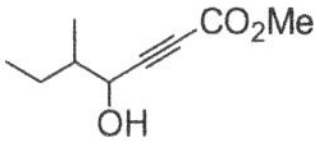 iPr-CH₂-CH(OH)-C≡C-CO₂Me	75
12	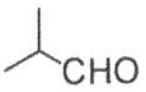 iPr-CHO	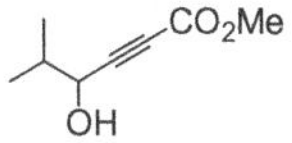 iPr-CH(OH)-C≡C-CO₂Me	74

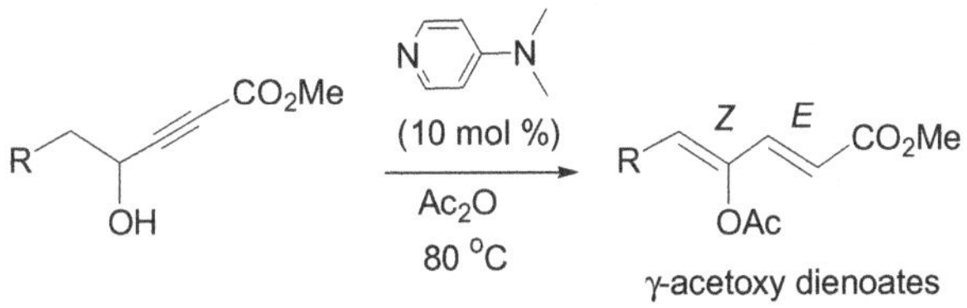

Scheme 4.6. Based-catalyzed conversion of γ-hydroxy-α,β-acetylenic esters to γ-acetoxy dienoates.

When the racemic γ-hydroxy-α,β-acetylenic esters are treated with a catalytic amount of N,N-dimethylaminopyridine (DMAP) (10 mol%) in acetic anhydride at 80°C, these compounds are converted to γ-acetoxy dienoates (Scheme 4.6).[26] The reaction is highly stereoselective with the formation of the (2E,4Z)-diene as the major product. As shown in Table 4.12, the reaction works well for the substrates derived from aliphatic aldehydes. The compounds with less sterically hindered alkyl groups give higher yields than those with more bulky alkyl groups as shown in entries 11 and 12. One of the γ-acetoxy dienoates is found to undergo an intramolecular Diels–Alder cycloaddition to generate a bicyclic γ-ketone ester (Scheme 4.7). Previously, Koide found that DBACO can catalyze the isomerization of the γ-aryl-substituted γ-hydroxy-α,β-acetylenic esters to γ-oxo-α,β-enoates in DMSO (Scheme 4.8).[27,28] When the γ-aryl group is replaced with an alkyl group, almost no reaction was observed.

We have conducted an NMR spectroscopic study of the DMAP-catalyzed reaction. When an γ-hydroxy-α,β-acetylenic ester is treated with DMAP in CDCl₃ at room temperature, the ¹H NMR spectrum shows a

Table 4.12. Conversion of various γ-hydroxy-α,β-acetylenic esters to γ-acetoxy dienoates in the presence of DMAP and acetic anhydride.

Entry	γ-Hydroxy-α,β-acetylenic esters	γ-Acetoxy dienoate	Yield (%)
1			78
2			62
3			73
4			65
5			80
6			65
7			53
8			59
9			67
10			63
11			52
12			41

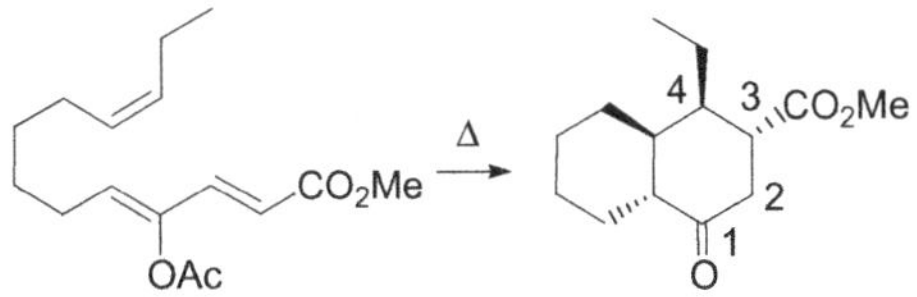

Scheme 4.7. An intramolecular Diels–Alder reaction of a γ-acetoxy dienoates.

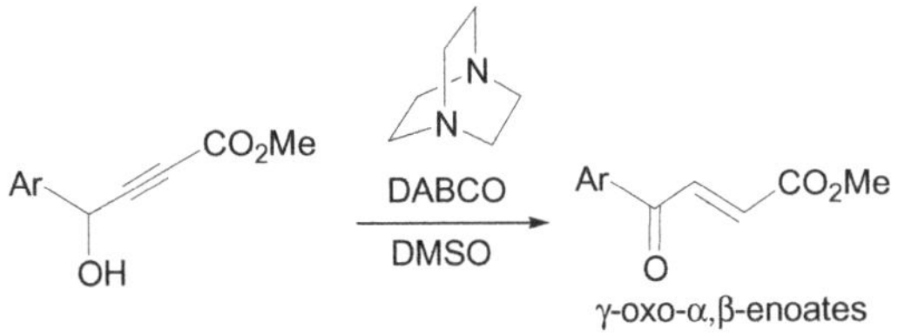

Scheme 4.8. Base-catalyzed conversion of γ-hydroxy-α,β-acetylenic esters to γ-oxo-α,β-enoates.

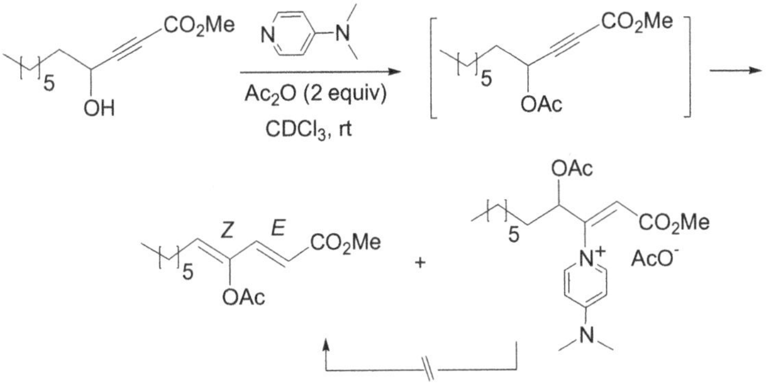

Scheme 4.9. NMR experiment for the reaction of an γ-hydroxy-α,β-acetylenic ester with DMAP and acetic anhydride at room temperature.

rapid formation of a ester, which is slowly converted to a mixture of an γ-acetoxy dienoate and a zwitterionic DMAP adduct in 1:3 ratio at room temperature (Scheme 4.9). The DMAP adduct does not move on silica gel. The isolated DMAP adduct cannot be converted to the γ-acetoxy dienoate even at high temperature. This indicates that this adduct is not an intermediate to the γ-acetoxy dienoate. No γ-oxo-α,β-enoate as shown in Scheme 4.8 is generated.

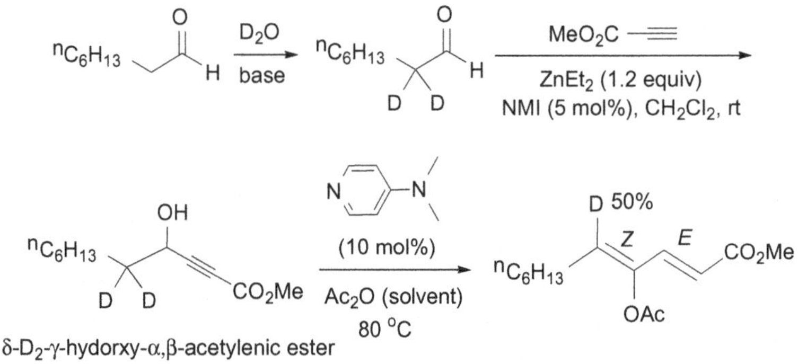

Scheme 4.10. A δ-isotope labeling experiment.

A α-deuterium labeled aldehyde is prepared by treatment of octanal with D_2O in the presence of base (Scheme 4.10). Reaction of the deuterium-labeled aldehyde with methyl propiolate in the presence of $ZnEt_2$ and N-methyl imidazole gives the δ-D_2-γ-hydroxy-α,β-acetylenic ester. Reaction of this compound with DMAP in acetic anhydride at 80°C gives a δ-D-γ-acetoxy dienoate product. No deuterium incorporation at the α and β positions is observed. At the δ position, there are 50% hydrogen and 50% deuterium. This indicates a partial H–D exchange during the reaction.

Oxidation of a γ-hydroxy-α,β-acetylenic ester with MnO_2 followed by reduction with $LiAlD_4$ in the presence of (S)-BINOL and MeOH gives a deuterated γ-hydroxy-α,β-acetylenic ester (Scheme 4.11). In the presence of DMAP and acetic anhydride, this compound is converted to the γ-acetoxy dienoate product. No deuterium incorporation in the product is observed.

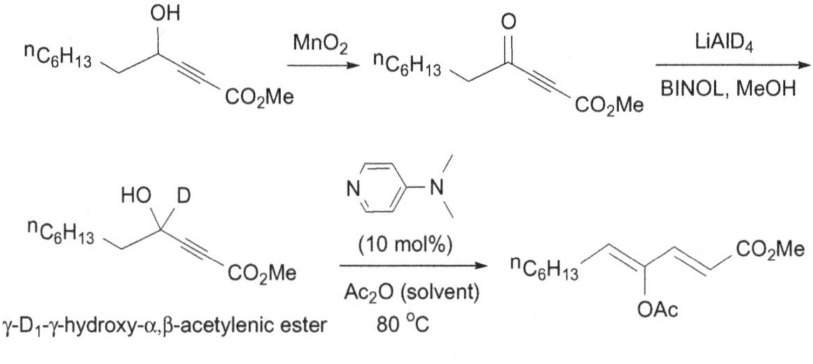

Scheme 4.11. A γ-isotope labeling experiment.

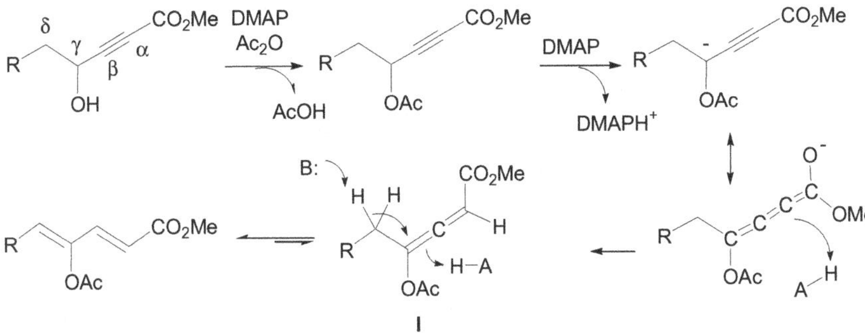

Scheme 4.12. A proposed mechanism for the base-catalyzed conversion of a γ-hydroxy-α,β-acetylenic ester to a γ-acetoxy dienoate.

On the basis of the above studies, a mechanism for the base-catalyzed conversion of a γ-hydroxy-α,β-acetylenic ester to a γ-acetoxy dienoate is proposed in Scheme 4.12. After the base-catalyzed formation of the acetate in the first step, DMAP can remove the γ-proton to form a resonance-stabilized anion. Protonation at the α-position of the anion should give the intermediate **I**. The AH could be acetic acid or water that are normally present in acetic anhydride. Deprotonation at the δ position of this intermediate together with protonation at the β position will give the γ-acetoxy dienoate product. This final step is probably reversible, leading to the observed partial H–D exchange at the δ position shown in Scheme 4.10. We have also dried and distilled acetic anhydride over Na to remove acetic acid and water. When the γ-D_1-γ-hydroxy-α,β-acetylenic ester is treated with DMAP and acetic anhydride under the rigorously dry conditions, still no deuterium incorporation in the dienoate product is observed. This indicates that the α-protons of acetic anhydride could also serve as the proton source in the reaction.

4.2.4. Catalysis by 3,3′-Bisanisyl-Substituted BINOLs[29,30]

In addition to the use of BINOL for the asymmetric alkyne addition to aldehydes, we have also explored the application of functionalized BINOL ligands. In Section 3.7.3 of Chapter 3, we have described the synthesis of ligand (R)-**4.1** [= (R)-**3.30**] and its use in the $ZnEt_2$ addition to aldehydes. This ligand and its various derivatives as shown in Fig. 4.1 are used to catalyze the asymmetric alkyne addition to aldehydes. However, all of these 3,3′-anisyl substituted BINOL compounds give only 0–67% *ee* for the

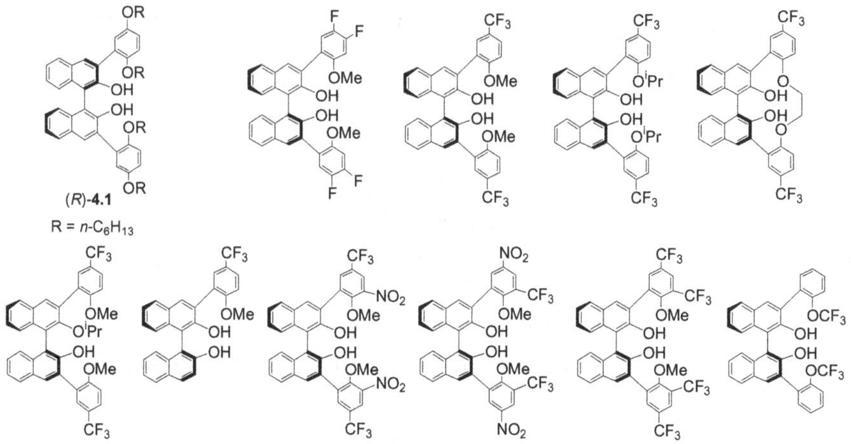

Figure 4.1. Various 3,3'-bisanisyl BINOL ligands.

phenylacetylene addition to benzaldehyde in the presence of ZnEt$_2$ with or without Ti(O^{i}Pr)$_4$.

Ligand (S)-**4.2** is then prepared and it is found that this compound (20 mol%) catalyzes the reaction of phenylacetylene (two equivalents) with benzaldehyde in the presence of ZnEt$_2$ (two equivalents) and Ti(O^{i}Pr)$_4$ (one equivalent) at room temperature to give the desired propargylic alcohol product with 80% ee. Since the study of the ligands shown in Fig. 4.1 indicates that neither the electron-donating groups nor the electron-withdrawing groups on the 3,3'-anisyl substituents of these ligands can give high enantioselectivity, we suspect that the increased enantioselectivity of (S)-**4.2** may be due to the increased size of the phenyl substituents. In order to further improve the enantioselectivity, ligand (S)-**4.3** that contains t-butyl groups at the para-position of the 3,3'-anisyl substituents of BINOL is prepared. Table 4.13 gives the results when (S)-**4.3** is used to catalyze the phenylacetylene addition to benzaldehyde in the presence of Ti(O^{i}Pr)$_4$ and ZnEt$_2$. Entry 1 uses 10 mol% of (S)-**4.3**, 20 mol% of Ti(O^{i}Pr)$_4$ and two equivalents of ZnEt$_2$ for the reaction of phenylacetylene (2.1 equivalents) with benzaldehyde in THF. In this reaction, the solution of ZnEt$_2$ and phenylacetylene is heated at reflux under nitrogen for 5 h before the addition of (S)-**4.3**, Ti(O^{i}Pr)$_4$ and benzaldehyde at room temperature. This gives the propargylic alcohol product with 85% ee. Other solvents lead to lower enantioselectivity (entries 2–4). Lower temperature in the

Table 4.13. Reaction of phenylacetylene with benzaldehyde catalyzed by (S)-**4.3** in the presence of Ti(O^{i}Pr)$_4$ and diethylzinc.

Entry	Solvent	Temp (second step)	Isolated yield (%)	ee (%)	t (h)
1	THF	rt	~70	85	12
2	CH$_2$Cl$_2$	rt	~70	83	12
3	Toluene	rt	~70	40	12
4	Ether	rt	~70	40	12
5	THF	0°C	50	88	35

second step gives a slightly better *ee* but with a significantly lower yield (entry 5).

(S)-**4.2**

(S)-**4.3** or (R)-**4.3**

The use of ligand (R)-**4.3**, the enantiomer of (S)-**4.3**, is further studied. It is found that stirring a THF solution of (R)-**4.3** (10 mol%), phenylacetylene (2.1 equivalents) and ZnEt$_2$ (2 equivalents) at room temperature for 12 h can avoid the refluxing condition in the first step. This indicates that the zinc complex generated from the reaction of (R)-**4.3** with ZnEt$_2$ can catalyze the formation of the alkynylzinc and reduce the reaction temperature. After this step, the mixture is stirred with Ti(O^{i}Pr)$_4$ (0.5 equivalents) for 1 h and then with benzaldehyde for 3 h. This gives the propargylic alcohol product with 89% *ee* and 80% yield.

Another ligand (S)-**4.4** (see Scheme 3.21 for preparation) containing a smaller methyl group is also tested. The catalytic properties of (S)-**4.4** are compared with those of (R)-**4.3**. As shown in Table 4.14, in the reaction of phenylacetylene with aromatic aldehydes, the enantioselectivity of (R)-**4.3**

Table 4.14. Asymmetric addition of phenylacetylene to aromatic aldehydes catalyzed by (R)-**4.3** and (S)-**4.4**.

Entry[a]	Aldehyde	Catalyst	ee (%)	Configuration
1	3-Cl-C₆H₄-CHO	(R)-**4.3**	84	R
2	4-F-C₆H₄-CHO	(R)-**4.3**	84	R
3	4-Me-C₆H₄-CHO	(R)-**4.3**	85	R
4	3-Me-C₆H₄-CHO	(R)-**4.3**	89	R
5	3-Cl-C₆H₄-CHO	(S)-**4.4**	56	S
6	4-F-C₆H₄-CHO	(S)-**4.4**	73	S
7	4-Me-C₆H₄-CHO	(S)-**4.4**	81	S
8	3-Me-C₆H₄-CHO	(S)-**4.4**	83	S

[a] Reaction carried out at room temperature with 10 mol% catalyst, 25 mol% Ti(O^{i}Pr)$_4$, two equivalents of ZnEt$_2$ and 2.1 equivalents of phenylacetylene in 5 mL of THF at room temperature.

is generally much better than that of (S)-**4.4**. The catalytic activity of (R)-**4.3** is also significantly higher than that of (S)-**4.4**.

(S)-**4.4**

Ligand (S)-**4.7** is further synthesized from the Suzuki coupling of (S)-**4.5** (see Scheme 3.32 for preparation) with the aryl bromide **4.6** followed by hydrolysis as shown in Scheme 4.13. This ligand contains two very bulky and rigid adamantanyl groups. In combination with ZnEt$_2$ this ligand can catalyze the reaction of phenylacetylene with benzaldehyde even *without* Ti(O^{i}Pr)$_4$. Table 4.15 shows the conditions explored for this reaction. Entry 3 is chosen as the optimized condition because it provides both good yield and good enantioselectivity (75% yield and 84% *ee*). Although entry 7 gives a higher *ee*, it is conducted at 0°C and requires much longer reaction time. In entry 3, a THF solution of (S)-**4.7** (10 mol%) and ZnEt$_2$ (two equivalents) is stirred at room temperature for 1 h and then combined with phenylacetylene (1.5 equivalents) for another hour. It is then treated with benzaldehyde for 18 h to give the desired propargylic alcohol. The configuration of the product is assigned to be *S*.

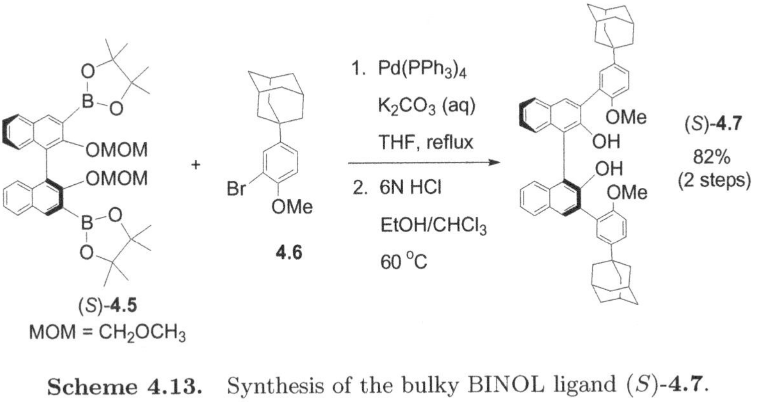

Scheme 4.13. Synthesis of the bulky BINOL ligand (S)-**4.7**.

Table 4.15. Reaction of phenylacetylene with benzaldehyde catalyzed by (S)-**4.7** in the presence of ZnEt$_2$.

Entry	ZnEt$_2$ (equiv)	(S)-**4.7** (mol%)	Solvent	Temp	Time (h)	ee (%)	Yield (%)
1	2.0	10	Toluene	rt	5	22	—
2	2.0	10	CH$_2$Cl$_2$	rt	5	13	—
3	2.0	10	THF	rt	18	84	75
4	2.0	10	THF	0°C	36	88	42
5	2.0	20	THF	rt	16	76	—
6	4.0	10	THF	rt	9	83	57
7	4.0	10	THF	0°C	36	92	48

Table 4.16. Reaction of phenylacetylene with aromatic aldehydes in the presence of (S)-**4.7** and ZnEt$_2$.[a]

Entry	Aldehyde	Product	Isolated yield (%)	ee (%)[b]
1	Ph-CHO		75	84
2[c]	Ph-CHO		42	88
3[c,d]	Ph-CHO		48	92
4	2-Cl-C$_6$H$_4$-CHO		74	94
5	4-Cl-C$_6$H$_4$-CHO		75	85
6	2-Me-C$_6$H$_4$-CHO		72	84
7	3-Me-C$_6$H$_4$-CHO		63	84
8	2-OMe-C$_6$H$_4$-CHO		64	91
9	3-OMe-C$_6$H$_4$-CHO		71	85
10[c]	1-naphthyl-CHO		45	80

[a] The reactions were carried out with 10 mol% of (S)-**4.7**, two equivalents of ZnEt$_2$ and 1.5 equivalents of phenylacetylene in THF at room temperatures unless otherwise noted.
[b] Determined by HPLC analysis on a Chiralcel OD column.
[c] 0°C.
[d] Four Equivalents of ZnEt$_2$ were used.

Ligand (S)-**4.7** is used to catalyze the reaction of phenylacetylene with other aromatic aldehydes under the conditions of entry 3 in Table 4.15 and the results are shown in Table 4.16. Enantioselectivities in the range of 80–90% ee are obtained for benzaldehydes containing *ortho-*, *meta-* and *para-*electron-donating or electron-withdrawing substituents. This demonstrates that the BINOL-based ligand can catalyze the asymmetric alkynylzinc addition to aldehydes without using an additional Lewis acid metal such as $Ti(O^iPr)_4$.

4.2.5. Catalysis by 3,3'-Bis(diphenylmethoxy)methyl Substituted BINOLs[31]

In order to further improve the catalytic properties of ligand (S)-**4.7** for the asymmetric alkyne addition to aldehydes, we have prepared compound (R)-**4.9** that contains two bulky substituents closer to the BINOL center in collaboration with Yu (Scheme 4.14). Treatment of the MOM protected (R)-BINOL with ^{n}BuLi followed by the addition of diphenyl ketone and hydrolysis gives (R)-**4.8**. Reaction of (R)-**4.8** with methanol in the presence of HCl gives (R)-**4.9**. We find that (R)-**4.9** in combination with ZnEt$_2$ can catalyze the reaction of phenylacetylene with benzaldehyde without $Ti(O^iPr)_4$. Table 4.17 gives the various conditions tested for this reaction. THF is found to be the best solvent and entry 11 gives the optimized conditions. In the presence of (R)-**4.9** (30 mol%) and ZnEt$_2$ (two equivalents) at 0°C, phenylacetylene (1.5 equivalents) reacts with benzaldehyde to give the R propargylic alcohol product with 91% ee.

Under the same conditions as entry 11 of Table 4.17, (R)-**4.9** is used to catalyze the reaction of phenylacetylene with other aromatic aldehydes. As the results summarized in Table 4.18 show, high enantioselectivities

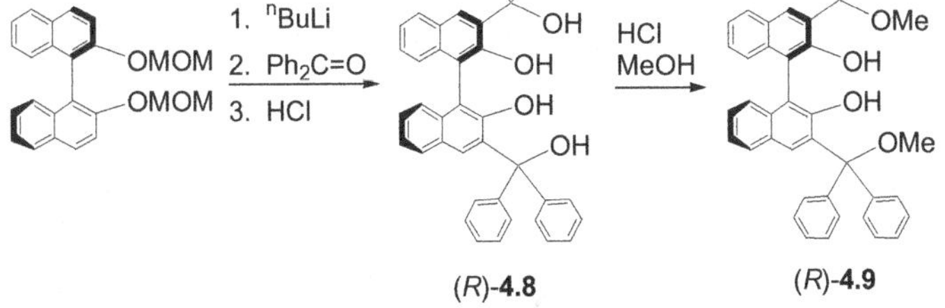

Scheme 4.14. Synthesis of the bulky BINOL ligand (R)-**4.9**.

Table 4.17. Results for the reaction of phenylacetylene with benzaldehyde catalyzed by (R)-**4.9**.[a]

Entry	(R)-**4.9** (mol %)	Solvent	$T(°C)$	ee (%)
1[b]	20	CH_2Cl_2	0	3
2[b]	20	toluene	0	10
3[b]	20	ether	0	47
4[b]	20	THF	0	74
5[c]	20	THF	0	53
6	20	THF	0	85
7	20	THF	30	55
8	20	THF	−21	89
9	15	THF	0	80
10	25	THF	0	89
11	**30**	**THF**	**0**	**91**
12	35	THF	0	91

[a] Unless otherwise indicated the following procedure was used. (R)-**4.9** and $ZnEt_2$ (two equivalents) in THF (3 mL) were stirred at rt for 1 h, and then phenylacetylene (1.5 equivalents) was added. After an additional hour, benzaldehyde was added at 0°C.
[b] (R)-**4.9**, phenylacetylene and $ZnEt_2$ in solvent were stirred at rt for 2 h. Then benzaldehyde was added at 0°C.
[c] Phenylacetylene and $ZnEt_2$ in THF (1 mL) were stirred at rt for 1 h, and then (R)-**4.9** in THF (2 mL) was added. After an additional hour, benzaldehyde was added at 0°C.

Table 4.18. Results for phenylacetylene addition to aldehydes catalyzed by (R)-**4.9**.

Entry	Aldehyde	Product	Isolated yield (%)	ee (%)
1	Ph-CHO	Ph-C≡C-CH(OH)-Ph	81	91
2	2-MeO-C₆H₄-CHO	2-MeO-C₆H₄-CH(OH)-C≡C-Ph	76	87
3	3-MeO-C₆H₄-CHO	3-MeO-C₆H₄-CH(OH)-C≡C-Ph	75	94
4	4-MeO-C₆H₄-CHO	4-MeO-C₆H₄-CH(OH)-C≡C-Ph	70	86
5	2-Me-C₆H₄-CHO	2-Me-C₆H₄-CH(OH)-C≡C-Ph	67	90

(Continued)

Table 4.18. (Continued)

Entry	Aldehyde	Product	Isolated Yield (%)	ee (%)
6	3-Me-C₆H₄-CHO	3-Me-C₆H₄-CH(OH)-C≡C-Ph	71	90
7	4-Me-C₆H₄-CHO	4-Me-C₆H₄-CH(OH)-C≡C-Ph	77	88
8	3-Cl-C₆H₄-CHO	3-Cl-C₆H₄-CH(OH)-C≡C-Ph	80	86
9	4-Cl-C₆H₄-CHO	4-Cl-C₆H₄-CH(OH)-C≡C-Ph	75	92
10	4-F-C₆H₄-CHO	4-F-C₆H₄-CH(OH)-C≡C-Ph	80	90
11	4-Br-C₆H₄-CHO	4-Br-C₆H₄-CH(OH)-C≡C-Ph	75	92
12	1-naphthyl-CHO	1-naphthyl-CH(OH)-C≡C-Ph	70	90

(86–92% ee) are observed for the reaction of various aromatic aldehydes including o-, m- and p-substituted benzaldehydes and 1-naphthaldehyde. The enantioselectivity of (R)-**4.9** in most cases is higher than that of (S)-**4.7**, though their reaction conditions are quite different.

Several analogs of (R)-**4.9** including (R)-**4.8**, (R)-**4.10** and (R)-**4.11** are tested for the reaction of phenylacetylene with benzaldehyde in the presence of ZnEt$_2$. The results are compared in Table 4.19. It shows that the tetrahydroxyl compound (R)-**4.8** gives a greatly reduced and opposite enantioselectivity in comparison with (R)-**4.9**. This indicates very different structures of their catalytic sites. Without the bulk phenyl substituents of (R)-**4.9**, ligand (R)-**4.11** gives very low enantioselectivity. Ligand (R)-**4.10** without the more acidic BINOL hydroxyl groups is inactive for the alkyne addition. This ligand probably cannot catalyze the

Table 4.19. Results for the reaction of phenylacetylene with benzaldehyde in the presence of chiral ligands and ZnEt$_2$.

Entry	Ligand	ee (%)	Configuration
1	(R)-**4.9**	91	R
2	(R)-**4.8**	28	S
3	(R)-**4.10**	No reaction	
4	(R)-**4.11**	32	R

reaction of phenylacetylene with ZnEt$_2$ to form the nucleophilic alkynylzinc reagent for the subsequent addition to aldehydes.

(R)-**4.10**   (R)-**4.11**

One mechanism is proposed for the reaction catalyzed by (R)-**4.9** (Scheme 4.15). When (R)-**4.9** is treated with excess ZnEt$_2$, complex **4.13** could be generated via complex **4.12**. The central ZnEt$_2$ unit in **4.13** is probably activated by the two Lewis basic oxygen atoms of the ligand and it can deprotonate phenylacetylene to form the alkynylzinc unit in **4.14**. An aldehyde molecule can be activated by coordination with two of the zinc centers of **4.14** to form complex **4.15**. In **4.15**, migration of the alkynyl group to the carbonyl followed by displacement with ZnEt$_2$ can generate the zinc propargyloxide aggregates, which upon aqueous work-up will give the propargylic alcohol product.

Unlike (R)-**4.9**, it should be much harder for (R)-**4.10** to react with excess ZnEt$_2$ to form a complex such as **4.13** because after deprotonation of the two hydroxyl groups of (R)-**4.10** with two equivalents of ZnEt$_2$, the central two oxygen atoms of the resulting complex **4.16** are both sterically and electronically less favorable to simultaneously coordinate to a ZnEt$_2$ molecule. Thus, it should be harder for this complex to activate ZnEt$_2$ to deprotonate the terminal alkyne to generate the nucleophilic alkynylzinc

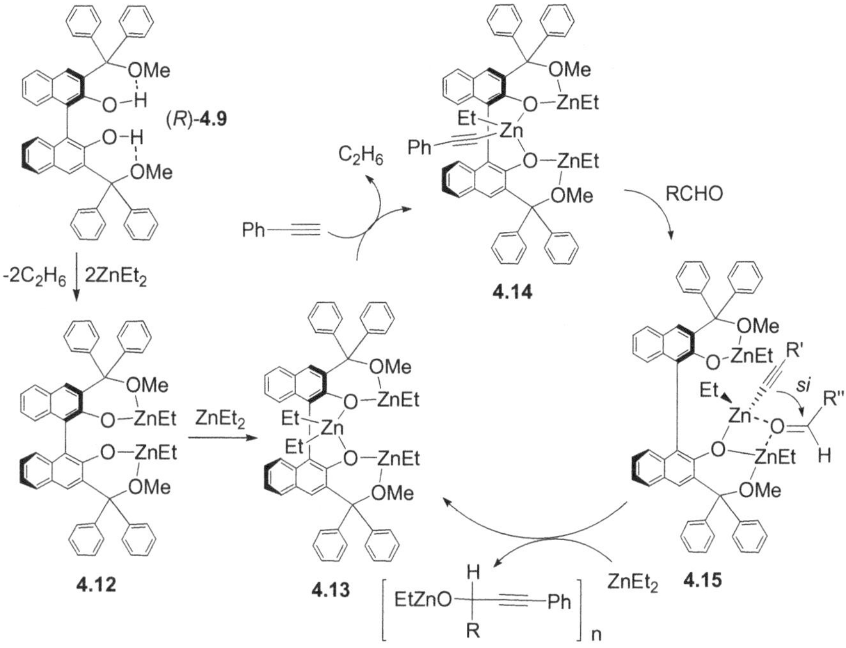

Scheme 4.15. A proposed mechanism for the catalytic asymmetric reaction of phenylacetylene with aldehydes in the presence of (R)-**4.9** and ZnEt$_2$.

reagent. This could explain as to why (R)-**4.10** cannot catalyze the alkyne addition.

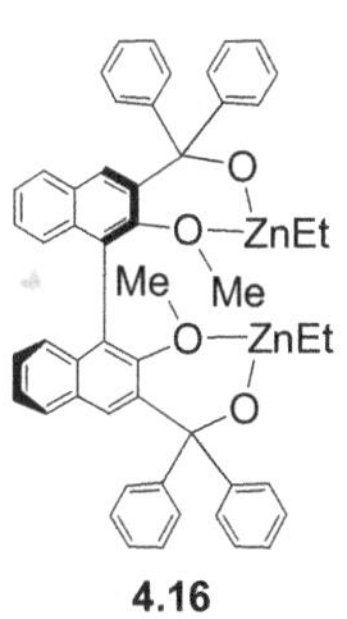

4.16

Although the tetrahydroxyl ligand (R)-**4.8** gives only very low enantioselectivity for the reaction of phenylacetylene with benzaldehyde in the presence of ZnEt$_2$, we find that addition of a Lewis base can improve the catalytic process. Table 4.20 lists the effects of various Lewis base additives on the reaction catalyzed by (R)-**4.8**. As entry 9 shows, in the presence

Table 4.20. Reaction of phenylacetylene with benzaldehyde catalyzed by using (R)-**4.8**, ZnEt$_2$ and a Lewis Base.[a]

Entry	(R)-**4.8** (equiv)	Additive[b]	ee (%)
1	0.2	HMPA (2 eq)	15(S)
2	0.2	DMSO (2 eq)	7(S)
3	0.2	Pyridine (2 eq)	9(S)
4	0.2	Ph$_3$PO (2 eq)	7(S)
5	0.2	IM (2 eq)	22(S)
6	0.2	DMBA (2 eq)	66(S)
7	0.2	DMA (2 eq)	76(S)
8	0.2	NMM (2 eq)	77(S)
9	**0.2**	**Et$_3$N (2 eq)**	**80(S)**
10	0.2	Et$_3$N (1 eq)	77(S)
11	0.2	Et$_3$N (4 eq)	77(S)

[a] ZnEt$_2$ (2.0 equiv), (R)-**4.8** (0.2 equiv), THF (3 mL), and a base was stirred at room temperature for 1 h. Phenylacetylene (1.5 equivalents) was added and stirred for 1 h. Then, an aldehyde (0.25 mmol) was added at 0°C, and the reaction mixture was stirred for 36–40 h.
[b] IM, imidazole; DMBA, N,N-dimetthylbenzylamine; DMA, N,N-dimethylaniline; NMM, N-methylmorpholine.

of two equivalents of Et$_3$N, the enantioselectivity is enhanced to be 80%. The reaction is conducted at room temperature in the presence of two equivalents of ZnEt$_2$. The product configuration is S, the same as that obtained without using Et$_3$N.

Under the same conditions as entry 9 of Table 4.20, (R)-**4.8** is used to catalyze the reaction of phenylacetylene with other aromatic aldehydes, giving 50–86% ee (Table 4.21).

4.2.6. Catalysis by Acyclic and Macrocyclic Binaphthyl Salens[32,33]

BINOL-based salen ligands have been used in a number of Lewis-acid-catalyzed asymmetric reactions.[34] We have prepared a few acyclic and macrocyclic BINOL-salen ligand and explored their application in the asymmetric alkyne addition to aldehydes. As shown in Scheme 4.16, the enantiomerically pure acyclic BINOL-salen ligands are synthesized from the condensation of 3-formyl BINOL with 1,2-cyclohexanediamine or 1,2-diphenylethylenediamine. Among these ligands, (S)-**4.17** and (R)-**4.17** are a pair of enantiomers, and they are diastereomers of (S)-**4.18**. Ligands (S)-**4.19** and (S)-**4.20** are diastereomers of each other. The macrocyclic

Table 4.21. Results for phenylacetylene addition to aldehydes catalyzed by using (R)-**4.8**, ZnEt$_2$ and Et$_3$N.

Entry	Aldehyde	Product	Yield (%)	ee (%)
1	Ph-CHO		83	80 (S)
2	2-MeO-C$_6$H$_4$-CHO		76	86
3	3-MeO-C$_6$H$_4$-CHO		74	76
4	4-MeO-C$_6$H$_4$-CHO		65	66
5	3-Me-C$_6$H$_4$-CHO		76	70
6	4-Me-C$_6$H$_4$-CHO		84	69
7	3-Cl-C$_6$H$_4$-CHO		83	50
8	4-Cl-C$_6$H$_4$-CHO		75	57
9	4-F-C$_6$H$_4$-CHO		78	51
10	2-furyl-CHO		83	54
11	PhCH=CH-CHO		75	65

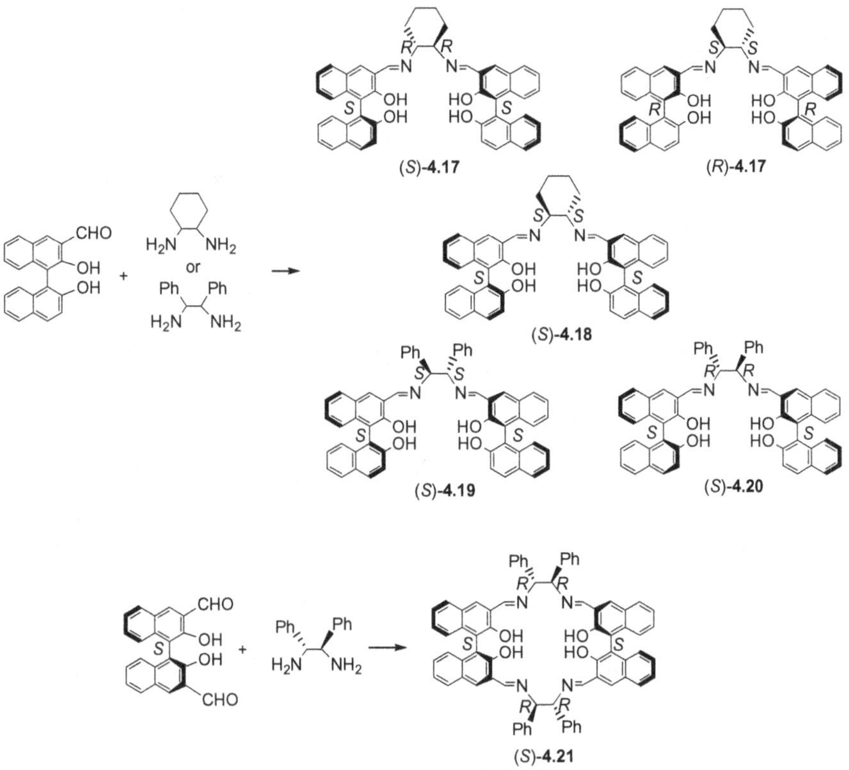

Scheme 4.16. Synthesis of BINOL-salen ligands.

salen (S)-**4.21** is synthesized from the condensation of (S)-3,3'-diformylBINOL with (R, R)-1,2-diphenyl ethylenediamine.

Table 4.22 gives the results for the reaction of phenylacetylene with benzaldehyde catalyzed by the BINOL-salen ligands. In these reactions, a toluene solution of phenylacetylene is sequentially treated with ZnEt$_2$, a chiral ligand and then benzaldehyde at room temperature. Among the chiral ligands, compound (S)-**4.17** gives the highest enantioselectivity (entry 1). No Ti(O^{i}Pr)$_4$ is needed. The low enantioselectivity of (S)-**4.18** shows that the matching of the chirality of the BINOL unit with that of the amine unit is important.

We have varied the reaction conditions in order to further improve the enantioselectivity of (S)-**4.17**. Table 4.23 gives the experiments tested for the reaction of phenylacetylene with benzaldehyde catalyzed by (S)-**4.17**. Among these, the optimized conditions are shown in entry 11. In

Table 4.22. Results for the reaction of phenylacetylene with benzaldehyde in the presence of the BINOL-salen ligands in toluene at room temperature.

Entry	Ligand (mol%)	Phenylacetylene (equiv)	ZnEt$_2$ (equiv)	ee (%)
1	(S)-**4.17** (15)	2.1	2.0	79
2	(S)-**4.18** (15)	2.1	2.0	<10
3	(S)-**4.19** (15)	2.1	2.0	13
4	(S)-**4.20** (15)	2.1	2.0	<10
5	(S)-**4.21** (15)	2.1	2.0	17

Table 4.23. Conditions for the reaction of phenylacetylene with benzaldehyde in the presence of (S)-**4.17**.

Entry	(S)-**4.17** (mol%)	Temp	Solvent	Phenylacetylene (equiv)	ZnR$_2$ (equiv)	ee (%)
1	15	rt	Toluene	2.1	ZnEt$_2$ (2.0)	79
2	15	rt	THF	2.1	ZnEt$_2$ (2.0)	57
3	15	rt	ether	2.1	ZnEt$_2$ (2.0)	39
4	15	rt	CH$_2$Cl$_2$	2.1	ZnEt$_2$ (2.0)	29
5	15	rt	Toluene	3.0	ZnEt$_2$ (2.0)	45
6	11	rt	Toluene	2.0	ZnEt$_2$ (2.0)	71
7	22	rt	Toluene	2.0	ZnEt$_2$ (2.0)	87
8	30 mol%	rt	Toluene	2.0	ZnEt$_2$ (2.0)	87
9	30 mol%	rt	Toluene	3.0	ZnEt$_2$ (3.0)	87
10	22 mol%	0°C	Toluene	2.0	ZnEt$_2$ (2.0)	80
11	**22 mol%**	**rt**	**Toluene**	**2.0**	**ZnMe$_2$ (2.0)**	**92**

this experiment, ZnMe$_2$ is used to replace ZnEt$_2$, which gives excellent enantioselectivity (92% ee). The reaction is carried out at room temperature in toluene in the presence of 22 mol% of (S)-**4.17**. The product configuration is S.

We have also tested Jacobsen's salen ligand **4.22** for the phenylacetylene addition to benzaldehyde using the conditions in entry 11 in Table 4.23, which gives only 39% ee. Thus, the chiral BINOL unit in (S)-**4.17** is necessary for its high enantioselectivity.

4.22

Ligand (S)-**4.17** is found to be generally enantioselective for the reaction of various terminal alkynes with aromatic aldehydes. Under the above optimized conditions, excellent results are obtained which are summarized in Table 4.24.

Table 4.24. Results for the alkyne additions to aldehydes catalyzed by (S)-**4.17**.

Entry	Alkyne	Aldehyde	Isolated yield (%)	ee (%)
1	PhC≡CH	Ph-CHO	–	92
2[a]	PhC≡CH	Ph-CHO	84	92
3	PhC≡CH	H₃CO-C₆H₄-CHO	85	94
4	PhC≡CH	H₃C-C₆H₄-CHO	82	91
5	PhC≡CH	Cl-C₆H₄-CHO	61	89
6	PhC≡CH	3-H₃C-C₆H₄-CHO	80	92
7	PhC≡CH	3-H₃CO-C₆H₄-CHO	72	89
8	PhC≡CH	3-Cl-C₆H₄-CHO	67	86
9	PhC≡CH	2-CH₃-C₆H₄-CHO	80	87
10	PhC≡CH	iPr-CHO	75	94
11	PhC≡CH	1-naphthyl-CHO	87	97
12	PhC≡CH	2-naphthyl-CHO	61	91
13	PhC≡CH	2-furyl-CHO	90	89

(*Continued*)

Table 4.24. (*Continued*)

Entry	Alkyne	Aldehyde	Isolated yield (%)	ee (%)
14	Cl(CH$_2$)$_3$C≡CH	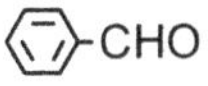 ⟨⟩-CHO	60	95
15	Cl(CH$_2$)$_3$C≡CH	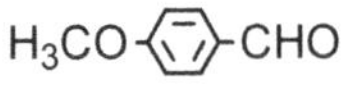 H$_3$CO-⟨⟩-CHO	72	95
16	Cl(CH$_2$)$_3$C≡CH	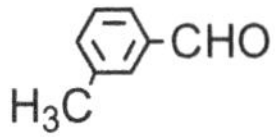 H$_3$C(⟨⟩-CHO)	72	90
17	Cl(CH$_2$)$_3$C≡CH	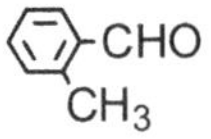 ⟨⟩-CHO, CH$_3$	76	89
18	Cl(CH$_2$)$_3$C≡CH	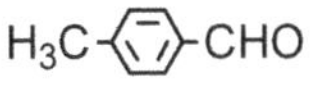 H$_3$C-⟨⟩-CHO	74	96
19	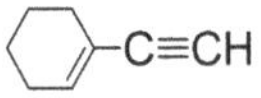 ⟨⟩-C≡CH	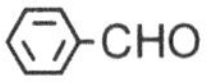 ⟨⟩-CHO	64	86

[a] (*R*)-**4.17** is used.

Although (*S*)-**4.17** is highly enantioselective for the reaction of various alkynes with aromatic aldehydes, when it is used to catalyze the reaction of phenylacetylene with octyl aldehyde, an aliphatic aldehyde, it gives only 61% ee. The use of macrocycles (*S*)-**4.21** and (*S*)-**4.23** is then tested for the asymmetric alkyne addition to octyl aldehydes. Both (*S*)-**4.21** and (*S*)-**4.23** give improved enantioselectivity (71 and 74% ee, respectively).

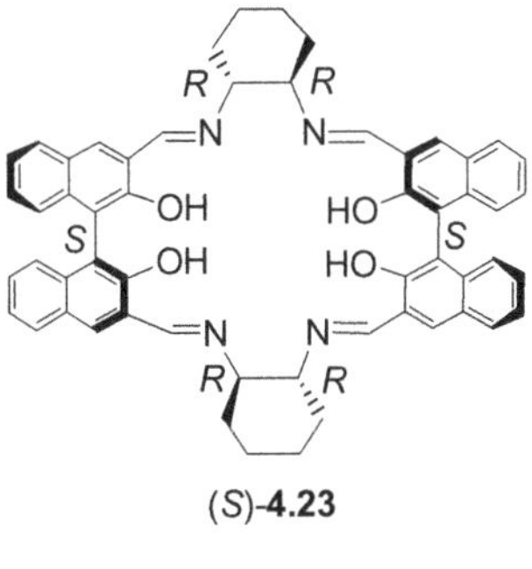

(*S*)-**4.23**

Because of the enhanced enantioselectivity of the macrocycles for the alkyne addition to aliphatic aldehydes, the reaction conditions using these macrocycles are further explored. Table 4.25 summarizes the results obtained for the reaction of phenylacetylene with *iso*-valeraldehyde in

Table 4.25. Attempted reactions of phenylacetylene with *iso*-valeraldehyde using the chiral macrocycles.[a]

Entry	Solvent (mL)	ZnMe$_2$ (equiv)	Ligand (mol %)	ee (%)
1	THF (2)	1.6	(S)-**4.23** (20)	83
2	THF (2)	2	(S)-**4.23** (20)	87
3	THF (2)	2.5	(S)-**4.23** (20)	87
4[b]	THF (2)	2	(S)-**4.23** (20)	85
5	THF (2)	2	(S)-**4.23** (10)	82
6	THF (2)	2	(S)-**4.21** (20)	84
7	Toluene (2)	2	(S)-**4.23** (20)	13
8	Et$_2$O (2)	2	(S)-**4.23** (20)	31
9	CH$_2$Cl$_2$ (2)	2	(S)-**4.23** (20)	83
10	THF (2.5)	2	(S)-**4.23** (20)	89
11	THF (3.5)	2	(S)-**4.23** (20)	91
12	THF (4.5)	2	(S)-**4.23** (20)	91.5
13	**THF (6)**	**2**	**(S)-4.23 (20)**	**93**
14	THF (10)	2	(S)-**4.23** (20)	92
15	THF (20)	2	(S)-**4.23** (20)	NR
16	THF (10)	2	(S)-**4.23** (10)	91
17	CH$_2$Cl$_2$ (6)	2	(S)-**4.23** (20)	94

[a] Two equivalents of phenylacetylene were used at room temperature.
[b] 4 Å MS were added.

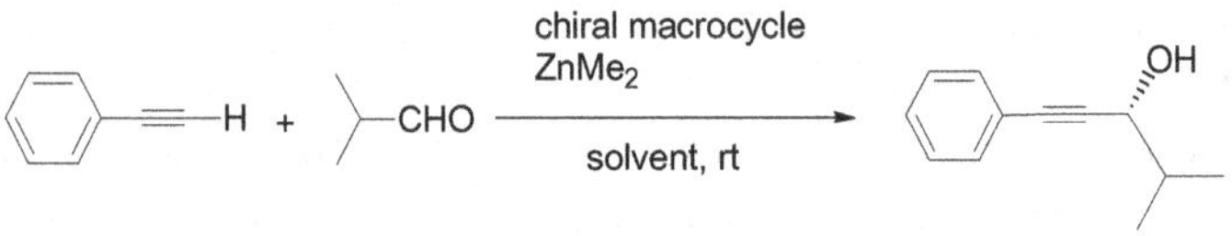

Scheme 4.17. Reaction of phenylacetylene with *iso*-valeraldehyde in the presence of the chiral macrocycles and ZnMe$_2$.

the presence of (S)-**4.21** and (S)-**4.23** in combination with ZnMe$_2$ (Scheme 4.17). Entry 13 gives the optimized conditions. The reaction is conducted in THF at room temperature using two equivalents of phenylacetylene, 20 mol% of (S)-**4.23**, two equivalents of ZnMe$_2$, which gives the propargylic alcohol product with 93% ee.

The conditions in entry 13 of Table 4.25 are applied for the reaction of phenylacetylene with various aldehydes. As shown in Table 4.26, (S)-**4.23** in combination with ZnMe$_2$ catalyzes the reaction of the alkyne with linear and branched aliphatic aldehydes with high enantioselectivity (89–95% ee, entries 1–8). The product configuration for the reaction of phenylacetylene

Table 4.26. Results for the addition of phenylacetylene to aldehydes catalyzed by (S)-**4.23**.

Entry	Aldehydes	Yield (%)	ee (%)
1	⟩−CHO	72	93
2[a]	⟩−CHO	–	93
3	Ph−CH$_2$−CHO	76	91
4	Cy−CHO	68	89
5	⟩−CHO	54	95
6	Me(CH$_2$)$_3$−CHO	67	93
7	Me(CH$_2$)$_6$−CHO	64	93
8	Me(CH$_2$)$_7$−CHO	76	93
9	⟩=CHO	52	91
10	Ph−CH=CH−CHO	79	96
11	CH=CH−CHO	76	96
12	C≡C−CHO	67	87

[a] (R)-**4.23**, the enantiomer of (S)-**4.23** made of (R)-binaphthyl and (S,S)-cyclohexane-1,2-diamine, was used for the reaction.

with nonyl aldehyde (entry 8) is determined to be R. The reactions of the vinyl aldehydes give high ee's (87–96% ee, entries 9–12). Macrocycle (S)-**4.23** also shows good enantioselectivity (81% ee) for the phenylacetylene addition to benzaldehyde.

The *ee* of the macrocycle (*S*)-**4.23** is found to be linearly related with the *ee* of the product in the reaction of phenylacetylene with *iso*-valeraldehyde. Thus, in this catalytic process, (*S*)-**4.23** probably acts as a monomeric catalyst rather than an intermolecular aggregate. This is consistent with the observation that lowering the concentration of (*S*)-**4.23** gives better enantioselectivity.

4.2.7. *Catalysis by 3,3′-Bismorpholinomethyl* $H_8 BINOL$[34,35]

We have developed a very efficient method to synthesize another functional BINOL ligand for asymmetric catalysis by starting with the partially hydrogenated BINOL, H$_8$BINOL, obtained from the reaction of BINOL with H$_2$ in the presence of PtO$_2$. As shown in Scheme 4.18, when the optically pure (*S*)-H$_8$BINOL is treated with morpholinomethanol in dioxane at 60°C, it gives (*S*)-3,3′-bismorpholinomethyl H$_8$BINOL, (*S*)-**4.24**, in 95% yield and with over 99% *ee*. The enantiomeric ligand (*R*)-**4.24** is obtained from the reaction of (*R*)-H$_8$BINOL.

Compound (*S*)-**4.24** is used to catalyze the reaction of phenylacetylene with benzaldehyde in the presence of ZnEt$_2$ and Ti(O^{i}Pr)$_4$. As shown in Table 4.27, unlike the other BINOL-based ligands, the enantioselectivity of (*S*)-**4.24** is not very sensitive to solvent, temperature, concentration and the catalyst loading. It gives generally good enantioselectivity under all the tested conditions (81–86% *ee*). The configuration of the propargylic alcohol product is *R*.

The conditions in entry 3 of Table 4.27 are applied to the phenylacetylene addition to a variety of aldehydes. The results summarized in Table 4.28 show that (*S*)-**4.24** gives generally good enantioselectivity for the reaction of aromatic aldehydes (89–98% *ee*). Particularly, its enantioselectivity for

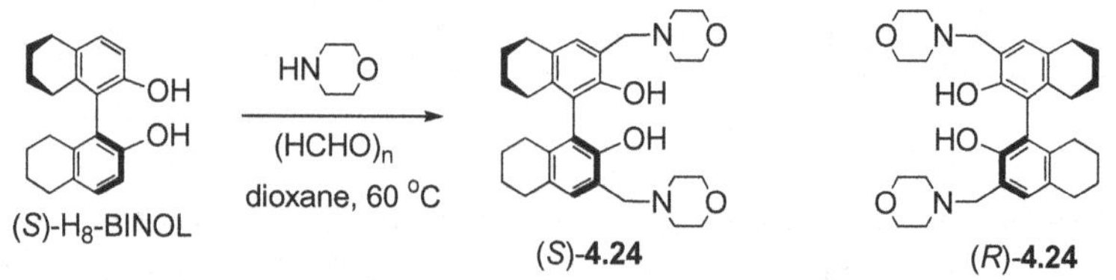

Scheme 4.18. Synthesis of enantiomerically pure 3,3′-bismorpholinylmethyl H$_8$-BINOL.

Table 4.27. Results for the reaction of phenylacetylene with benzaldehyde in the presence of (S)-**4.24**, Ti(O^{i}Pr)$_4$ and ZnEt$_2$.

Entry	(S)-**4.24** (mol%)	PhCCH	ZnEt$_2$	Ti(O^{i}Pr)$_4$ (mol%)	Solvent	T (°C)	Yield (%)	ee (%)
1	2	4 eq.	4 eq.	100	THF (3 mL)	rt	25	81
2	5	4 eq.	4 eq.	100	THF (3 mL)	rt	64	83
3	10	4 eq.	4 eq.	100	THF (3 mL)	rt	93	83
4	20	4 eq.	4 eq	100	THF (3 mL)	rt	95	84
5	40	4 eq.	4 eq.	100	THF (3 mL)	rt	90	85
6	20	4 eq.	4 eq.	100	CH$_2$Cl$_2$ (3 mL)	rt	96	82
7	20	4 eq.	4 eq.	100	Ether (3 mL)	rt	60	84
8	20	4 eq.	4 eq.	100	Toluene (3 mL)	rt	85	84
9	10	4 eq.	4 eq.	100	THF (3 mL)	0	85	83
10	10	2 eq.	2 eq.	100	THF (3 mL)	rt	63	86
11	10	1.5 eq.	1.5 eq.	100	THF (3 mL)	rt	58	83
12	20	2 eq.	2 eq.	50	THF (3 mL)	rt	45	83
13	20	2 eq.	2 eq.	100	THF (3 mL)	rt	77	84
14	20	2 eq.	2 eq.	200	THF (3 mL)	rt	89	84
15	20	2 eq.	2 eq.	100	THF (1 mL)	rt	38	83
16	20	2 eq.	2 eq.	100	THF (6 mL)	rt	65	84

Table 4.28. Asymmetric reactions of phenylacetylene with aromatic aldehydes in the presence of (S)-**4.24**, ZnEt$_2$ and Ti(O^{i}Pr)$_4$.[a]

Entry	Aldehyde	Isolated yield (%)	ee (%)
1	Ph–CHO	93	83
2	4-Cl-C$_6$H$_4$–CHO	88	97
3	4-OCH$_3$-C$_6$H$_4$–CHO	76	93
4	4-OEt-C$_6$H$_4$–CHO	83	92
5	4-CH$_3$-C$_6$H$_4$–CHO	88	89
6[b]	4-NO$_2$-C$_6$H$_4$–CHO	62	98

(Continued)

Table 4.28. (Continued)

Entry	Aldehyde	Isolated yield (%)	ee (%)
7	2-Br-C6H4-CHO	84	97
8	2,4,5-(CH3)3-C6H2-CHO	68	96
9	2-H3CO-C6H4-CHO	93	86
10	4-Br-C6H4-CHO	86	87
11	4-F-C6H4-CHO	50	86
12	3-H3C-C6H4-CHO	98	91
13	4-H3C-C6H4-CHO	86	83
14	4-H3CO-C6H4-CHO	99	85
15	1-naphthyl-CHO	85	83
16[c]	2,6-Cl2-C6H3-CHO	62	86
17[d]	CH3(CH2)6CHO	76	67

[a] Unless otherwise indicated, reactions were conducted by stirring (S)-**4.24**:ZnEt$_2$:PhCCH:Ti(O^{i}Pr)$_4$:RCHO = 0.1:4:4:1:1 at room temperature in THF for 4 h.
[b] Two equivalents of ZnEt$_2$ and two equivalents of PhCCH were used.
[c] The reaction time was 6 h.
[d] 20 mol% (S)-**4.24** was used.

ortho-substituted benzaldehydes is very high. For the reaction of octyl aldehyde, the enantioselectivity is significantly lower (67% ee, entry 17).

We have also tested the use of this ligand to catalyze the alkyne addition to ketones, a reaction that is much less developed. This ligand

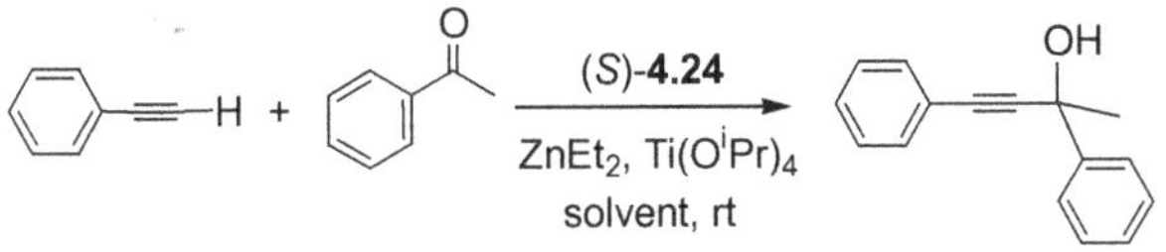

Scheme 4.19. Reaction of phenylacetylene with acetophenone catalyzed by (S)-4.24.

Table 4.29. Reaction of phenylacetylene with acetophenone in the presence of (S)-4.24 under various conditions.

Entry	(S)-4.24	Ti(O^{i}Pr)$_4$	Solvent	ee (%)
1	10 mol%	100 mol%	toluene	0
2	10 mol%	100 mol%	CH$_2$Cl$_2$	34
3	10 mol%	100 mol%	Et$_2$O	19
4	10 mol%	100 mol%	THF	63 (60)[a]
5	10 mol%	40 mol%	THF	60
6	10 mol%	100 mol%	THF[b]	56
7	10 mol%	40 mol%	1,4-dioxane	69

[a] Isolated yield in the parentheses.
[b] 3 mL THF was used.

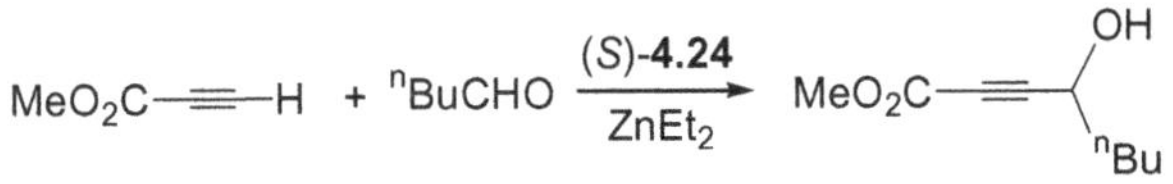

Scheme 4.20. Asymmetric reaction of methyl propiolate with valeraldehyde.

is found to be an active catalyst for the reaction of phenylacetylene with acetophenone at room temperature (Scheme 4.19). In the experiments tested, (S)-4.24 (10 mol%) is mixed with ZnEt$_2$ (four equivalents) and phenylacetylene (four equivalents) in a solvent at room temperature for 2 h and then combined with Ti(O^{i}Pr)$_4$ (one equivalent) and acetophenone. The results are summarized in Table 4.29. The highest enantioselectivity is obtained in 1,4-dioxane (69% ee, entry 7).

Ligand (S)-4.24 is also used to catalyze the reaction of methyl propiolate with valeraldehyde. We find that this ligand in combination with ZnEt$_2$ is highly active even *without Ti(O^{i}Pr)$_4$* (Scheme 4.20). Table 4.30 shows that the highest enantioselectivity is 70% ee obtained by using 10 mol% of (S)-4.24 in diethyl ether at −20°C (entry 4). The yields for these experiments are generally above 70%.

Table 4.30. Results for the reaction of methyl propiolate with valeraldehyde in the presence of (S)-**4.24**.

Entry	(S)-**4.24**	Solvent	ZnR$_2$	T (°C)	ee (%)
1	10 mol%	Toluene	1.2 eq. ZnEt$_2$	25	28
2	10 mol%	Toluene	1.2 eq. ZnEt$_2$	−20	43
3	10 mol%	CH$_2$Cl$_2$	1.2 eq. ZnEt$_2$	−20	28
4	10 mol%	Et$_2$O	1.2 eq. ZnEt$_2$	−20	70
5	10 mol%	Et$_2$O	1.2 eq. ZnEt$_2$	−40	62
6	10 mol%	Et$_2$O	1.2 eq. ZnMe$_2$	−40	50
7	20 mol%	Et$_2$O	2.4 eq. ZnEt$_2$	−20	69

4.3. Asymmetric Arylzinc Addition to Aldehydes

Arylzinc addition to aldehydes can generate chiral α-substituted benzyl alcohols that exist in many organic structures. We have studied the use of two types of BINOL derivatives to catalyze the reaction of arylzincs with aldehydes, that is the 3,3′-bisanisyl-substituted BINOLs[36,37] and the 3,3′-bismorpholinomethyl-substituted BINOLs.[35,38] During this period, other catalysts have also been developed for the enantioselective arylzinc addition to aldehydes and ketones.[39]

4.3.1. Catalysis by the 3,3′-Bisanisyl-Substituted BINOLs[36,37]

We have tested the use of ligand (R)-**4.1** for the diphenylzinc addition to aldehydes (Scheme 4.21). In the presence of 10 mol% of (R)-**4.1**,

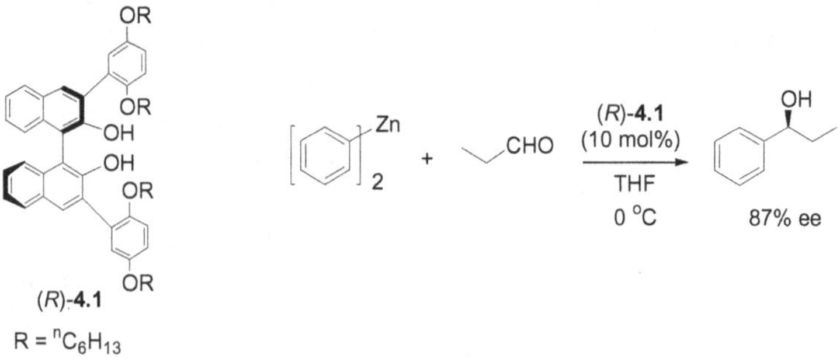

Scheme 4.21. Asymmetric diphenylzinc addition to propionaldehyde catalyzed by (R)-**4.1**.

diphenylzinc reacts with propionaldehyde in toluene at 0°C to give (S)-1-phenylpropanol with 87% ee. This is the first highly enantioselective diphenylzinc addition to aldehydes.

We have further studied the use of (R)-**4.1** to catalyze the reaction of diphenylzinc with aromatic aldehydes. Table 4.31 gives the conditions for the reaction of a variety of aromatic aldehydes. As shown in entry 2 of Table 4.31, even without the catalyst, diphenylzinc can still react with p-anisylaldehyde at 0°C. We find that in order to achieve high enantioselectivity, $ZnEt_2$ needs to be added, which probably reacts with (R)-**4.1** to form a better catalyst for the diphenylzinc addition to the aldehyde (93% ee and 84% yield, entry 5) than that formed from the reaction of (R)-**4.1** with diphenylzinc. For different aldehydes, the reaction conditions such as solvent, temperature, concentration and additive need to be varied in order to obtain high enantioselectivity. For example, for the reaction of cinnamaldehyde, besides $ZnEt_2$, methanol also needs to be added and the reaction mixture needs to be heated at reflux, which gives the diphenylzinc addition product with 83% ee (entry 11).

For the asymmetric diphenylzinc addition to aldehydes, one of the main challenges is the uncatalyzed reaction which is often in competition with the chiral catalyst-catalyzed reaction. This leads to reduced enantioselectivity. In order to increase the catalytic activity of (R)-**4.1**, electron-withdrawing groups are introduced to the 3,3'-ansiyl substituents. This strategy could increase the Lewis acidity of the corresponding zinc complex and thus increase the catalytic activity.

A series of 3,3'-bisanisyl BINOL ligands (S)-**4.26**–(S)-**4.31** are synthesized from the Suzuki coupling of (S)-**4.5** [=(S)-**3.45**, see preparation in Section 3.32] with the corresponding aryl bromides **4.25** (Scheme 4.22). Compounds (S)-**4.26**–(S)-**4.29** contain 1–3 electron-withdrawing fluorine atoms on one of their 3,3'-anisyl substituents. Compound (S)-**4.30** contains two additional bromine atoms at the 6,6'-positions of the BINOL unit. In compound (S)-**4.31**, the hexyl substituents of (S)-**4.28** are replaced with two methyl groups. Compounds (S)-**4.27** and (S)-**4.30** have a methyl group at the *ortho* position of the anisyl substituents. These two ligands should have both electronically and sterically modified environments in comparison with (R)-**4.1**.

Ligands (S)-**4.26**–(S)-**4.31** are tested for the reaction of diphenylzinc with cinnamaldehyde without using methanol as additive. Table 4.32 compares these ligands with (R)-**4.1**. In these experiments, the ligands are first treated with two equivalents of $ZnEt_2$ in methylene chloride to

Table 4.31. Asymmetric diphenylzinc addition to aldehydes catalyzed by (R)-**4.1**.

Aldehyde	ZnPh$_2$ (equiv.)	(R)-Ligand or +additive (mol%)	Aldehyde concentration (mM)	Solv.	T (°C)	Time (h)	Isolated yield (%)	ee (%)	Conf.[a]
propanal-CHO	1.2	10	100	Tol	0	20	90	87[b]	S
MeO-C$_6$H$_4$-CHO	1	0	50	Tol	0	10	61	0	
	2	0	50	Tol	0	10	76	54[b]	R
	1	5 + 10 ZnEt$_2$	50	Tol	0	10	87	77[b]	R
	1	20 + 40 ZnEt$_2$	50	Tol	−30	24	84	93[b]	R
Cl-C$_6$H$_4$-CHO	1	20 + 40 ZnEt$_2$	50	Et$_2$O	rt	10	70	57[c]	R
	1	20 + 40 ZnEt$_2$	5	Et$_2$O	rt	10	86	94[c]	R
naphthyl-CHO	1	20 + 40 ZnEt$_2$	5	THF	−10	96	66	87[b]	R
cinnamaldehyde	1	20 + 40 ZnEt$_2$	5	CH$_2$Cl$_2$	rt	28	98	50[b]	S[5]
	1	20 + 80 ZnEt$_2$ + 40 MeOH	5	CH$_2$Cl$_2$	rt	22	92	77[b]	S
	1	20 + 80 ZnEt$_2$ + 40 MeOH	5	CH$_2$Cl$_2$	Reflux	10	94	83[b]	S

[a] Determined by comparing the optical rotation with the literature data.
[b] Determined by HPLC-Chiracel-OD column.
[c] Determined by analyzing the acetate derivative of the alcohol product on a HPLC-Chiracel-OD column.

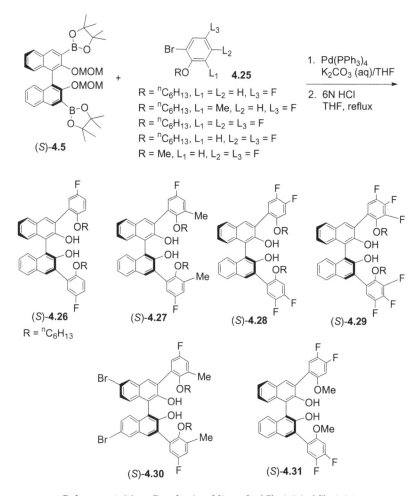

Scheme 4.22. Synthesis of ligands (S)-**4.26**–(S)-**4.31**.

generate the corresponding zinc complexes. A 20 mol% of the zinc complex is then used to catalyze the diphenylzinc addition to cinnamaldehyde. As shown in Table 4.32, all these electronically and sterically modified ligands exhibit improved enantioselectivity over (R)-**4.1**. Among these, ligand (S)-**4.28** gives the best result with up to 87% ee (entry 4). Addition of methanol cannot further increase the enantioselectivity of (S)-**4.28**.

Scheme 4.23 shows a proposed mechanism for the diphenylzinc addition catalyzed by (S)-**4.28**. Treatment of the ligand with two equivalents of ZnEt$_2$ followed by coordination of diphenylzinc can generate the zinc

Table 4.32. Reaction of cinnamaldehyde with diphenylzinc in the presence of the chiral ligands.[a]

Entry	Ligand	Isolated yield (%)	ee (%)	Configuration
1	(R)-**4.1**	88	50	S
2	(S)-**4.26**	90	73	R
3	(S)-**4.27**	88	81	R
4	(S)-**4.28**	92	87	R
5	(S)-**4.29**	90	81	R
6	(S)-**4.30**	88	70	R

[a] The reaction was carried out under nitrogen at room temperature in CH_2Cl_2 in the presence of 20 mol% of the chiral ligand and 40 mol% of $ZnEt_2$. The concentration of aldehyde was 5 mM. The reaction was quenched in 5 h. The ee was determined by HPLC-Chiracel-OD column. The absolute configuration of the product was determined by comparing the optical rotation with the literature data.

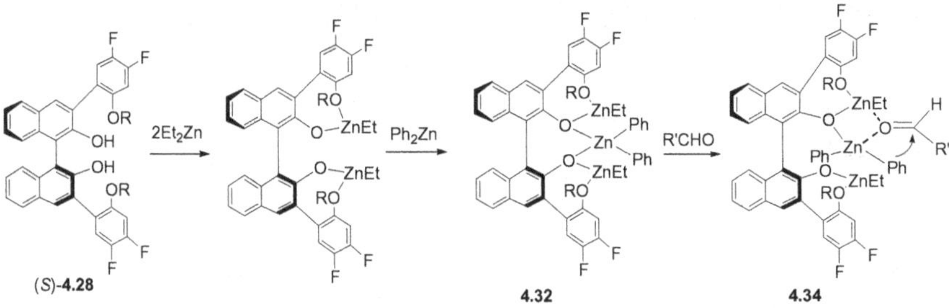

Scheme 4.23. A proposed mechanism for the catalytic asymmetric diphenylzinc addition by (S)-**4.28**.

complex **4.32**. An aldehyde molecule can coordinate to the zinc centers to generate **4.34**. In **4.34**, the electron-withdrawing fluorine atoms on the anisyl substituent can make the Zn center of the [ZnEt] unit more Lewis acidic and thus provide more catalyst control for the diphenylzinc addition to aldehydes, leading to the much higher enantioselectivity of (S)-**4.28** than (R)-**4.1**. However, the additional fluorine atoms in (S)-**4.29** cannot further increase the enantioselectivity. The fluorine atom adjacent to the alkoxyl group may have an unfavorable interaction with the [ZnEt] unit, which could reduce the enantioselectivity. The significantly lower enantioselectivity of (S)-**4.30** than (S)-**4.27** shows that the electron-withdrawing bromine atoms on the BINOL unit are not favorable for the diphenylzinc addition. As shown in intermediate **4.34**, coordination of $ZnPh_2$ with the Lewis basic central oxygen atoms of the BINOL unit should have activated the subsequent phenyl addition to the aldehyde. When the bromine atoms are introduced to the 6,6′-positions of the BINOL unit as in ligand (S)-**4.30**, it should reduce the Lewis basicity of the central two oxygen atoms. This should provide a reduced activation effect for the coordinated ZnPh$_2$ unit and decrease the catalyst-controlled diphenylzinc addition. The lower catalyst control means more of the uncatalyzed diphenylzinc addition. This could explain the lower enantioselectivity of (S)-**4.30** than (S)-**4.28**.

Ligand (S)-**4.28** is used to catalyze the reaction of diphenylzinc with several aromatic aldehydes and high enantioselectivity is observed

Table 4.33. Synthesis of chiral diarylcarbinols by the diphenylzinc addition to aryl aldehydes catalyzed by (S)-**4.28**.[a]

Entry	Aldehyde	Isolated yield (%)	ee (%)	Configuration[b]
1	Ph-CH=CH-CHO	92	87[c]	R
2	Cl-C$_6$H$_4$-CHO	92	95[d]	S
3	2-CH$_3$-C$_6$H$_4$-CHO	87	91[c]	S
4	2-naphthyl-CHO	90	88[c]	S
5	pyridyl-CHO	86	80[c]	(+)
6	pyridyl-CHO	90	70[c,e]	(+)
7	pyridyl-CHO	89	86[c,f]	(+)

[a] The reaction was carried out under nitrogen at room temperature in CH$_2$Cl$_2$ in the presence of 20 mol% of (S)-**4.28** and 40 mol% of ZnEt$_2$ unless otherwise indicated. The concentration of aldehyde was 5 mM. The reaction was quenched in 5 h.
[b] Determined by comparing the optical rotation with the literature data.
[c] Determined by HPLC-Chiracel-OD column.
[d] Determined by analyzing the acetate derivative of the alcohol product by HPLC-Chiracel-OD column.
[e] (S)-**4.30** was used in place of (S)-**4.28**.
[f] Et$_3$B-pretreated aldehyde was used.

(Table 4.33). This ligand possesses better catalytic properties than (R)-**4.1**. As shown earlier in Table 4.31, when (R)-**4.1** is used to catalyze the reaction of diphenylzinc with aldehydes, it often requires longer reaction time, lower or higher temperature and very different conditions for different substrates. However, when (S)-**4.28** (20 mol%) is used in combination with ZnEt$_2$ (40 mol%), it catalyzes the diphenylzinc addition to various aldehydes in methylene chloride at room temperature in 5 h with high enantioselectivity. For example, the reaction of 2-naphthaldehyde in the presence of (R)-**4.1** takes four days at −10°C with incomplete conversion and 66% isolated yield. Using (S)-**4.28** for this reaction only takes 5 h to give the product in 90% yield and similar ee. When ligand (S)-**4.30** where the hexyl groups of (S)-**4.28** are replaced with methyl groups is used, it gives lower ee than (S)-**4.28** for the reaction of 3-pyridinecarboxaldehyde (entries 5 and 6).

4.3.2. Catalysis by 3,3′-Bismorpholinomethyl-Substituted BINOL and H$_8$BINOL Ligands[35,38]

In Section 4.1.6, we have shown that the 3,3′-bismorpholinomethyl H$_8$BINOL ligand (S)-**4.24**, prepared from the reaction of H$_8$BINOL with morpholine and paraformaldehyde in dioxane at 60°C in high yield and high enantiomeric purity, in combination with ZnEt$_2$ and Ti(O^{i}Pr)$_4$ can catalyze the alkyne addition to aromatic aldehydes with high enantioselectivity. We have explored the use of this ligand for the asymmetric arylzinc addition to aldehydes. Besides (S)-**4.24**, we have also studied the synthesis of the 3,3′-bismorpholinomethyl BINOL (S)-**4.34** and examined its catalytic properties for this reaction (Scheme 4.24).

We find that the conditions for the synthesis of (S)-**4.24** are not applicable for the synthesis of (S)-**4.34**. No reaction occurs when BINOL is treated with morpholinomethanol at 60°C. Apparently BINOL is much less reactive than H$_8$BINOL because its aromatic rings are less electron rich

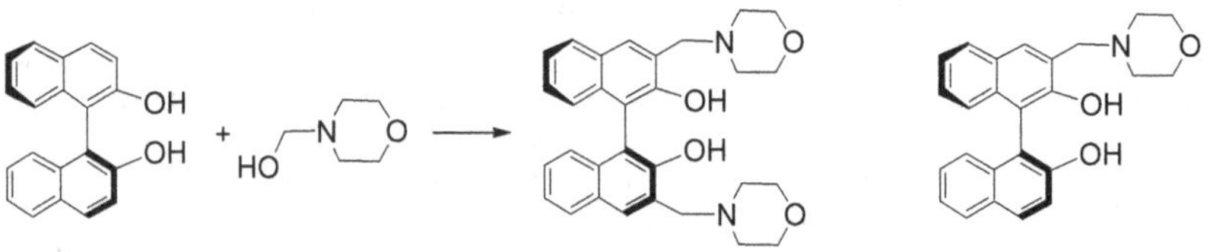

Scheme 4.24. Preparation of the BINOL derivative (S)-**4.34**.

Table 4.34. Reaction of (S)-BINOL with morpholinomethanol under various conditions (under 30 psi nitrogen).

Entry	Temp (°C)	Additive	Time (h)	Yield of (S)-**4.34** (%)	ee of (S)-**4.34** (%)
1	160	None	24	94	0
2	95–100	None	60	~5	>99
3	90	NaBH$_4$ (1 eq)	24	~0	—
4	90	P$_2$O$_5$ (1 eq)	24	~0	—
5	90	Et$_2$Zn (4–6 eq)	24	~0	—
6	130	None	48	90	<30
7	120	None	72	63	<50
8	110	None	72	55	75

in comparison with those in H$_8$BINOL. At 95–100°C, the main product is the monosubstituted BINOL (S)-**4.35** obtained in 60% yield and 99% ee. At 160°C, (S)-**4.34** is obtained in >90% yield but as a completely racemized product. Table 4.34 shows various conditions explored for the synthesis of (S)-**4.34**. Entry 8 gives the best condition, which produces (S)-**4.34** at 110°C with 75% ee and 55% yield. Recrystallization gives (S)-**4.34** with >99% ee. In this experiment, (S)-**4.35** is also obtained in 30% yield and >87% ee. Higher temperature gives more racemized product and lower temperature gives low yield. Using various additives shown in entries 3–5 cannot generate the product. Addition of a range of Lewis acid complexes, including CeCl$_3$, Zn(OTf)$_2$, TbCl$_3$, InCl$_3$, ZnI$_2$, LiCl and VO(acac)$_2$, also cannot improve this reaction.

The single crystal X-ray structures of (S)-**4.34** and (S)-**4.24** are obtained. Strong O–H...N intramolecular hydrogen bonds are found in both compounds. In these two compounds, their BINOL units have a transoid conformation with the binaphthyl dihedral angle of 113.7° and 100.0° for (S)-**4.34** and (S)-**4.24**, respectively. The dihedral angle of (S)-**4.24** is closer to the 90° of the orthogonal conformation because of its larger steric interaction between the partially hydrogenated six-membered rings.

Compounds (S)-**4.34** and (S)-**4.24** are used to catalyze the reaction of diphenylzinc with valeraldehyde (Scheme 4.25) and the results are given in Table 4.35. The optimized conditions in entry 4 give the desired diphenylzinc addition product with 92% ee with the use of (S)-**4.24**. Under the same conditions, the BINOL derivative (S)-**4.34** gives a slightly reduced ee (87% ee, entry 5). Thus, the H$_8$BINOL derivative forms a better catalyst

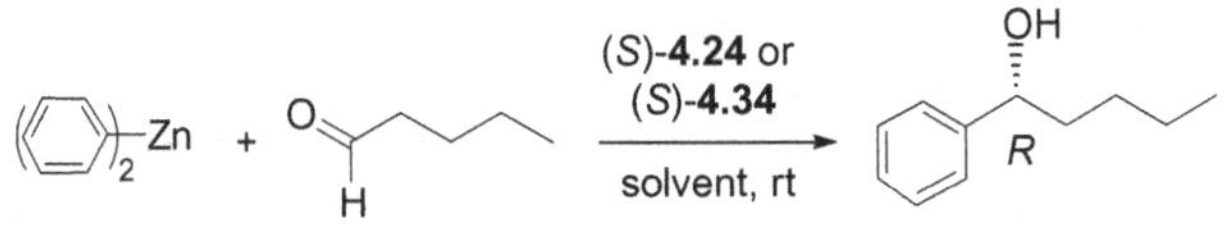

Scheme 4.25. Asymmetric diphenylzinc addition to valeraldehyde.

Table 4.35. Results for the diphenylzinc addition to valeraldehyde under various conditions catalyzed by (S)-**4.34** and (S)-**4.24**.[a]

Entry	Ligand	Solvent	ZnPh$_2$ (equiv.)	ee (%)[b,c]
1	(S)-**4.24** (10 mol%)	CH$_2$Cl$_2$	1.2	35
2	(S)-**4.24** (10 mol%)	Toluene	1.2	38
3	(S)-**4.24** (10 mol%)	Et$_2$O	1.2	39
4	(S)-**4.24** (10 mol%)	**THF**	**1.2**	**92**
5	(*S*)-**4.34** (10 mol%)	THF	1.2	87
6[d]	(S)-**4.24** (10 mol%)	THF	1.2	93
7[e]	(S)-**4.24** (10 mol%)	THF	1.0	91
8[f]	(S)-**4.24** (10 mol%)	THF	1.6	89
9[e,f]	(S)-**4.24** (10 mol%)	THF	1.6	91
10	(S)-**4.24** (5 mol%)	THF	1.2	84
11	(S)-**4.24** (20 mol%)	THF	1.4	87

[a] Unless otherwise indicated, the following procedure was used: Under nitrogen to a flask containing (S)-**4.24** (12.3 mg, 0.025 mmol) and diphenylzinc (66.0 mg, 0.3 mmol), a solvent (2 mL, dried) was added and the solution was stirred at room temperature for 1 h. Valeraldehyde (0.25 mmol) was then added and the resulting solution was stirred for 12 h. Aqueous work up and column chromatography on silica gel gave 1-phenyl-1-pentanol.
[b] Isolated yields of all these reactions were 82–89%.
[c] ee's were determined by HPLC-Chiral OD column.
[d] Valeraldehyde was added dropwise over 2 h.
[e] (S)-**4.24** was pretreated with diethylzinc (20 mol%) for 1 h followed by the addition of diphenylzinc and aldehyde.
[f] MeOH (40 mol%) was added after (S)-**4.24** was treated with diphenylzinc or diethylzinc.

than the BINOL derivative probably for both steric and electronic reasons. In addition, the synthesis of the H$_8$BINOL is of much higher yield, milder condition, and free of racemization.

We have used (S)-**4.24** to catalyze the diphenylzinc addition to a variety of aliphatic aldehydes using the conditions in entry 4 of Table 4.35. Very high enantioselectivity has been achieved for the reaction of both linear and branched aliphatic aldehydes (Table 4.36). In these reactions, R alcohols are produced.

Table 4.36. Asymmetric diphenylzinc addition to aliphatic aldehydes catalyzed by (S)-**4.24**.

Entry	Aldehyde	Product	Time (h)	Yield (%)[a]	ee (%)[b,c]
1	n-C$_3$H$_7$CHO	n-C$_3$H$_7$CH(OH)Ph	12	87	92
2	n-C$_6$H$_{13}$CHO	n-C$_6$H$_{13}$CH(OH)Ph	12	78	93
3	CH$_2$=CHCH$_2$CH$_2$CHO	CH$_2$=CHCH$_2$CH$_2$CH(OH)Ph	12	75	92
4	cyclohexyl-CHO	cyclohexyl-CH(OH)Ph	8	96	98
5	t-BuCHO	t-BuCH(OH)Ph	6	93	98
6	i-PrCH$_2$CHO	i-PrCH$_2$CH(OH)Ph	16	82	92
7	MeO$_2$C(CH$_2$)$_3$CHO	MeO$_2$C(CH$_2$)$_3$CH(OH)Ph	12	80	81

[a] Isolated yield.
[b] Determined by HPLC-chiral column or GC-chiral column.
[c] All racemic compounds were prepared by mixing the aldehydes with diphenylzinc.

Ligand (S)-**4.24** is also used to catalyze the diphenylzinc addition to aromatic aldehydes. As shown in Table 4.37, high enantioselectivity is achieved for the reaction of *para*-substituted benzaldehydes (89–96% ee, entries 1–5). For the reactions of the *ortho*- and *meta*-substituted benzaldehydes, the enantioselectivity is lower (51–78% ee, entries 7–10). The configuration of the diarylcarbinol products is S.

The diphenylzinc addition to α,β-unsaturated aldehydes in the presence of (S)-**4.24** is further investigated and the results are summarized in Table 4.38. Using the conditions of entry 4 in Table 4.35 for the reaction of 2-methyl-but-2-enal with diphenylzinc gives very high enantioselectivity

Table 4.37. Diphenylzinc addition to various aromatic aldehydes catalyzed by (S)-**4.24**.

Entry	Aldehyde	Product	Time (h)	Yield (%)[a]	ee (%)[b,c]
1	4-Me-C6H4-CHO	4-Me-C6H4-CH(OH)-Ph	16	90	89
2	4-Br-C6H4-CHO	4-Br-C6H4-CH(OH)-Ph	16	91	89
3	4-Cl-C6H4-CHO	4-Cl-C6H4-CH(OH)-Ph	16	90	89
4	4-F-C6H4-CHO	4-F-C6H4-CH(OH)-Ph	16	92	94
5	4-MeO-C6H4-CHO	4-MeO-C6H4-CH(OH)-Ph	16	91	91
6	2-naphthyl-CHO	2-naphthyl-CH(OH)-Ph	16	97	89
7	2-Me-C6H4-CHO	2-Me-C6H4-CH(OH)-Ph	16	80	78
8	2-Cl-C6H4-CHO	2-Cl-C6H4-CH(OH)-Ph	16	95	51
9	2-MeO-C6H4-CHO	2-MeO-C6H4-CH(OH)-Ph	16	94	60
10	3-Br-C6H4-CHO	3-Br-C6H4-CH(OH)-Ph	16	78	68

[a] Isolated yield.
[b] Determined by HPLC-chiral column or GC-chiral column.
[c] All racemic compounds were prepared by mixing the aldehydes with diphenylzinc.

Table 4.38. Reaction of diphenylzinc with α,β-unsaturated aldehydes in the presence of (S)-**4.24**.

Entry	Aldehyde	(S)-**4.24** (mol%)	Additive	T	THF (mL)	ZnPh$_2$	ee (%)
1	(CH$_3$)C=CHCHO	10	None	rt	2	1.2 equiv.	96 (88)[a]
2	PhCH=CHCHO	10	None	rt	2	1.2 equiv.	77 (98)[a]
3	PhCH=CHCHO	10	None	0°C	2	1.2 equiv.	70
4	PhCH=CHCHO	20	None	rt	2	2.4 equiv.	78
5	PhCH=CHCHO	20	40 mol% MeOH	rt	2	2.4 equiv.	73
6	PhCH=CHCHO	10	20 mol% ZnEt$_2$	rt	2	1.2 equiv.	78
7	PhCH=CHCHO	10	None	rt	5	1.2 equiv.	75
8	PhCH=CHCHO	10	None	rt	1	1.0 equiv.	78

[a]The isolated yield is given in the parentheses and the reaction time was 12 h.

and high yield (96% ee and 88% yield, entry 1). The reaction of cinnamaldehyde shows 77% ee (entry 2), which is not improved even with varying reaction conditions.

The ee of (S)-**4.24** is found to be linearly related with the ee of the products generated from the diphenylzinc addition to both aromatic and aliphatic aldehydes. This indicates that the catalyst in this reaction probably contains the monomeric chiral ligand. NMR study shows that when (S)-**4.24** is treated with one equivalent of ZnPh$_2$, it probably generates an intermediate **4.36** (Scheme 4.26). Addition of 2,2-dimethylpropanal to **4.36** shows no reaction. That is, the phenyl group on the zinc center in **4.36** cannot add to the aldehyde. Further reaction of **4.36** with 0.5

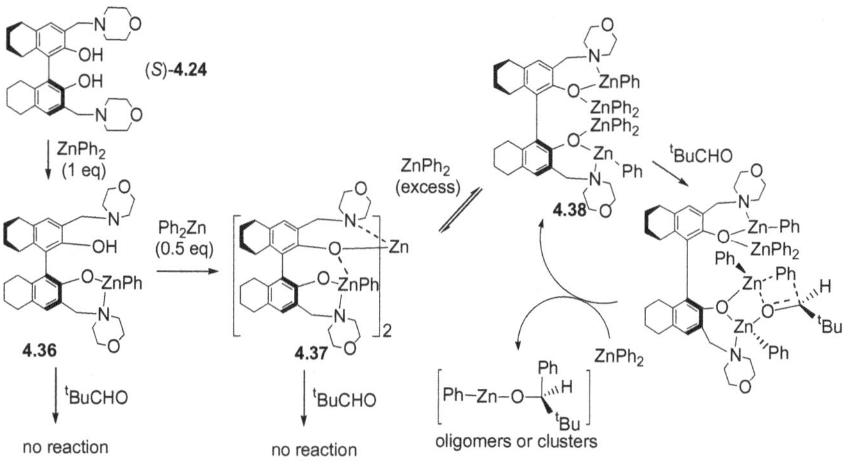

Scheme 4.26. A proposed mechanism for the catalytic diphenylzinc addition.

equivalent of ZnPh$_2$ gives a (2 + 3) complex **4.37**. Complex **4.37** also cannot react with the aldehyde. When excess ZnEt$_2$ is added, **4.37** probably dissociates to form the C_2 symmetric monomeric complex **4.38**. Addition of 2,2-dimethylpropanal leads to the phenyl addition product. After the consumption of the aldehyde, complex **4.38** is regenerated. Thus, the monomeric C_2 symmetric complex **4.38** may be the catalytically active species in this reaction.

4.3.3. Asymmetric Functional Arylzinc Addition Catalyzed by the 3,3′-Substituted BINOL and H$_8$BINOL Ligands[40]

Diphenylzinc is the only commercially available diarylzinc reagent. In order to expand the application of the catalytic asymmetric diphenylzinc addition described above to other functional arylzincs, we have studied the reaction of the *in situ* generated diarylzincs with aldehydes. Scheme 4.27 shows the reaction of *m*-iodoanisole with ZnEt$_2$, which presumably produces a diarylzinc complex as reported by Knochel.[41] When this arylzinc is treated with cyclohexanecarboxaldehyde at room temperature, the corresponding alcohol is isolated in only 30% yield after 19 h.

The use of ligand (*S*)-**4.24** is tested for the reaction of the functional arylzincs with aldehydes. It is found that in the presence of 10 mol% of (*S*)-**4.24**, the arylzinc generated *in situ* from *m*-iodoanisole reacts with

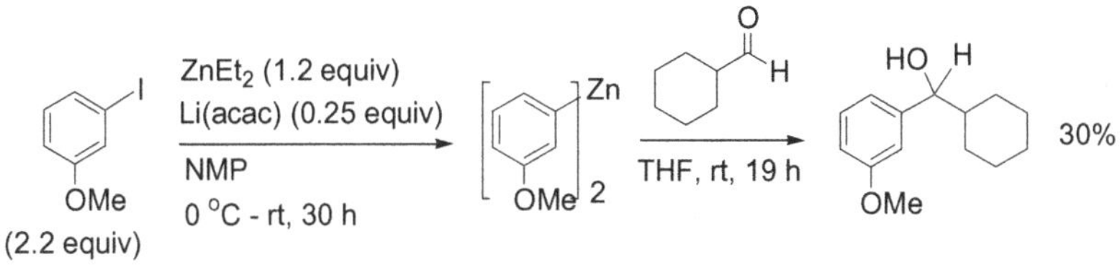

Scheme 4.27. Preparation of a substituted arylzinc and its addition to an aldehyde.

cyclohexanecarboxaldehyde to give the corresponding alcohol not only in greatly increased yield (94%) but also with very high enantioselectivity (>99% ee). Thus, (S)-**4.24** not only activates the nucleophilic arylzinc addition to the aldehyde but also provides excellent stereocontrol.

Table 4.39 summarizes the reaction of the *in situ* prepared (3-MeO-$C_6H_4)_2$Zn with various aldehydes catalyzed by (S)-**4.24**. In the presence of 10 mol% of (S)-**4.24**, the arylzinc additions to aromatic and aliphatic aldehydes are activated. Without (S)-**4.24**, the reactions are much slower and give very low yields of the alcohol products. For example, almost no product is obtained for the arylzinc addition to *m*-nitrobenzaldehyde in the *absence* of (S)-**4.24** (entry 4). However, using (S)-**4.24** leads to 93% yield of the desired diarylcarbinol. In addition, high enantioselectivity (83–>99% ee) is achieved for the arylzinc addition to both aromatic and aliphatic aldehydes. In entry 8, the addition to a chiral aldehyde generates both diastereomers in 1:1 ratio, each with over 99% ee. The following outlines the general procedure used for the reactions shown in Table 4.39. Under nitrogen at 0°C, $ZnEt_2$ (1.2 equivalent) is added dropwise to a mixture of Li(acac) (0.25 equivalent), 3-iodoanisole (2.2 equivalent) and NMP (1.5 mL) to produce the diarylzinc (the first step). Then a solution of (S)-**4.24** (10 mol%) in THF (5 mL) is added (the second step). Subsequently at room temperature, an aldehyde is added (the third step), which gives the chiral alcohol product upon work-up and purification. The *ee*'s of the products are determined by using HPLC-Chiralcel OD-H column (solvent: hexanes:isopropanol). The absolute configuration for *m*-methoxyphenyl phenyl carbinol generated from (S)-**4.24** is determined to be *R*.

Using a procedure similar to that described above, we have conducted the reaction of benzaldehyde with the arylzinc generated *in situ* from methyl *p*-iodobenzoate and $ZnEt_2$ (Scheme 4.28) As shown in entry 1 of Table 4.40, this gives only low yield (47%) and ee (67%) of the corresponding

Table 4.39. Addition of the arylzinc generated from *m*-iodoanisole to aldehydes in the presence of (S)-**4.24**.

Entry	Product	t (h)[a]	Yield (%)	ee (%)	Yield without (S)-**4.24**
1	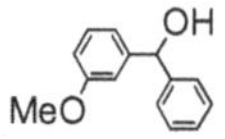 (MeO-C6H4)CH(OH)(Ph)	7	93	91	56
2	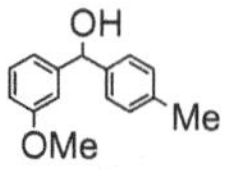 (3-MeO-C6H4)CH(OH)(4-Me-C6H4)	11	95	91	43
3	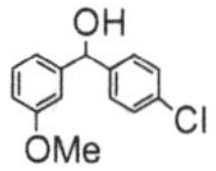 (3-MeO-C6H4)CH(OH)(4-Cl-C6H4)	10	85	90	41
4	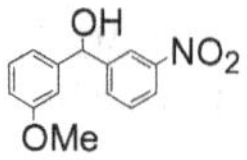 (3-MeO-C6H4)CH(OH)(3-NO2-C6H4)	8	93	83	Trace
5	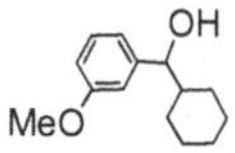 (3-MeO-C6H4)CH(OH)(cyclohexyl)	16	93	>99	30
6	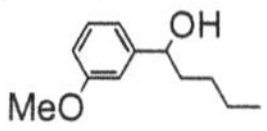 (3-MeO-C6H4)CH(OH)(n-Pr)	14	85	96	10
7	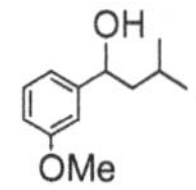 (3-MeO-C6H4)CH(OH)(i-Bu)	16	85	93	17
8	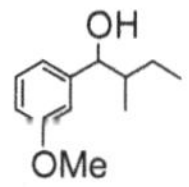 (3-MeO-C6H4)CH(OH)(sec-Bu)	10	90[b]	>99; 99	25

[a]Third step.
[b]The combined yield of both diastereomers.

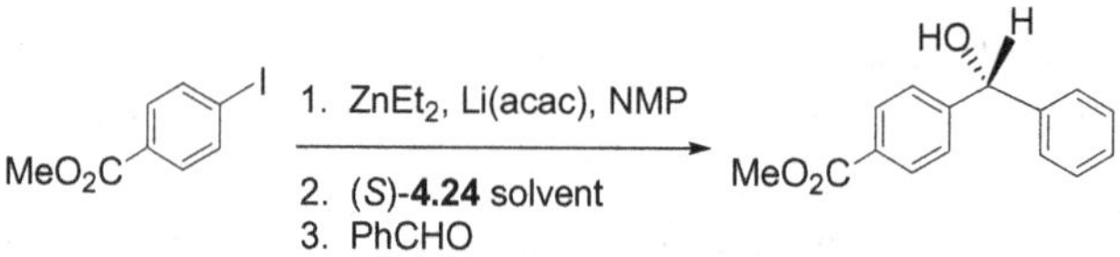

Scheme 4.28. Reaction of methyl *p*-iodobenzoate with benzaldehyde in the presence of ZnEt$_2$ and (S)-**4.24**.

Table 4.40. Conditions for the reaction of methyl p-iodobenzoate with benzaldehyde in the presence of $ZnEt_2$ and (S)-**4.24**.[a,b]

Entry	Reaction time 1st step (h)	(S)-**4.24** (mol%)	Solvent 2nd step (mL)	Reaction Temp 2nd step (°C)	Reaction Temp 3rd step (°C)	Reaction Time 3rd step (h)	Isolated yield (%)	ee (%)
1[c]	12	10	THF (5)	rt	rt	27	47	67
2[d]	12	10	THF (5)	rt	rt	27	63	71
3[e]	18	10	THF (5)	rt	rt	22	75	54
4[f]	16	10	THF (5)	rt	rt	27	90	52
5	18.5	20	THF (5)	0	rt	17.5	86	65
6[g]	21	10	Toluene (5)	0	rt	23	Quant.	45
7	21	10	Et_2O (5)	0	rt	23	80	55
8	18.5	10	CH_2Cl_2 (5)	0	rt	16	88	79
9	12	10	CH_2Cl_2 (10)	0	rt	9	93	84
10	17	20	CH_2Cl_2 (5)	0	rt	9	99	84
11	**18**	**20**	**CH_2Cl_2 (20)**	**0**	**0**	**17**	**97**	**94**

[a] Unless otherwise indicated, the following conditions were employed: 2.0 mmol methyl p-iodobenzoate (2.2 equiv), 0.24 mmol Li(acac) (0.25 equiv), 1.1 mmol $ZnEt_2$ (1.2 equiv), NMP (1.5 mL), and 0.91 mmol benzaldehyde. Preparation of the arylzinc in the first step was conducted at 0°C. Upon addition of a solution of (S)-**4.24** to the arylzinc solution, the reaction mixture was stirred for 1 h prior to addition of benzaldehyde.
[b] ee's were measured on Chiralpak HPLC AD column (2% IPA:98% hexanes, 1 mL/min).
[c] Arylzinc was prepared at 0°C for 6 h and then at room temperature for 6 h.
[d] Arylzinc was prepared at 0°C for 0.5 h and then at room temperature for 11.5 h.
[e] (S)-**4.24** was mixed with $ZnPh_2$ (2.0 equiv versus the ligand) at room temperature for 1 h and then combined with benzaldehyde before addition to the arylzinc solution.
[f] (S)-**4.24** was mixed with $ZnEt_2$ (2.0 equiv versus the ligand) at room temperature for 1 h and then combined with benzaldehyde before addition to the arylzinc solution.
[g] 3.2 equiv of aryl iodide was used.

diarylcarbinol product. Increasing the time for the arylzinc formation step at room temperature increases the yield to 63% and the ee to 71% (entry 2). Using $ZnPh_2$ or $ZnEt_2$ to prepare the zinc complex of (S)-**4.24** as the catalyst increases the yield but decreases the ee (entries 3 and 4). Increasing the amount of (S)-**4.24** to 20 mol% does not improve the enantioselectivity (entry 5). Changing the solvent in the second step from THF to toluene or Et_2O decreases the ee (entries 6 and 7). Using CH_2Cl_2 as the solvent in the second step significantly improves both the yield and ee (entry 8). Increasing the amount of CH_2Cl_2 further improves the enantioselectivity (entry 9). Finally, by using 20 mol% of (S)-**4.24** in CH_2Cl_2 at 0°, up to 97% yield and 94% ee are obtained for this reaction (entry 11).

The optimized conditions of entry 11 in Table 4.40 are applied to the reaction of various aldehydes with the arylzinc generated *in situ* from methyl *p*-iodobenzoate and ZnEt$_2$, and the results are summarized in Table 4.41. High yield and enantioselectivity are observed for a variety of aliphatic, aromatic and α, β-unsaturated aldehydes. Using (S)-**4.24** has

Table 4.41. Addition of the arylzinc generated from methyl *p*-iodobenzoate to aldehydes in the presence of (S)-**4.24**.[a]

Entry	Product	T (h)[b]	Yield (%)	ee (%)	Yield without (S)-**4.24**
1	MeO$_2$C–C$_6$H$_4$–CH(OH)–cyclohexyl	16	86	96	4
2	MeO$_2$C–C$_6$H$_4$–CH(OH)–iPr	15	91	94	28
3	MeO$_2$C–C$_6$H$_4$–CH(OH)–(3-NO$_2$-C$_6$H$_4$)	13	84	92	37
4	MeO$_2$C–C$_6$H$_4$–CH(OH)–Ph	17	97	94	
5	MeO$_2$C–C$_6$H$_4$–CH(OH)–(3-Cl-C$_6$H$_4$)	15	93	88	
6	MeO$_2$C–C$_6$H$_4$–CH(OH)–(4-OMe-C$_6$H$_4$)	15	52	89	
7	MeO$_2$C–C$_6$H$_4$–CH(OH)–(2-naphthyl)	15	89	92	
8	MeO$_2$C–C$_6$H$_4$–CH(OH)–CH=CH–Ph	20	90	87	

[a]Conditions: 2.0 mmol methyl *p*-iodobenzoate (2.2 equiv), 0.23 mmol Li (acac) (0.25 equiv), 1.1 mmol ZnEt$_2$ (1.2 equiv), NMP (1.5 mL), (S)-**4.24** (20 mol%) and 0.91 mmol aldehyde. Preparation of the arylzinc in the first step was conducted at 0°C for 18 h. Upon addition of a CH$_2$Cl$_2$ (20 mL) solution of (S)-**4.24** to the arylzinc solution (second step), the reaction mixture was stirred at 0°C for 1 h prior to addition of aldehyde (third step).
[b]Third step.

greatly activated the arylzinc addition with high stereocontrol. For example, without (S)-**4.24**, there is little addition of the arylzinc generated from methyl p-iodobenzoate to cyclohexanecarboxaldehyde (4% yield). In the presence of 20 mol% of (S)-**4.24**, the desired product is isolated in 86% yield and 96% ee (entry 1).

The reaction of benzaldehyde with the arylzinc generated from m-iodobenzonitrile in the presence of (S)-**4.24** is investigated (Scheme 4.29). Table 4.42 summarizes the conditions for this reaction. It is found that THF is a better solvent than CH_2Cl_2, toluene, or Et_2O (entries 1–11). The optimized conditions involve the use of 40 mol% of (S)-**4.24** in THF at 0°C, which gives the product in 87% yield and 84% ee (entry 14). The conditions in entry 14 of Table 4.42 are applied to the reactions of several aldehydes with the arylzinc generated in situ from m-iodobenzonitrile and $ZnEt_2$. As shown in Table 4.43, good yields and enantioselectivity are observed for the addition to several aromatic and aliphatic aldehydes.

The H_8BINOL-based ligands (S)-**4.39**, (S)-**4.40** and (S)-**4.41** are prepared in ways similar to the preparation of (S)-**4.24**. These ligands are used to catalyze the reaction of m-iodobenzonitrile with p-methoxybenzaldehyde under the same conditions as those used in Table 4.43. As the results listed in Table 4.44 show, these ligands exhibit improved enantioseletivity over (S)-**4.24**.

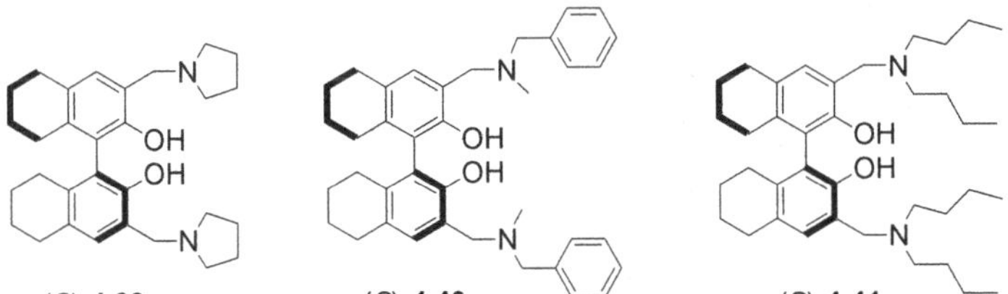

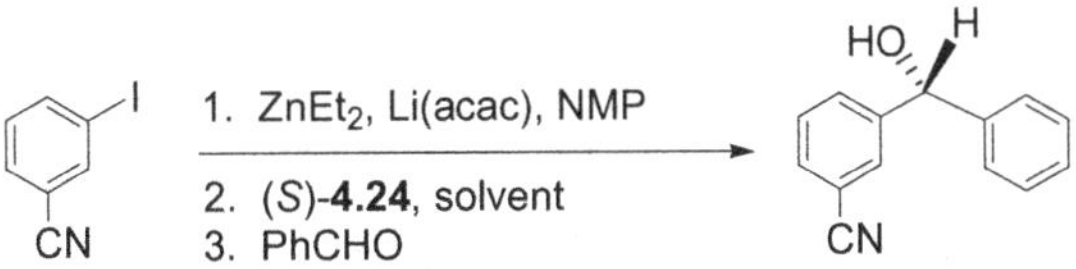

Scheme 4.29. Reaction of m-iodobenzonitrile with benzaldehyde in the presence of $ZnEt_2$ and (S)-**4.24**.

Table 4.42. Conditions for the reaction of m-iodobenzonitrile with benzaldehyde in the presence of $ZnEt_2$ and (S)-**4.24**[a,b]

Entry	m-IC_6H_4CHO (equiv versus PhCHO)	(S)-**4.24** (mol%)	Solvent 2nd step (mL)	Reaction temp 2nd step (°C)	Reaction temp 3rd step (°C)	Reaction time 3rd step (h)	Isolated yield (%)	ee (%)
1[c]	2.2	10	THF (5)	0	rt	30	77	66
2[c,d]	2.2	10	CH_2Cl_2 (5)	0	rt	20	93	35
3	2.2	10	CH_2Cl_2 (5)	0	rt	20	87	52
4	2.2	10	Tol (5)	0	rt	20	80	25
5[e]	2.2	10	Et_2O (5)	0	rt	20	80	42
6	2.2	20	THF (20)	0	rt	24	61	77
7	2.6[f]	20	CH_2Cl_2 (10)	0	0	22	65	52
8	2.6[f]	20	THF (10)	0	0	22	27	38
9[g]	2.2	20	THF (20)	rt	rt	24	70	71
10[g]	2.2	20	CH_2Cl_2 (20)	rt	rt	24	40	73
11[g,h]	3.2[i]	20	Tol (20)	rt	rt	36	70	17
12	2.2	20	THF (20)	0	11–13	44	50	79
13[j]	3.2[i]	40	THF (25)	rt	0	36	60	88
14[k]	**3.2**[i]	**40**	**THF (25)**	**rt**	**0**	**48**	**87**	**84**

[a] Unless otherwise indicated, the following conditions were employed: Li(acac) (0.25 equiv), $ZnEt_2$ (1.2 equiv), NMP (1.5 mL), and 0.91 mmol benzaldehyde. Arylzinc preparation proceeded at 0 °C for 18 h. A solution of (S)-**4.24** was added to the arylzinc solution and stirred for 1 h before the addition of benzaldehyde.
[b] ee's were measured on Chiralpak HPLC AD column (solvent: isopropanol/hexanes).
[c] Arylzinc was prepared at 0 °C for 17 h.
[d] 1.1 mmol m-iodobenzonitrile and 0.5 mmol benzaldehyde were used.
[e] (S)-**4.24** was sonicated in Et_2O for 45 m.
[f] Li(acac) (0.31 equiv) and Et_2Zn (1.4 equiv).
[g] (S)-**4.24** was mixed with $ZnPh_2$ (two equiv versus the ligand) at room temperature for 1 h and then combined with benzaldehyde before addition to the arylzinc solution.
[h] Arylzinc was prepared at 0 °C for 12 h.
[i] Li(acac) (0.38 equiv) and $ZnEt_2$ (1.8 equiv).
[j] Arylzinc was prepared at 0 °C for 14 h, and then at room temperature for 4 h.
[k] Arylzinc was prepared at 0 °C for 2 h, and then at room temperature for 34 h.

4.4. Asymmetric Alkylzinc Addition to Aldehydes

4.4.1. Using the 3,3'-Diaryl-Substituted BINOL Ligands

In 1986, Noyori reported that (-)-3-exo-dimethylaminoisobornenol catalyzed the reaction of $ZnMe_2$ with benzaldehyde to generate α-methyl benzyl alcohol with 95% ee.[42] Since then, a great number of catalysts have

Table 4.43. Addition of the arylzinc generated from m-iodobenzonitrile to aldehydes in the presence of (S)-**4.24**.[a]

Entry	Product	t (h)[b]	Yield (%)	ee (%)
1	(3-CN-C6H4)CH(OH)(C6H5)	48	87	84
2	(3-CN-C6H4)CH(OH)(4-OMe-C6H4)	48	74	79
3	(3-CN-C6H4)CH(OH)(iPr)	48	82	91
4	(3-CN-C6H4)CH(OH)(cyclohexyl)	48	77	>98

[a]Conditions: Li (acac) (0.38 equiv), ZnEt$_2$ (1.8 equiv), NMP (1.5 mL) and 0.91 mmol aldehyde. Arylzinc preparation proceeded at 0°C for 2 h and room temperature for 34 h (first step). A THF (25 mL) solution of (S)-**4.24** (40 mol%) was added to the arylzinc solution (second step) and stirred at 0°C for 1 h before the addition of aldehyde (third step).
[b]Third step.

Table 4.44. Addition of the arylzinc generated from m-iodobenzonitrile to p-methoxybenzaldehyde in the presence of ligands (S)-**4.39**, (S)-**4.40** and (S)-**4.41**.

Entry	Ligand	Product	Yield (%)	ee (%)
1	(S)-**4.39**	(3-CN-C6H4)CH(OH)(4-OMe-C6H4)	97	88
2	(S)-**4.40**	(3-CN-C6H4)CH(OH)(4-OMe-C6H4)	80	91
3	(S)-**4.41**	(3-CN-C6H4)CH(OH)(4-OMe-C6H4)	~60	85
4	(S)-**4.24**	(3-CN-C6H4)CH(OH)(4-OMe-C6H4)	74	79

been developed for the asymmetric alkylzinc addition to aldehydes. Among these catalysts, many are found to be good for the reaction of aromatic aldehydes, less are good for the reaction of aliphatic aldehydes and few are good for the reaction of α,β-unsaturated aldehydes. No catalyst is found to be generally enantioselective for all three types of aldehydes previously.

As described in Section 3.7.3 in Chapter 3, the BINOL derivative (R)-**4.1** [= (R)-**3.30**] is found to be the most generally enantioselective catalyst for the alkylzinc addition to aldehydes.[43] It catalyzes the reaction of $ZnEt_2$ with a variety of aromatic, aliphatic and α,β-unsaturated aldehydes with 90->99% ee. It also shows high enantioselectivity for the $ZnMe_2$ addition to aldehydes. At the 3,3'-positions of the BINOL unit in (R)-**4.1**, the two aryl groups can generate three chiral conformations as shown by **4.42**–**4.44**. These three diastereomeric structures may contribute differently for the catalytic properties of (R)-**4.1**. At room temperature, these three conformations interconvert rapidly because of the low rotation barrier around the aryl–aryl bonds at the 3,3'-positions. These different conformations can be observed at low temperature by NMR spectroscopy.

In order to study the influence of the chiral conformation of the 3,3'-diaryl-substituted BINOL ligand on the asymmetric alkylzinc addition to aldehydes, we have synthesized compounds (S,S,S)-**4.45**, (S,S,R)-**4.46** and (R,S,R)-**4.47** (Scheme 4.30).[44] From the Suzuki coupling of 2-methoxy-naphth-1-yl boronic acid with the MOM protected 3,3'-diiodoBINOL followed by hydrolysis, the three diastereomeric compounds (S,S,S)-**4.45**, (S,S,R)-**4.46** and (R,S,R)-**4.47** are obtained. Since these tetranaphthalene compounds have greatly increased rotation barrier around the 3,3'-aryl–aryl bonds, they can be separated at room temperature by column chromatograph. On silica gel column, it is found that the least

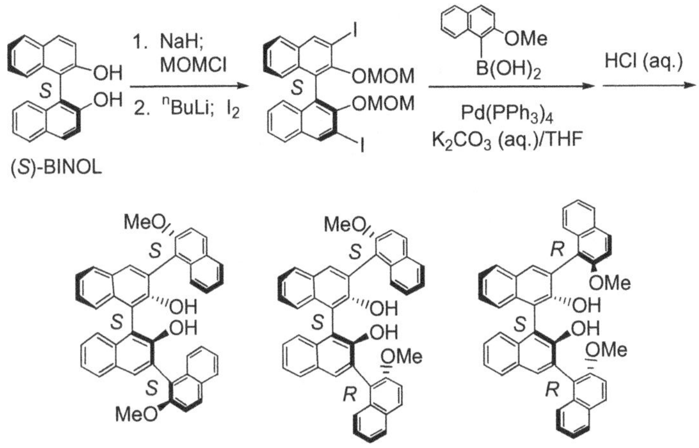

Scheme 4.30. Synthesis of diastereomeric BINOL derivatives (S,S,S)-**4.45**, (S,S,R)-**4.46** and (R,S,R)-**4.47**.

polar isomer is (S,S,S)-**4.45** and the most polar isomer is (R,S,R)-**4.47** when eluted with ethyl acetate and hexane.

The ^{1}H NMR spectra of compounds (S,S,S)-**4.45** and (R,S,R)-**4.47** show a singlet methoxyl signal at δ 3.95 (s, 6H), and 3.83 (s, 6H), respectively, consistent with the expected C_2 symmetry. The ^{1}H NMR spectrum of (S,S,R)-**4.46** shows two methoxyl signals at δ 3.81 (s, 3H) and 3.94 (s, 3H), indicating a C_1 symmetric compound. A single crystal X-ray analysis of the C_2 symmetric diastereomer (S,S,S)-**4.45** in collaboration with Sabat establishes its absolute structure. As shown in Fig. 4.2, the configurations of the tetranaphthalene units are SSS. The dihedral angle of the top binaphthyl unit is 73°, the middle binaphthyl unit 69°, and the bottom binaphthyl 74°. The three enantiomers of (S,S,S)-**4.45**, (S,S,R)-**4.46** and (R,S,R)-**4.47** are also obtained by starting with (R)-BINOL. They give the same spectroscopic characteristics as their corresponding enantiomers except their opposite optical rotations.

When the isolated diastereomers are stored at $-5°C$, no isomerization is observed after a long period of time. A solution of (R,S,R)-**4.47** in chloroform-d is found to slowly isomerize back to a mixture of the three diastereomers at room temperature when monitored by ^{1}H NMR spectrometry. An equilibrium is reached after 30 days, which contains compounds

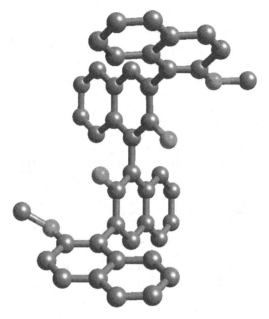

Figure 4.2. The X-ray structure of diastereomer (S, S, S)-**4.45**.

(S, S, S)-**4.45**, (S, S, R)-**4.46** and (R, S, R)-**4.47** in a 1.1:2.2:1.0 ratio. This demonstrates that there is essentially no thermodynamic preference for any one of the three diastereomers in solution.

Compounds (S, S, S)-**4.45**, (S, S, R)-**4.46** and (R, S, R)-**4.47** are used to catalyze the reaction of diethylzinc with p-anisaldehyde in toluene-d_8 at room temperature and the reactions are monitored by using ^{1}H NMR spectroscopy (Scheme 4.31). As shown in Table 4.45, the three diastereomeric compounds exhibit catalytic properties very different from each other. Compound (S, S, S)-**4.45** has the highest catalytic activity as well as enantioselectivity. It catalyzes the diethylzinc addition to p-anisaldehyde to give (S)-1-(4-methoxy-phenyl)-propan-1-ol with 87% ee and a reaction half-life of 95 min. Both compounds (S, S, R)-**4.46** and (R, S, R)-**4.47** show much lower reaction rates as well as much lower enantioselectivity. The two C_2 symmetric ligands (S, S, S)-**4.45** and (R, S, R)-**4.47** show better enantioselectivity than the C_1 symmetric (S, S, R)-**4.46**. All the three diastereomers produce (S)-1-(4-methoxy-phenyl)-propan-1-ol as the major enantiomer. This indicates that the central BINOL unit controls the chiral configuration of the product.

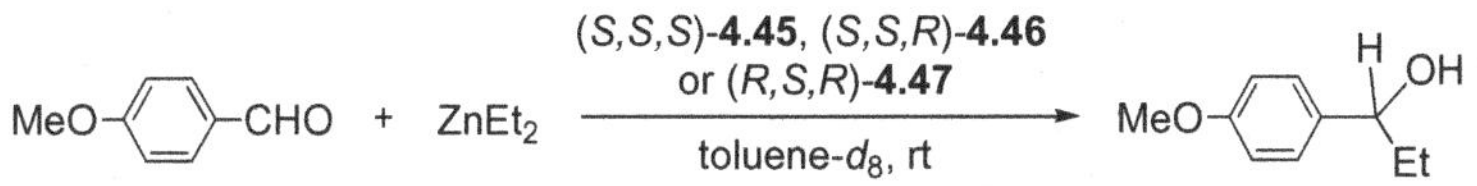

Asymmetric Catalysis by BINOL and Its Non-polymeric Derivatives

Scheme 4.31. Catalytic asymmetric reaction of *p*-anisaldehyde with diethylzinc in the presence of compounds (S, S, S)-**4.45**, (S, S, R)-**4.46** and (R, S, R)-**4.47**.

Table 4.45. Reaction of diethylzinc with *p*-anisaldehyde catalyzed by compounds (S, S, S)-**4.45**, (S, S, R)-**4.46** and (R, S, R)-**4.47**.[a]

Entry	ZnEt$_2$ (M)	*p*-Anisaldehyde (M)	Compound	$t_{1/2}$ (min)	ee (%)	Configuration
1	0.267	0.133	(S, S, S)-**4.45**	95	87	S
2	0.270	0.130	(S, S, R)-**4.46**	340	14	S
3	0.263	0.133	(R, S, R)-**4.47**	1440	61	S

[a] 5 mol% of compound (S, S, S)-**4.45**, (S, S, R)-**4.46** or (R, S, R)-**4.47** is used in each reaction.

The X-ray structure of (S, S, S)-**4.45** indicates that a zinc center can interact only with no more than two of the adjacent oxygen atoms in this compound. The molecular modeling (MacSpartanPlus-AM1) suggests that all the four oxygen atoms in (R, S, R)-**4.47** could coordinate to a single zinc center and up to three of the oxygen atoms in (S, S, R)-**4.46** could coordinate to a zinc center. Such differences in coordination ability may be responsible for the observed differences in catalytic activity. Because of the least number of oxygen coordination, the proposed zinc catalyst (S, S, S)-**4.48** generated from the reaction of (S, S, S)-**4.45** with diethylzinc may have the highest Lewis acidity, leading to the highest catalytic activity among the three diastereomeric catalysts.

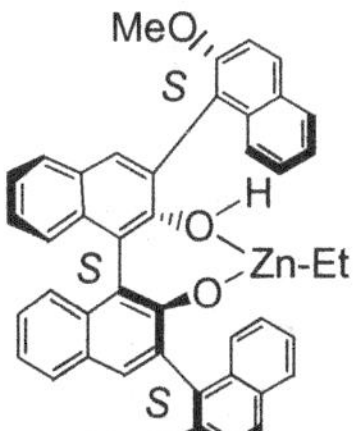

(S,S,S)-**4.48**

Table 4.46. Enantioselectivity of the catalytic asymmetric reaction of aldehydes with diethylzinc in the presence of compound (S, S, S)-**4.45**.

Entry	Aldehyde	ee (%)	Configuration
1	H₃CO–C₆H₄–CHO	93	S
2	C₆H₅–CHO	93	S
3	2-Cl-C₆H₄–CHO	89	S
4	3-Cl-C₆H₄–CHO	96	S
5	4-Cl-C₆H₄–CHO	92	S
7	1-naphthyl–CHO	91	S
8	2-naphthyl–CHO	93	S
9	n-alkyl–CHO	94	S

We have also studied the use of (S, S, S)-**4.45** to catalyze the reaction of diethylzinc with other aldehydes. At 0°C in toluene, 5 mol% of (S, S, S)-**4.45** shows 89–96% ee for the diethylzinc addition to both aromatic and aliphatic aldehydes (Table 4.46). The high enantioselectivity of (S, S, S)-**4.45** is similar to (R)-**4.1** but with the opposite alcohol product configuration because of their opposite 1,1′-binaphthyl configuration. This indicates that the catalytically active species generated from (S, S, S)-**4.45** may be structurally similar to that generated from (S)-**4.1**, the enantiomer of (R)-**4.1**.

The results described above demonstrate that the 3,3′-aryl conformation of the 1,1′-binaphthyl-based catalysts is very important for both the stereoselectivity and catalytic activity. The most favorable catalyst should have an (S, S)-3,3′-aryl conformation for the (S)-BINOL ligand. Correspondingly, the (R, R)-3,3′-aryl conformation should be most favorable for the (R)-BINOL ligand.

4.4.2. Using the 3,3'-Morpholinomethyl-Substituted H_8BINOL Ligand[35b]

The catalytic properties of the 3,3'-morpholinomethyl-substituted H_8BINOL (S)-**4.24** in the ZnMe$_2$ and ZnEt$_2$ additions to aldehydes are examined. For the reaction of ZnMe$_2$ and ZnEt$_2$ with valeraldehyde in THF at room temperature, (S)-**4.24** cannot give the desired product. For the reaction of ZnEt$_2$ with benzaldehyde in THF at room temperature, (S)-**4.24** gives the addition product 1-phenyl-propyl-1-ol with 73% *ee* and 78% yield. Thus, although (S)-**4.24** is highly enantioselective for the arylzinc and alkynylzinc additions to aldehydes, its enantioselectivity for ZnEt$_2$ addition is lower.

4.4.3. Using a Dendritic BINOL Ligand[45]

The BINOL-centered aryleneethynylene dendrimer (S)-**4.49** [= (S)-**5.5**; for the synthesis of this compound and its analogs, see Section 5.2 in Chapter 5] is used to catalyze the asymmetric reaction of benzaldehyde with diethylzinc in the absence of Ti(O^{i}Pr)$_4$. We find that the nanoscale dendritic material behaves very differently from the small BINOL molecule. Dendrimer (S)-**4.49** shows much higher catalytic activity than (S)-BINOL and also generates the opposite enantiomeric product. In the presence of (S)-**4.49** (5 mol%) in toluene solution, a 98.6% conversion of benzaldehyde is observed in 24 h at room temperature. However, only 37% conversion is observed under the same condition when BINOL is used. Both the enantioselectivities of (S)-**4.49** and BINOL are very low. The increased catalytic activity of (S)-**4.49** over BINOL indicates that the zinc complex generated from the reaction of (S)-**4.49** with diethylzinc may have much higher Lewis acidity than the zinc complex generated from BINOL. The zinc complex formed from the reaction of BINOL with diethylzinc probably exists as aggregates in solution through intermolecular Zn–O–Zn bonds, which should greatly reduce the Lewis acidity of the zinc center. It is expected that such aggregate cannot be generated in the case of (S)-**4.49** due to the large dendritic arms. These bulky and rigid dendritic arms prevent (S)-**4.49** from forming oligomers through Zn–O–Zn bonds, yet still allow small molecules such as benzaldehyde and diethylzinc to approach the chiral core for the catalytic reaction. The opposite enantioselectivity of (S)-**4.49** *versus* (S)-BINOL also indicates that the *in situ* generated catalytically active species when these ligands are used are very different. It may be monomeric for (S)-**4.49** but oligomeric for BINOL.

(S)-**4.49**

In the presence of Ti(O^{i}Pr)$_4$, (S)-**4.49** becomes highly enantioselective for the reaction of aldehydes with diethylzinc. In toluene, (S)-**4.49** (20 mol%) and Ti(O^{i}Pr)$_4$ (1.4 equivalent) catalyze the reaction of 1-naphthaldehyde with diethylzinc with 100% conversion and 90% *ee* in 5 h. No side product is observed. Under similar condition, (S)-**4.49** shows 100% conversion and 89% *ee* for the reaction of benzaldehyde with diethylzinc in the presence of Ti(O^{i}Pr)$_4$. These catalytic properties of (S)-**4.49** are very similar to those of BINOL in the presence of the titanium complex as reported by Chan and coworkers.[46] The same enantiomeric product is produced by using either (S)-**4.49** or (S)-BINOL in the presence of Ti(O^{i}Pr)$_4$. This study indicates that unlike the reaction in the absence of Ti(O^{i}Pr)$_4$, the asymmetric reaction of aldehydes with diethylzinc in the presence of the titanium complex may involve structurally similar catalytically active species when either the chiral dendrimer (S)-**4.49** or the small molecule BINOL is used. Since the titanium complex of (S)-**4.49** is not expected to generate any dimeric or oligomeric structure through the binaphthyl core, the catalytically active species for the diethylzinc addition to aldehydes formed from the interaction of BINOL with Ti(O^{i}Pr)$_4$ may also be monomeric. In these catalytic reactions, (S)-**4.49** can be easily recovered from the reaction mixture by precipitation with methanol due to the large size difference between this dendritic molecule and the products and reagents.

4.5. Asymmetric TMSCN Addition to Aldehydes

Cyanohydrins are very useful synthetic precursors to organic compounds such as α-hydroxyacids, α-hydroxyketones, α-amino acids, β-amino alcohols and 1,2-diols. A number of catalysts have been developed for the asymmetric reaction of aldehydes with TMSCN to generate chiral cyanohydrins.[47] We have studied the use of two types of ligands to catalyze this reaction. One of the catalysts is based on the 3,3′-bismorpholinomethyl BINOL and H$_8$BINOL ligands, and the other is based on bisBINOL salen ligands.

4.5.1. Catalysis by the 3,3′-Bismorpholinomethyl BINOL and H$_8$BINOL Ligands[35b,48]

The morpholinomethyl-substituted BINOL and H$_8$BINOL ligands (S)-**4.34**, (S)-**4.35** and (S)-**4.24** are used in combination with Me$_2$AlCl to catalyze the TMSCN addition to benzaldehyde (Scheme 4.32). The reactions are conducted by stirring a ligand (10 mol%), Me$_2$AlCl (10 mol%) and 4 Å molecular sieves in a solvent (1 mL) at room temperature for 1 h. Ph$_3$P=O is then added and the mixture is cooled to −20°C. Both benzaldehyde and TMSCN are added in one portion. After 48 h, the reaction is quenched with HCl (1 N) and the product is converted to an O-acetyl cyanohydrin before isolation and characterization. Table 4.47 gives the results for the reactions catalyzed by (S)-**4.34**, (S)-**4.35** and (S)-**4.24** in various solvents. The best result is obtained by using (S)-**4.34** as the chiral ligand in toluene solution, which gives the product with 96% *ee* (entry 2). Both the monosubstituted BINOL ligand (S)-**4.35** and the H$_8$BINOL-based ligand (S)-**4.24** gives much lower enantioselectivity.

When Ti(O^{i}Pr)$_4$ is used in place of Me$_2$AlCl for the reaction catalyzed by (S)-**4.34**, only moderate *ee* is observed. Switching Ph$_3$P=O with HMPA reduces the reaction time and slightly increases the yield. Therefore, the

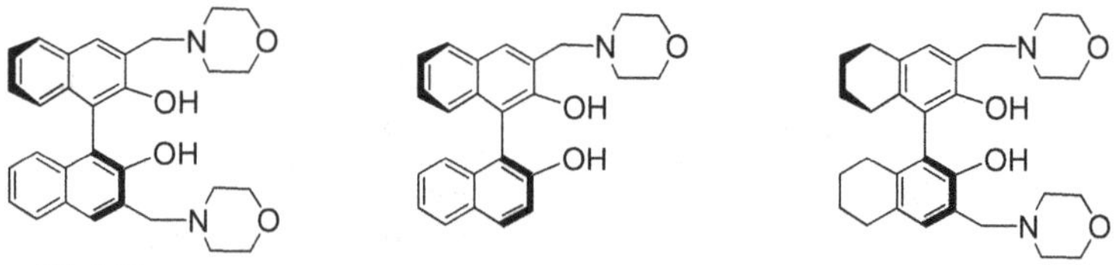

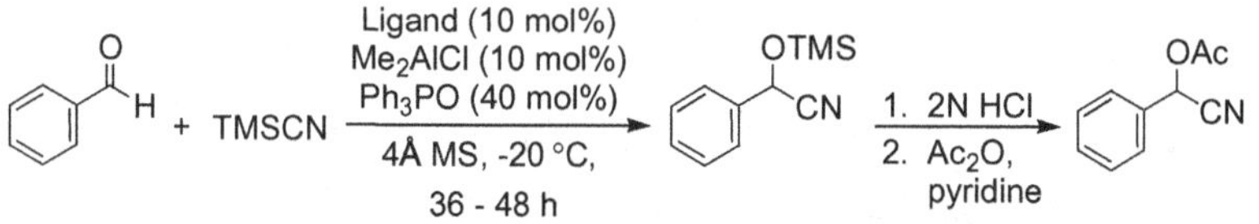

Scheme 4.32. Catalytic TMSCN addition to benzaldehyde in the presence of (S)-**4.34**, (S)-**4.35** and (S)-**4.24**.

Table 4.47. Results for the TMSCN addition to benzaldehyde in the presence of the BINOL ligands.

Entry	Ligand	Solvent	ee (%)
1	(S)-**4.34**	THF	72
2	**(S)-4.34**	**Toluene**	**96**
3	(S)-**4.34**	Et$_2$O	73
4	(S)-**4.35**	THF	7
5	(S)-**4.35**	Toluene	20
6	(S)-**4.35**	Et$_2$O	11
7	(S)-**4.35**	CH$_2$Cl$_2$	7
8	(S)-**4.24**	Toluene	53

optimized conditions are identified as using (S)-**4.34** in combination with Me$_2$AlCl, HMPA and 4 Å MS in toluene at −20°C. Under these conditions, the reaction of TMSCN with benzaldehyde gives the cyanohydrin product with 92% yield and 92% ee. The reactions of TMSCN with other aromatic aldehydes under the optimized conditions are conducted. As shown in Table 4.48, good enantioselectivity is observed for various aromatic aldehydes.

Since very few good catalysts are found to catalyze the reaction of TMSCN with aliphatic aldehydes, we have investigated the use of (S)-**4.34** for this reaction. Table 4.49 summarizes the conditions tested for the reaction of octyl aldehyde with TMSCN in the presence of (S)-**4.34**, Me$_2$AlCl, HMPA and 4 Å MS. It is found that changing the

Table 4.48. Results of TMSCN addition to aromatic aldehydes.

Entry	Aldehyde	Product	Yield (%)	ee (%)
1	PhCHO	Ph-CH(OAc)CN	92	94
2	3-MeC₆H₄CHO	3-MeC₆H₄-CH(OAc)CN	90	93
3	3-ClC₆H₄CHO	3-ClC₆H₄-CH(OAc)CN	77	94
4	3-BrC₆H₄CHO	3-BrC₆H₄-CH(OAc)CN	88	82
5	3-MeOC₆H₄CHO	3-MeOC₆H₄-CH(OAc)CN	82	80
6	4-MeC₆H₄CHO	4-MeC₆H₄-CH(OAc)CN	86	75
7	4-FC₆H₄CHO	4-FC₆H₄-CH(OAc)CN	85	90
8	4-MeOC₆H₄CHO	4-MeOC₆H₄-CH(OAc)CN	80	80
9	2-furyl-CHO	2-furyl-CH(OAc)CN	70	74
10	2-ClC₆H₄CHO	2-ClC₆H₄-CH(OAc)CN	94	80

Table 4.49. Asymmetric addition of TMSCN to octyl aldehyde in the presence of (S)-**4.34** and Me$_2$AlCl.[a]

Entry	Solvent	4 Å MS	t (h)	ee (%)
1	Toluene	5 mg	24	91
2	THF	5 mg	24	87
3	**Et$_2$O**	**5 mg**	**24**	**97**
4[b]	Toluene	5 mg	24	82
5	Toluene	None	24	46
6[c]	Et$_2$O	5 mg	3	76
7[c,d]	Et$_2$O	15 mg	4	31
8[c]	Hexane	5 mg	24	6

[a] The following procedure was used unless otherwise indicated: (S)-**4.34** (0.025 mmol, 10 mol%), 4 Å molecular sieves, and Me$_2$AlCl (10 mol%, 1 M in hexanes) in a solvent (1 mL) were stirred under nitrogen at room temperature for 3 h. It was then combined with the additive HMPA (40 mol%) and cooled down to −20°C. TMSCN (3.0 equivalent) and an aldehyde were added.
[b] No additive.
[c] Reaction at room temperature.
[d] 20 mol% (S)-**4.34** and 20 mol% Me$_2$AlCl.

solvent from toluene to diethyl ether leads to further enhancement of the enantioselectivity (97% ee, entry 3). Changing the temperature from −20°C to room temperature allows the reaction to proceed much faster with a reduced but still significant ee (76% ee, entry 6).

The conditions of entry 3 in Table 4.49 are applied to the TMSCN addition to other aliphatic aldehydes. As shown in Table 4.50, (S)-**4.34** (10 mol%) in combination with Me$_2$AlCl (10 mol%) gives very high enantioselectivity for all types of aliphatic aldehydes (92–99% ee). The reactions of the α,β-unsaturated aldehydes are also highly enantioselective. These reactions produce R products.

Ligand (S)-**4.34** exhibits a large positive non-linear effect in the catalytic reaction of TMSCN with octyl aldehyde with a 40% ee of (S)-**4.34** leading to 84% ee of the product. This indicates that the intermolecular aggregates of the ligand complex might be involved in the catalytic process. Although NMR experiments are conducted, no structural information can be obtained for the catalytically active species of this reaction. Structure **4.50** is a working hypothesis for the transition state of the TMSCN addition to an aldehyde catalyzed by (S)-**4.34**. In **4.50**, the aldehyde is activated by coordination to the Al(III) center and the TMSCN is activated by the Lewis basic HMPA and the morpholinyl nitrogen.

Table 4.50. Asymmetric addition of TMSCN to aliphatic aldehydes in the presence of (S)-**4.34** and Me$_2$AlCl.

Entry	Aldehyde[a]	Product	Yield (%)	ee (%)[b]
1	M$_6$CHO	M$_6$CH(OAc)CN	91	97
2	M$_7$CHO	M$_7$CH(OAc)CN	92	98
3	M$_3$CHO	M$_3$CH(OAc)CN	87	96
4	cyclohexyl-CHO	cyclohexyl-CH(OAc)CN	90	99
5	iPr-CHO	iPr-CH(OAc)CN	65	97
6	iBu-CHO	iBu-CH(OAc)CN	72	96
7	PhCH$_2$CH$_2$CHO	PhCH$_2$CH$_2$CH(OAc)CN	86	95
8	MeCH=C(Me)CHO	MeCH=C(Me)CH(OAc)CN	70	98
9	PhCH=CHCHO	PhCH=CHCH(OAc)CN	74	94
10	CH$_2$=CHCH$_2$CH$_2$CHO	CH$_2$=CHCH$_2$CH$_2$CH(OAc)CN	67	96
11	MeO$_2$C(CH$_2$)$_3$CHO	MeO$_2$C(CH$_2$)$_3$CH(OAc)CN	90	92

[a] Freshly distilled.

4.50

Both (S)-**4.34** and (S)-**4.24** in combination with Me₂AlCl are used to catalyze the reaction of acetophenone with TMSCN. Under the same conditions as the reaction of the aldehydes, very low yield of the product is obtained in each case with no enantioselectivity. When N-methylmorpholine-N-oxide is used in place of HMPA as the additive, the racemic product is obtained in high yield.

4.5.2. Asymmetric TMSCN Addition to Aldehydes Catalyzed by Binaphthyl Salens[49]

In Section 4.1.6, the use of the BINOL-Salen compound (S)-**4.17** to catalyze the highly enantioselective alkyne additions to aldehydes is described. We have also explored the application of this ligand in the reaction of TMSCN with aldehydes.

The metal complex of (S)-**4.17** is first isolated after it is treated with a Lewis acid metal precursor. Then this complex is used to catalyze the reaction of TMSCN with benzaldehyde at room temperature (Scheme 4.33). The trimethylsilylether product is converted to the corresponding acetate for *ee* determination. As shown in Table 4.51, many reaction conditions are examined by varying the amount of the ligand, the Lewis acid complexes and the solvents. Entry 7 gives the highest *ee* of 69%, which uses Ti(O^{i}Pr)₄ as the Lewis acid precursor and methylene chloride as the solvent.

In order to improve the enantioselectivity for this reaction, the structure of the salen ligand is modified. Compounds (S)-**4.51**–(S)-**4.53** are prepared and in these compounds one of the hydroxyl groups in the BINOL unit as in (S)-**4.17** is alkylated with various alkyl groups (Scheme 4.34). These compounds in combination with Ti(O^{i}Pr)₄ are used to catalyze the reaction of TMSCN with benzaldehyde at room temperature. Table 4.52

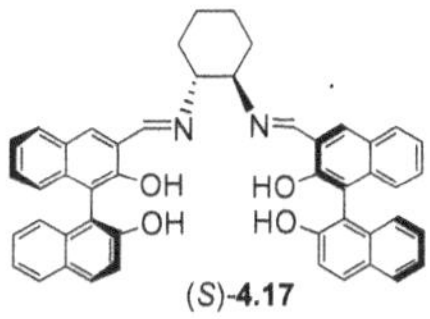

Asymmetric Catalysis by BINOL and Its Non-polymeric Derivatives   231

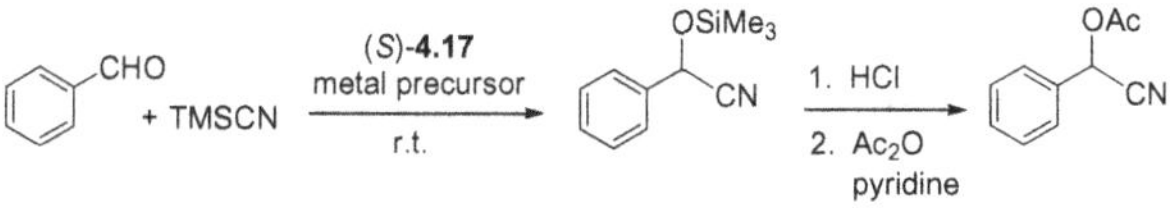

Scheme 4.33. Asymmetric cyanohydrin synthesis catalyzed by (S)-**4.17** in combined with Lewis acids.

Table 4.51. Asymmetric addition of trimethylsilyl cyanide to benzaldehyde in the presence of (S)-**4.17**.

Entry	(S)-**4.17** (%)	TMSCN (equiv)	Lewis acid complex	Solvent	Time (h)	ee (%)
1	20	4	Ti(O^{i}Pr)$_4$ (20%)	CH$_2$Cl$_2$	22	46
2	20	4	Ti(O^{i}Pr)$_4$ (20%)	Toluene	22	14
3	20	4	Ti(O^{i}Pr)$_4$ (20%)	THF	22	14
4	20	4	Ti(O^{i}Pr)$_4$ (20%)	Ether	22	15
5	10	4	Ti(O^{i}Pr)$_4$ (10%)	CH$_2$Cl$_2$	22	66
6	10	4	Ti(O^{i}Pr)$_4$ (10%)^a	CH$_2$Cl$_2$	22	48
7	0.5	4	Ti(O^{i}Pr)$_4$ (0.5%)	CH$_2$Cl$_2$	36	69
8	0.05	4	Ti(O^{i}Pr)$_4$ (0.05%)	CH$_2$Cl$_2$	60	26
9	10	4	TiCl$_4$ (10%)	CH$_2$Cl$_2$	22	45
10	0.5	4	TiCl$_4$ (0.5%)	CH$_2$Cl$_2$	36	34
11	10	4	AlMe$_2$Cl (10%)	CH$_2$Cl$_2$	22	15
12	0.5	4	AlMe$_2$Cl (0.5%)	CH$_2$Cl$_2$	36	40
13	10	4	ZrCl$_4$ (10%)	CH$_2$Cl$_2$	18	2
14	10	4	CeCl$_3$ (10%)	CH$_2$Cl$_2$	22	17
15	10	4	O=VSO$_4$ (10%)	CH$_2$Cl$_2$	20	52

^{a}10% O=P(Ph)$_3$ was added.

shows that (S)-**4.53** containing hexyl groups gives the highest ee for the reaction (84% ee, entry 3). Ligand (S)-**4.54** is prepared as the analog of (S)-**4.53** by replacing the cyclohexane-1,2-diamine unit with the 1,2-diphenylethylenediamine unit (Scheme 4.34). Ligand (S)-**4.54** in combination with Ti(O^{i}Pr)$_4$ catalyzes the reaction of TMSCN with benzaldehyde with greatly reduced enantioselectivity (entries 4 and 5). Thus, the more rigid cyclohexane-1,2-diamine-based ligand (S)-**4.53** is much better than the more flexible 1,2-diphenylethylenediamine-based ligand (S)-**4.54**.

The reaction conditions for the use of (S)-**4.53** are further explored. Simplification of the reaction procedure by directly using a mixture of (S)-**4.53** and Ti(O^{i}Pr)$_4$ in methylene chloride for the catalysis without isolating the chiral metal complex is attempted. In this procedure, benzaldehyde

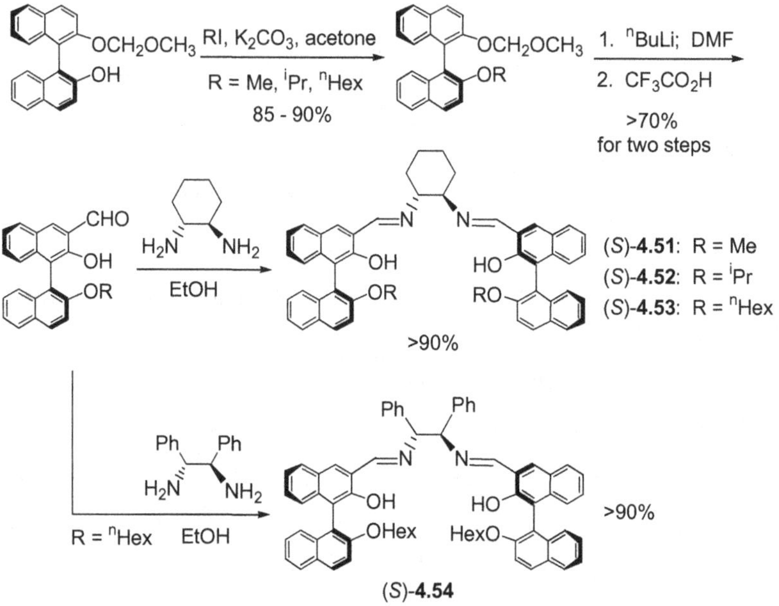

Scheme 4.34. Synthesis of BINOL-salen ligands (S)-**4.51**–(S)-**4.54**.

Table 4.52. Reactions of TMSCN with benzaldehyde in the presence of ligands (S)-**4.51**–(S)-**4.54**.

Entry	Ligand	TMSCN (equiv)	Ti(O^{i}Pr)$_4$ (%)	Solvent	Time (h)	ee (%)
1	(S)-**4.51** (12%)	4	10	CH$_2$Cl$_2$	22	52
2	(S)-**4.52** (12%)	4	10	CH$_2$Cl$_2$	4	66
3	(S)-**4.53** (12%)	4	10	CH$_2$Cl$_2$	4	84
4	(S)-**4.54** (12%)	4	10	CH$_2$Cl$_2$	4	32
5	(S)-**4.54** (12%)	4	10	Toluene	6	18

and TMSCN are added to the mixture of (S)-**4.53** and Ti(O^{i}Pr)$_8$ at room temperature. As shown in entry 2 of Table 4.53, when the simplified procedure is used, the enantioselectivity is almost the same as the one using the isolated chiral titanium complex. This simplified procedure is thus used for all the experiments in entries 3–11 in Table 4.53. No significant difference in enantioselectivity is observed under various conditions except a small ee increase at −20°C (entry 11). The configuration of the cyanahydrin product using (S)-**4.53** is determined to be S.

For convenience, the room temperature reaction conditions of entry 2 in Table 4.53 are chosen for the TMSCN addition to several aldehydes. In

Table 4.53. Reactions of TMSCN with benzaldehyde in the presence of (S)-**4.53** and Ti(O^{i}Pr)$_4$.

Entry	Ligand (%)	TMSCN (equiv)	Lewis acid	Solvent	Temperature	Time (h)	ee (%)
1	10	4	Ti(O^{i}Pr)$_4$ (10%)^a	CH$_2$Cl$_2$	rt	4	84
2	10	4	Ti(O^{i}Pr)$_4$ (10%)	CH$_2$Cl$_2$	rt	4	85
3	10	4	AlMe$_2$Cl (10%)	CH$_2$Cl$_2$	rt	18	3
4	10	4	Ti(O^{i}Pr)$_4$ (15%)	CH$_2$Cl$_2$	rt	2.5	55
5	10	4	Ti(O^{i}Pr)$_4$ (8%)	CH$_2$Cl$_2$	rt	3	82
6	25	4	Ti(O^{i}Pr)$_4$ (20%)	CH$_2$Cl$_2$	rt	4	80
7	12	4	Ti(O^{i}Pr)$_4$ (10%)	CH$_2$Cl$_2$	rt	3	85
8	12	4	Ti(O^{i}Pr)$_4$ (10%)^b	CH$_2$Cl$_2$	rt	3	85
9	12	4	Ti(O^{i}Pr)$_4$ (10%)^c	CH$_2$Cl$_2$	rt	3	86
10	12	2	Ti(O^{i}Pr)$_4$ (10%)	CH$_2$Cl$_2$	rt	4	85
11	12	2	Ti(O^{i}Pr)$_4$ (10%)^b	CH$_2$Cl$_2$	−20°C	4	89

^{a}The isolated BINOL–Salen–Ti(IV) complex was used.
^{b}10% O = PPh$_3$ was added.
^{c}4 Å molecular sieves were added.

the presence of (S)-**4.53** (10 mol%) and Ti(O^{i}Pr)$_4$ (10 mol%) in methylene chloride, the reaction of TMSCN with aldehydes gives the following results: for benzaldehyde 85% ee and 78% yield; for p-methoxybenzaldehyde 80% ee and 68% yield; for p-methylbenzaldehyde 85% ee and 70% yield; and for octyl aldehyde 75% ee and 64% yield.

Previously, a number of Ti(IV)–Salen complexes have been found to show good enantioselectivity for the reaction of TMSCN with aldehydes.[50] Most of these reactions need to pre-prepare the catalysts and often require low temperature for better enantioselectivity. As described above, when (S)-**4.53** is used, it does not require the pre-preparation of the catalyst and the reaction can be conducted at room temperature with good enantioselectivity. This makes this class of salen catalyst promising for practical application.

4.6. Asymmetric Hetero-Diels–Alder Reaction[51]

The Lewis-acid catalyzed hetero-Diels–Alder (HDA) reaction of Danishefsky's diene with the enamide aldehyde **4.55** can generate compounds **4.56** that is potentially useful in the natural product synthesis (Scheme 4.35). We have explored the use of the Lewis acid complexes of the BINOL ligands (R)-**4.57a-e** to catalyze the HDA reaction. Ligands (R)-**4.57a-d** were originally prepared by Yamamoto using a five-step synthesis.[52]

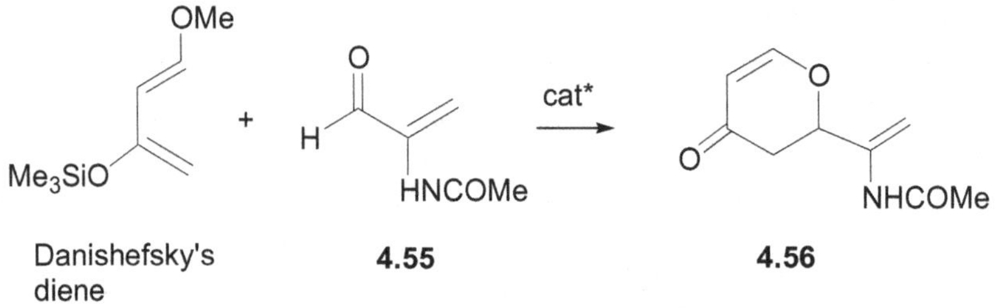

Scheme 4.35. The HDA reaction of Danishefsky's diene with **4.55**.

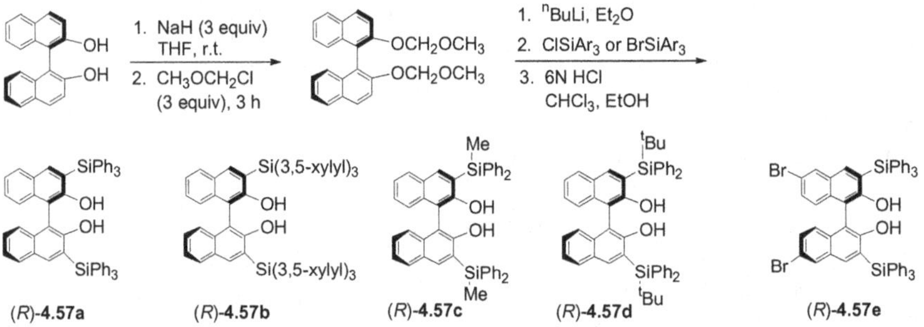

Scheme 4.36. Synthesis of chiral BINOL ligands containing bulky 3,3'-substituents.

Table 4.54. The asymmetric HDA of Danishefsky's diene with **4.55** to form **4.56** catalyzed by the complexes of (R)-**4.57a-e** and AlMe$_3$.

Entry	Ligand	Solvent	Yield (%)	ee (%)[a]
1	(R)-**4.57a**	Et$_2$O	0	
2	(R)-**4.57a**	THF	0	
3	(R)-**4.57a**	Toluene	40	42
4	(R)-**4.57a**	CH$_2$Cl$_2$	50	45
5	(R)-**4.57b**	Toluene	40	46
6	(R)-**4.57b**	CH$_2$Cl$_2$	45	73
7	(R)-**4.57b**	CH$_2$Cl$_2$	20	78[b]
8	(R)-**4.57c**	CH$_2$Cl$_2$	35	46
9	(R)-**4.57d**	CH$_2$Cl$_2$	30	25
10	(R)-**4.57e**	CH$_2$Cl$_2$	60	44

[a]The ee was determined on HPLC with a Chiracel OD column.
[b]Catalyzed by 10 mol% of AlMe$_3$ + 11 mol% of (R)-**4.57b**.

We have developed a simplified three-step synthesis by using Snieckus' ortho-aromatic metalation strategy as shown in Scheme 4.36. Ligand (R)-**4.57a** is also converted to (R)-**4.57e** in 70% yield by treatment with bromine in acetic acid and methylene chloride at $-20°$C.

The chiral Lewis acid complexes made from the combination of ligands (R)-**4.57a-e** with AlMe$_3$ are used to catalyze the asymmetric HDA of Danishefsky's diene with the enamide aldehyde **4.55**. The results are summarized in Table 4.54. All of these reactions are carried out at $-40°$C in the presence of 20 mol% of AlMe$_3$ and 22 mol% of chiral ligand for 36 h unless otherwise indicated. After hydrolysis and work-up, compound **4.56** is obtained and characterized. As shown in Table 4.54, these reactions are strongly influenced by solvents. In solvents containing coordinating oxygen atoms, such as THF and Et$_2$O, no reaction is observed. A polar solvent such as CH$_2$Cl$_2$ is found to be more favorable than a less polar toluene. The more sterically bulky ligand (R)-**4.57b** shows much higher enantioselectivity than (R)-**4.57a**. Up to 78% ee of **4.56** is obtained when the reaction is catalyzed by 10 mol% of (R)-**4.57b**+AlMe$_3$. Introduction of the 6,6'-bromine atoms in (R)-**4.57e** should increase the acidity of the hydroxyl groups, but shows no improvement on the enantioselectivity. Other chiral Lewis acids that are successfully used in different asymmetric HDA reactions, such as BINOL-Ti(O^{i}Pr)$_2$[53] and BINAP-Cu(OTf)$_2$,[54] could not catalyze this reaction.

References

1. Rosini C, Franzini L, Raffaelli A, Salvadori P, *Synthesis* 503, 1992.
2. Whitesell JK, *Chem Rev* **89**:1581, 1989.
3. Bringmann G, Walter R, Weirich R, *Angew Chem Int Ed Engl* **29**:977, 1990.
4. Pu L, *Chem Rev* **98**:2405–2494, 1998.
5. Chen Y, Yekta S, Yudin AK, *Chem Rev* **103**:3155–3211, 2003.
6. (a) Kočovský P, Vyskočil S, Smrčina M, *Chem Rev* **103**:3213–3245, 2003. (b) Brunel JM, *Chem Rev* **105**:857–898, 2005.
7. Brinkmeyer RS, Kapoor VM, *J Am Chem Soc* **99**:8339–8341, 1977.
8. (a) Midland MM, McLoughlin JI, Gabriel J, *J Org Chem* **54**:159, 1989. (b) Ramachandran PV, Teodorovic AV, Rangaishenvi MV, Brown HC, *J Org Chem* **57**:2379–2386, 1992.
9. (a) Brown HC, Pai GG, *J Org Chem* **50**:1384–1394, 1985. (b) Noyori R, Tomino I, Yamada M, Nishizawa M, *J Am Chem Soc* **105**:6717–6725, 1984.
10. (a) Parker KA, Ledeboer M, *J Org Chem* **61**:3214–3217, 1996. (b) Bach J, Berenguer R, Garcia J, Loscertales T, Vilarrasa J, *J Org Chem* **61**:9021–9025, 1996.

11. (a) Helal CJ, Margriotis PA, Corey EJ, *J Am Chem Soc* **118**:10938–10939, 1996. (b) Matsumura K, Hashiguchi S, Ikariya T, Noyori R, *J Am Chem Soc* **119**:8738–8739, 1997.
12. Corey EJ, Cimprich KA, *J Am Chem Soc* **116**:3151–3152, 1994.
13. Frantz DE, Fässler R, Tomooka CS, Carreira EM, *Acc Chem Res* **33**:373–381, 2000.
14. Moore D, Pu L, *Org Lett* **4**:1855–1857, 2002.
15. Gao G, Moore D, Xie R-G, Pu L, *Org Lett* **4**:4143–4146, 2002.
16. Lu G, Li X, Chan WL, Chan ASC, *J Chem Soc Chem Commun* 172–173, 2002.
17. Gao G, Xie R-G, Pu L, *Proc Natl Acad Sci* **101**:5417–5420, 2004.
18. Gao G, Wang Q, Yu X-Q, Xie R-G, Pu L, *Angew Chem Int Ed* **45**:122–125, 2006.
19. Rajaram AR, Pu L, *Org Lett* **8**:2019–2021, 2006.
20. Okhlobystin OY, Zakharkin LI, *J Organomet Chem* **3**:257–258, 1965.
21. (a) Midland MM, McDowell DC, Hatch RL, Tramontano A, *J Am Chem Soc* **102**:867–869, 1980. (b) Midland MM, Tramontano A, Kazubski A, Graham RS, Tsai DJS, Cardin DB, *Tetrahedron* **40**:1371–1380, 1984. (c) Noyori R, Yamada TM, Nishizawa M, *J Am Chem Soc* **106**:6717–6725, 1984.
22. (a) Duvold T, Rohmer M, *Tetrahedron Lett* **41**:3875–3878 2000. (b) Mulzer J, Csybowski M, Bats J-W, *Tetrahedron Lett* **42**:2961–2964, 2001. (c) Johansson M, Köpcke B, Anke H, Sterner O, *Tetrahedron* **58**:2523–2528, 2002.
23. Trost BM, Weiss AH, von Wangelin AJ, *J Am Chem Soc* **128**:8–9, 2006.
24. Zhou L-H, Yu XQ, Pu L, *J Org Chem* **74**:2013–2017, 2009.
25. Yang F, Xi P, Yang L, Lan J, Xie R, You JS, *J Org Chem* **72**:5457–5460, 2007.
26. Yue Y, Yu XQ, Pu L, *Chem European J* **15**:5104–5107, 2009.
27. Sonye JP, Koide K, *J Org Chem* **71**:6254–6257, 2006.
28. Sonye JP, Koide K, *Org Lett* **8**:199–202, 2006.
29. Moore D, Huang W-S, Xu M-H, Pu L, *Tetrahedron Lett* **43**:8831–8834, 2002.
30. Xu M-H, Pu L, *Org Lett* **4**:4555–4557, 2002.
31. Wang Q, Chen S-Y, Yu X-Q, Pu L, *Tetrahedron* **63**:4422–4428, 2007.
32. Li Z-B, Pu L, *Org Lett* **6**:1065–1068, 2004.
33. Li Z-B, Liu T-D, Pu L, *J Org Chem* **72**:4340–4343, 2007.
34. (a) Sasaki H, Irie R, Katsuki T, *Synlett* 300–302, 1993. (b) Katsuki T, *Coord Chem Rev* **140**:189–214, 1995. (c) DiMauro EF, Kozlowski MC, *Org Lett* **3**:1641–1644, 2001. (d) Annamalai V, DiMauro, EF, Carroll PJ, Kozlowski MC, *J Org Chem* **68**:1973–1981, 2003. (e) DiMauro EF, Kozlowski MC, *Organometallics* **21**:1454–1461, 2002.
35. (a) Liu L, Pu L, *Tetrahedron* **60**:7427–7430, 2004. (b) Qin Y-C, Liu L, Sabat M, Pu L, *Tetrahedron* **62**:9335–9348, 2006.
36. Huang W-S, Pu L, *J Org Chem* **64**:4222–4223, 1999.
37. Huang W-S, Pu L, *Tetrahedron Lett* **41**:145–149, 2000.
38. Qin Y-C, Pu L, *Angew Chem Int Ed* **45**:273–277, 2006. *Angew Chem* **118**:279–283, 2006.

39. (a) Schmidt F, Stemmler RT, Rudolph J, Bolm C, *Chem Soc Rev* **35**:454–470, 2006. (b) Dosa PI, Ruble JC, Fu GC, *J Org Chem* **62**:444–445, 1997. (c) Huang W-S, Pu L, *J Org Chem* **64**:4222–4223, 1999. (d) Bolm C, Muñiz K, *Chem Commun* 1295–1296, 1999.
40. DeBerardinis AM, Turlington M, Pu L, *Org Lett* **10**:2709–2712, 2008.
41. Kneisel FF, Dochnahl, M, Knochel, P, *Angew Chem Int Ed* **43**:1017–1021, 2004.
42. Kitamura M, Suga S, Kawai K, Noyori R, *J Am Chem Soc* **108**:6071, 1986.
43. Huang W-S, Hu Q-S, Pu L, *J Org Chem* **63**:1364–1365, 1998.
44. Simonson D, Kingsbury K, Xu M-H, Hu Q-S, Sabat M, Pu L, *Tetrahedron* **58**:8189–8193, 2002.
45. Hu Q-S, Pugh V, Sabat M, Pu L, *J Org Chem* **64**:7528–7536, 1999.
46. Zhang F-Y, Yip C-W, Cao R, Chan ASC, *Tetrahedron Asymmetry* **8**:585, 1997.
47. Reviews: (a) North M, *Tetrahedron Asymmetry* **14**:147–176, 2003. (b) Brunel J-M, Holmes IP, *Angew Chem Int Ed* **43**:2752–2778, 2004.
48. Qin Y-C, Liu L, Pu L, *Org Lett* **7**:2381–2383, 2005.
49. Li Z-B, Rajaram AR, Decharin N, Qin Y-C, Pu L, *Tetrahedron Lett* **46**:2223–2226, 2005.
50. Selected examples on using Salen–Ti complexes in the asymmetric cyanohydrin syntheses: (a) Hayashi M, Miyamoto Y, Inoue T, Oguni N, *J Org Chem* **58**:1515–1522, 1993. (b) Flores-Lope'z LZ, Parra-Hake M, Somanathan R, Walsh PJ, *Organometallics* **19**:2153–2160, 2000. (c) Jiang Y, Gong L, Feng X, Hu W, Pan W, Li Z, Mi A, *Tetrahedron* **53**:14327–14338, 1997. (d) Liang S, Bu XR, *J Org Chem* **67**:2702–2704, 2002. (e) Belokon YN, Green B, Ikonnikov NS, North M, Parsons T, Tararov VI, *Tetrahedron* **57**:771–779, 2001. (f) Yang Z-H, Wang L-X, Zhou Z-H, Zhou Q-L, Tang C-C, *Tetrahedron Asymmetry* **12**:1579–1582, 2001. (g) Zhou XG, Huang J-S, Ko P-H, Cheung K-K, Che CM, *J Chem Soc Dalton Trans* 3303–3309, 1999. (h) Belokon YN, Caveda-Cepas S, Green B, Ikonnikov NS, Khrustalev VN, Larichev VS, Moscalenko MA, North M, Orizu C, Tararov VI, Tasinazzo M, Timofeeva GI, Yashkina LV, *J Am Chem Soc* **121**:3968–3973, 1999.
51. Gong L-Z, Pu L, *Tetrahedron Lett* **41**:2327–2331, 2000.
52. Maruoka K, Itoh T, Araki Y, Shirasaka T, Yamamoto H, *Bull Chem Soc Jpn* **61**:2975–2976, 1988.
53. Keck GE, Li XY, Krishnamurthy D, *J Org Chem* **60**:5998–5999, 1995.
54. Yao S, Johannsen M, Hazell RG, Jørgensen KA, *Angew Chem Int Ed Engl* **37**:3121–3124, 1998.

Chapter 5

Enantioselective Fluorescent Sensors Based on 1,1′-Binaphthyl-Derived Dendrimers, Small Molecules and Macrocycles

5.1. Introduction

Because of the chirality of the 1,1′-binaphthyl structure and the good fluorescence property of the naphthalene rings, we have used these compounds to develop enantioselective fluorescent sensors for the recognition of chiral organic compounds. When appropriate functional groups are attached to 1,1′-binaphthyl, the resulting compounds can bind diverse organic molecules through non-covalent interactions such as hydrogen bonding and π–π attraction. The chirality of a 1,1′-binaphthyl compound allows its interaction with the two enantiomers of a chiral organic compound to generate two diastereomeric complexes, which should exhibit difference in their fluorescence, leading to enantioselective fluorescent response. We have developed both acyclic and macrocyclic 1,1′-binaphthyl-based fluorescent sensors for the enantioselective recognitions of chiral amines, amino alcohols, α-hydroxycarboxylic acids and amino acid derivatives. Dendritic chromophores have been attached to the binaphthyl molecules to amplify their fluorescent response and increase the fluorescent sensitivity. Enantioselective fluorescent sensors can potentially provide a rapid analytical tool to determine the enantiomeric composition of chiral compounds. These sensors could be used in the high-throughput screening of chiral catalysts. They also have potential in the detection of chiral organic molecules in biological system and could provide added sensitivity because of their capability for chiral recognition. We have demonstrated that enantioselective precipitation can also be used to develop sensors for chiral discrimination. References 1–10 are a few review articles and selected work of other research groups on the enantioselective fluorescent recognition.

5.2. Using Phenyleneethynylene-BINOL Dendrimers for the Recognition of Amino Alcohols[11–13]

Dendrimers that contain well-defined and highly branched structures have been extensively studied. The unique structures of the dendritic materials have led to many interesting properties and applications.[14–17] We are particularly interested in using the light harvesting properties of dendrimers to develop highly sensitive fluorescent sensors.

Previously, it was found that bromination of BINOL and its derivatives occurs specifically at the 6,6'-positions. We find that when an alkylated BINOL is treated with excess bromine in acetic acid at room temperature, there is a second-stage bromination occurring at the 4,4'-positions following the 6,6'-bromination to give (S)-**5.1** (Scheme 5.1). The two alkyl groups of (S)-**5.1** can be removed with the addition of BBr_3 followed by hydrolysis. The resulting tetrabromoBINOL is then treated with Ac_2O in the presence of pyridine to give (S)-**5.2**. From (S)-BINOL to the tetrabromide (S)-**5.2**, the overall yield is 62.5%. The much less electron-donating carboxylate groups in (S)-**5.2** make its bromines more reactive in the following Sonogashira coupling reactions than those in (S)-**5.1**.

As shown in Scheme 5.2, coupling of three different sizes of the aryleneethynylene-based dendrons[18] with (S)-**5.2** in the presence of $Pd(PPh_3)_4/CuI$ followed by basic hydrolysis generates the generation 0 (G0), generation 1 (G1) and generation 2 (G2) dendrimers (S)-**5.3**, (S)-**5.4** and (S)-**5.5**, respectively, in 82–90% yields. These molecules have been characterized by NMR spectroscopy and FAB/MALDI mass spectroscopies.

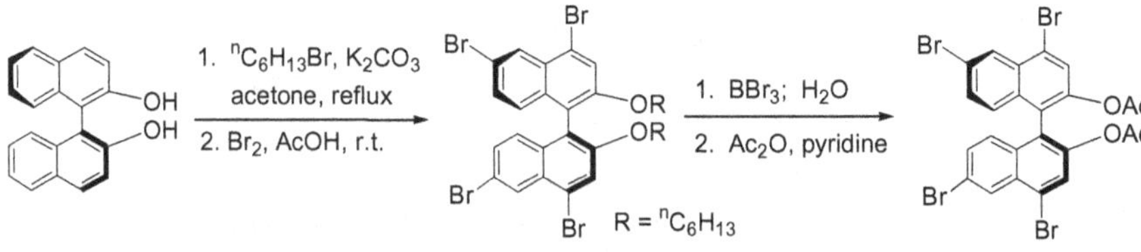

Scheme 5.1. Synthesis of 4,4',6,6'-tetrabrominated BINOL derivatives.

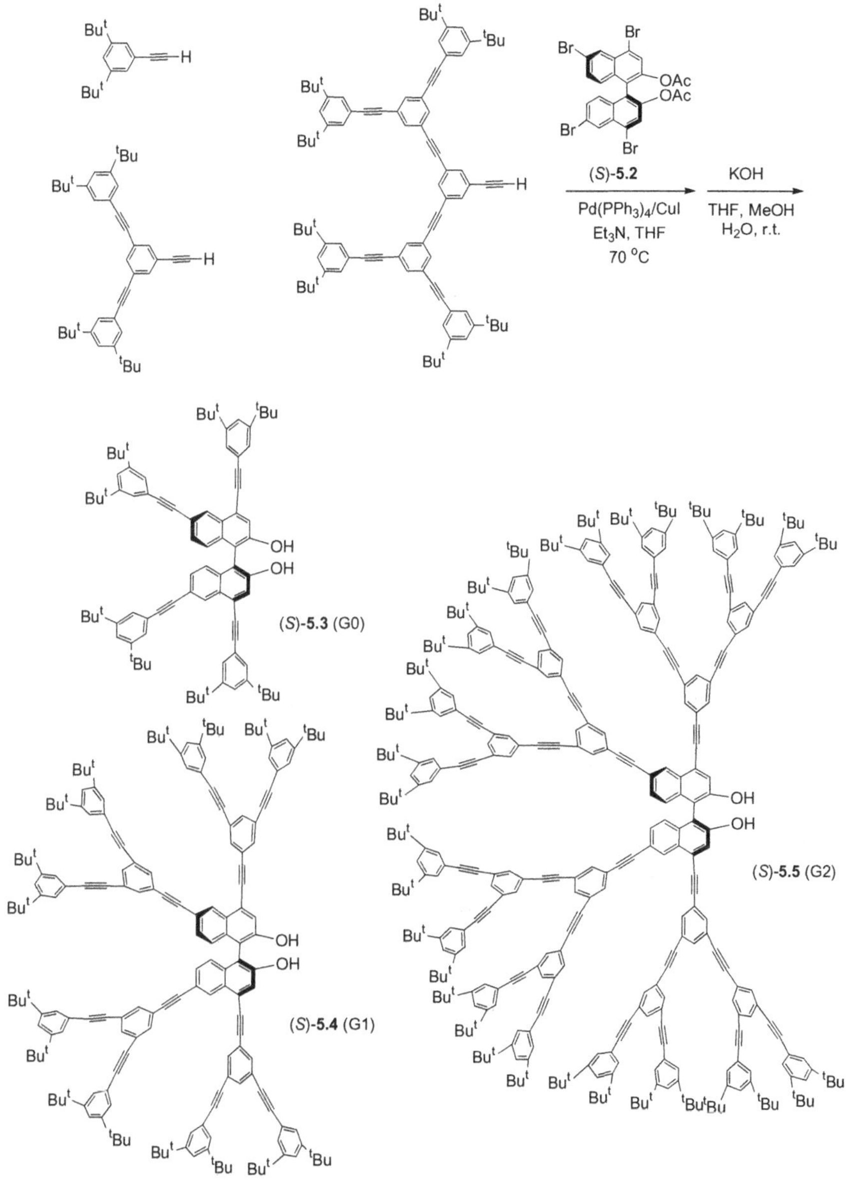

Scheme 5.2. Synthesis of aryleneethynylene–BINOL dendrimers.

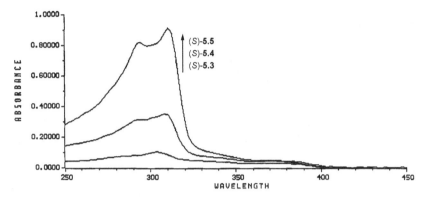

Figure 5.1. UV spectra of dendrimers (S)-**5.3**, (S)-**5.4** and (S)-**5.5** (CH$_2$Cl$_2$, 1.0×10^{-6} M).

These dendrimers exhibit two UV absorptions bands with a strong band around 250–330 nm and a weak band around 340–400 nm (Fig. 5.1). The short wavelength band can be attributed to the absorptions of the aryleneethynylene units and the naphthalene units, and the long wavelength band is attributed to the more conjugated system of the 4,4′,6,6′-tetraphenylethynyl-1,1′-binaphthyl core. The absorption in the short wavelength band increases significantly as the dendrimer generation increases because of the increased number of the arylenethynylene units. The molar extinction coefficient of the G2 dendrimer (S)-**5.5** at $\lambda_{max} = 310$ nm is nine times that of the G0 dendrimer (S)-**5.3** at $\lambda_{max} = 304$ nm.

The specific optical rotations $[\alpha]_D$ of the G0–G3 dendrimers are 114.4, 67.5 and 36.2, respectively. That is, as the size of the dendron increases, the optical rotation of the BINOL core is diluted. The CD spectrum of the G2 dendrimer gives two major Cotton effects: $[\theta](\lambda_{nm}) = -7.7 \times 10^4$ (363) and 4.2×10^5 (312), which are very close to those of the G0 and G1 dendrimers except a small blue shift from 312 nm of G2 to 305 nm of G0 at the major positive CD signal (Fig. 5.2). Thus, the dendrons in the G2 dendrimer do not achieve a chiral order to amplify the chiral optical property of the BINOL core.

The fluorescence spectrum of (S)-**5.5** (4.0×10^{-8} M in 1:4 benzene: hexane solution) is compared with that of (S)-BINOL under the same conditions. As shown in Fig. 5.3, there is a dramatic increase in the fluoresence intensity from (S)-BINOL to (S)-**5.5**.

In methylene chloride, when the G2 dendrimer (S)-**5.5** is excited at 310 nm, it shows an intense emission peak at 422 nm and a shoulder at

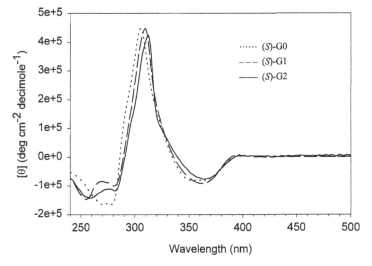

Figure 5.2. CD spectra of dendrimers (S)-**5.3**, (S)-**5.4** and (S)-**5.5**.

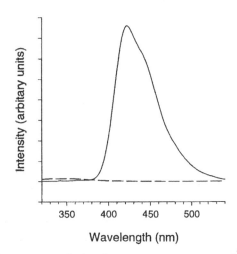

Figure 5.3. Comparison of the fluorescence spectrum of (S)-**5.5** (λ_{exc} = 310 nm) (solid line) and that of (S)-BINOL (λ_{exc} = 280 nm) (dashed line).

441 nm. The shape and position of the fluorescence signals of the G0 and G1 dendrimers are almost the same as those of the G2 dendrimer, but there is a large increase of intensity going from G0 to G2. As shown in Fig. 5.4, at the same molar concentration, the fluorescence intensity of the G2 dendrimer is about 12 times that of the G0. No separated emission

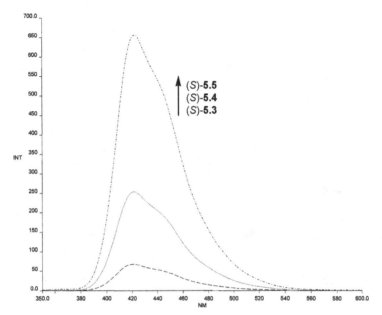

Figure 5.4. The fluorescence spectra of the dendrimers (S)-**5.3**, (S)-**5.4** and (S)-**5.5** (CH$_2$Cl$_2$, 4.0×10^{-8} M).

from the aryleneethynylene dendrons in these dendrimers is observed. This demonstrates that there is efficient energy migration from the dendritic arms to the more conjugated core in the dendrimers. The large number of light-absorbing phenyleneethynylene units in the G2 dendrimer serve as light harvesting antenna, which absorb light and funnel the energy to the center for emission. Despite the greatly increased fluorescence intensity from G0 to G2, the fluorescence quantum yields of these molecules do not change significantly and are 0.30, 0.32 and 0.40, respectively.

Previously, it was reported that the fluorescence of BINOL can be quenched by chiral amines and amino alcohols with a small enantioselectivity.[19] We find that the fluorescence intensity of the G2 dendrimer at 4.0×10^{-8} M is comparable to that of BINOL at a 170 times more concentrated sample. The much greater fluorescence intensity of the G2 dendrimer is expected to make a more sensitive fluorescent sensor when treated with a fluorescence quencher.

The fluorescence responses of the G0–G2 dendrimers towards chiral amino alcohols, including valinol, leucinol and phenylalaninol, are studied. All of these amino alcohols efficiently quench the fluorescence of the dendrimers. The fluorescence quenching follows the Stern–Volmer equation: $I_0/I = 1+K_{SV}[Q]$, where I_0 is the fluorescence intensity without a quencher and I the fluorescence in the presence of a quencher. $[Q]$ is the quencher concentration. K_{SV} is the Stern–Volmer constant, which measures the efficiency of quenching. Tables 5.1–5.3 give the Stern–Volmer constants for the chiral dendrimers (S)-**5.3**–(S)-**5.5** in the presence of the enantiomers of the amino alcohols. As the data in these tables show, there are measurable

Table 5.1. Stern–Volmer quenching constants for the dendrimers in the presence of (R)- and (S)-valinol in methylene chloride.

Dendrimer[a]	$K_{SV}^{(S_D,S_Q)}$	$K_{SV}^{(S_D,S_Q)}$	$K_{SV}^{(S_D,S_Q)} / K_{SV}^{(S_D,R_Q)}$
(S)-**5.3**	47	43	1.09
(S)-**5.4**	63	57	1.11
(S)-**5.5**	83	80	1.04

[a] Concentration = 4.0×10^{-8} M.

Table 5.2. Stern–Volmer quenching constants for the dendrimers in the presence of (R)- and (S)-leucinol in methylene chloride.

Dendrimer[a]	$K_{SV}^{(S_D,S_Q)}$	$K_{SV}^{(S_D,R_Q)}$	$K_{SV}^{(S_D,S_Q)} / K_{SV}^{(S_D,R_Q)}$
(S)-**5.3**	61	54	1.13
(S)-**5.4**	92	73	1.26
(S)-**5.5**	90	82	1.10

[a] Concentration = 4.0×10^{-8} M.

Table 5.3. Stern–Volmer quenching constants for the dendrimers and BINOL in the presence of (R)- and (S)-phenylalaninol in benzene:hexane (20:80).

Dendrimer	$K_{SV}^{(S_D,S_Q)}$	$K_{SV}^{(S_D,R_Q)}$	$K_{SV}^{(S_D,S_Q)} / K_{SV}^{(S_D,R_Q)}$
(S)-**5.3**[a]	305	253	1.21
(S)-**5.4**[a]	445	351	1.27
(S)-**5.5**[a]	520	426	1.22
(S)-**5.5**[b]	561	461	1.22
(S)-BINOL[c]	111	109	1.02

[a] Concentration = 1.0×10^{-6} M.
[b] Concentration = 4.0×10^{-8} M.
[c] Concentration = 1.0×10^{-4} M.

Table 5.4. Stern–Volmer quenching constants (M^{-1}) of (S)-**5.5** and (R)-**5.5** in the presence of (S)- and (R)-leucinol in methylene chloride.

Dendrimer	(S)-leucinol	(R)-leucinol
(S)-**5.5**	90	82
(R)-**5.5**	80	93

fluorescence differences when two enantiomers of an amino alcohol are interacted with these dendrimers. Generally, the S enantiomer of an amino alcohol quenches the fluorescence of a dendrimer more efficiently than the R enantiomer. The amino alcohol phenylalaninol contains an aromatic ring and shows much greater Stern–Volmer constant than the other two amino alcohols. This indicates stronger interaction between this amino alcohol and the dendrimers probably because of certain π–π attraction.

valinol   leucinol   phenylalaninol

Dendrimer (R)-**5.5** is prepared as the enantiomer of (S)-**5.5**. The interaction of (R)-**5.5** with the enantiomers of leucinol is compared with that of (S)-**5.5** and the Stern–Volmer constants are given in Table 5.4. There is a mirror image relationship between the fluorescence responses of (R)-**5.5** with those of (S)-**5.5**, consistent with an enantioselective response.

5.3. Using Phenylene-BINOL Dendrimers for the Recognition of Chiral Amino Alcohols

5.3.1. Synthesis and Study of Phenylene-BINOL Dendrimers[20]

BINOL dendrimers containing arylene dendrons are synthesized. As shown in Scheme 5.3, the Suzuki coupling of the G0–G2 aryl boronic acids[21] with the alkylated 4,4′,6,6′-tetrabromoBINOL compound (S)-**5.1** in the presence of $Pd(PPh_3)_4$ (3 mol%), and K_2CO_3 (2 M) in refluxing THF, followed by treatment with BBr_3 and hydrolysis, gives the G0–G2 dendrimers (S)-**5.6**, (S)-**5.7** and (S)-**5.8**, respectively. These compounds are pale yellow solid and soluble in common organic solvents. The molar optical rotations of the

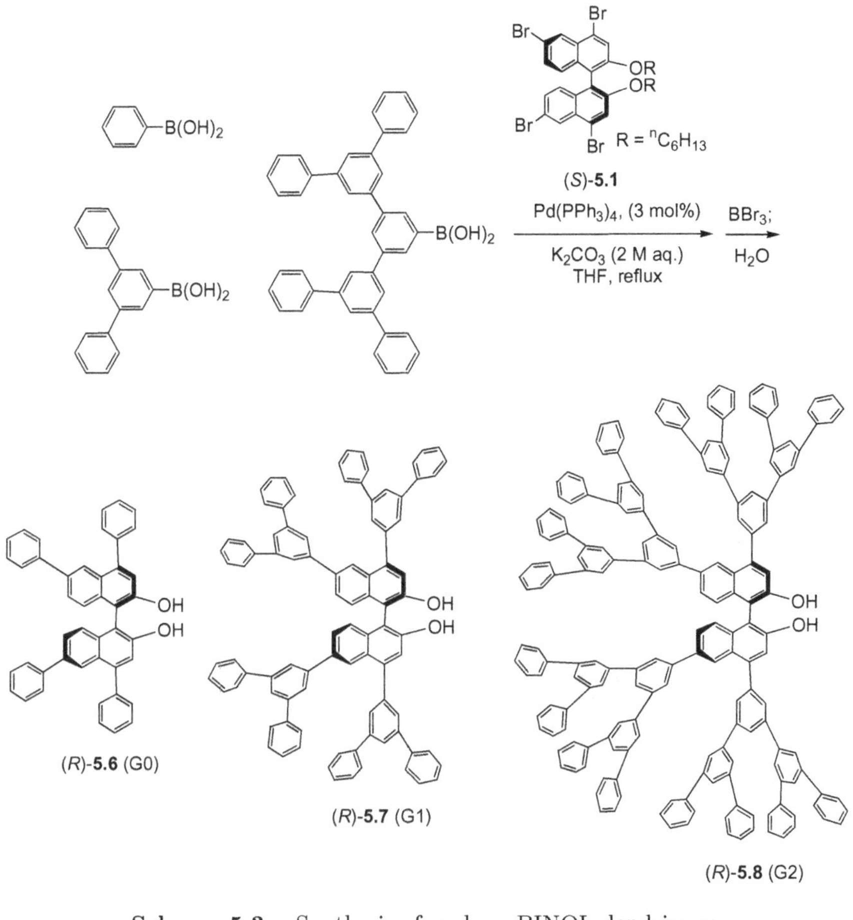

Scheme 5.3. Synthesis of arylene–BINOL dendrimers.

G0–G2 dendrimers are −391, −704, −754, respectively. The molar optical rotation of the G2 dendrimer (S)-**5.8** is about twice as much as that of the G0 dendrimer (S)-**5.6**. This indicates that there may be induced chiral conformation in the phenylene units of the higher generation dendrimers, which could contribute to the increased optical rotation. This is quite different from the aryleneethynylene-based dendrimers described in the previous section where there is only about 30% increase in the molar optical rotation from the G0 dendrimer (S)-**5.3** (1299) to the G2 dendrimer (S)-**5.5** in the molar optical rotation (1769). This is probably because there is more steric interaction between the *ortho* C–H bonds of the biaryl units in the

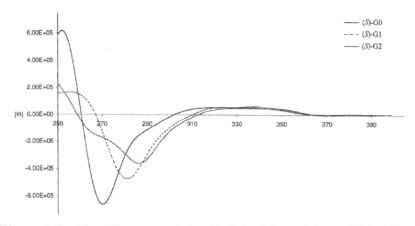

Figure 5.5. The CD spectra of the chiral dendrimers (*S*)-**5.6** (G0), (*S*)-**5.7** (G1) and (*S*)-**5.8** (G2) (CH$_2$Cl$_2$, 2.0×10^{-6} M).

Table 5.5. CD spectral data of the dendrimers (*R*)-**5.6**–(*R*)-**5.8**.

Dendrimers	(*R*)-**5.6** (G0)	(*R*)-**5.7** (G1)	(*R*)-**5.8** (G2)
CD [Θ] (λ nm) (CH$_2$Cl$_2$)	6.37×10^{-5} (254) -6.79×10^{-5} (272)	-4.75×10^{-5} (282)	2.72×10^{-5} (244) -3.64×10^{-5} (287)

phenylene-based dendrimers. This steric interaction could allow the chiral BINOL core to induce a preferred chiral conformation for the dendritic arms and amplify the chiral optical property.

The CD spectra of the chiral dendrimers (*S*)-**5.6**, (*S*)-**5.7** and (*S*)-**5.8** are shown in Fig. 5.5, and the data are given in Table 5.5. For these dendrimers, there is a small red shift in the major negative CD signal with reduced intensity from (*S*)-**5.6** to (*S*)-**5.8**. This is also quite different from the phenyleneethynylene-based BINOL dendrimers, which do not show obvious change in their CD spectra as the dendritic generation increases. It further indicates that the steric interaction among the phenyl rings of the phenylene-based BINOL dendrimers should have contributed to their optical properties.

The UV spectra of these molecules show two major absorption bands including a strong band at 250–300 nm and a weak band at 300–370 nm. The short wavelength strong band shows large increase from (*S*)-**5.6** to (*S*)-**5.8** due to the absorption of the increased number of phenyl rings. The absorption of (*S*)-**5.8** at $\lambda_{\max} = 256$ nm is five times that of (*S*)-**5.6** at

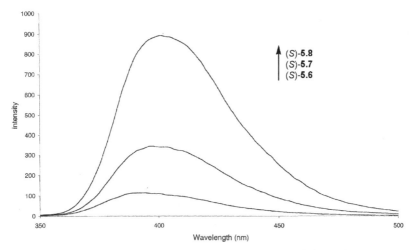

Figure 5.6. Fluorescence spectra of dendrimers (S)-**5.6**–(S)-**5.8** (CH_2Cl_2, 2.0×10^{-7} M).

the same location, whereas the long wavelength weak band does not change with the generation of the dendrimers and it is attributed to the absorption of the extended π system of the 4,4′,6′,-tetraphenylbinaphthyl core.

When the methylene chloride solutions of dendrimers (S)-**5.6**–(S)-**5.8** are irradiated at 256 nm, they exhibit emissions at 395, 402 and 405 nm, respectively (Fig. 5.6). The fluorescence intensity increases as the dendrimer generation increases. The emission intensity of (S)-**5.8** is about 2.5 times that of (S)-**5.7**, which is about three times that of (S)-**5.6**. No separated emission from the phenylene dendrons is observed, which indicates efficient energy migration from the dendrons to the more conjugated core. The fluorescence intensity of the (S)-**5.8** dendrimer is about 100 times stronger than that of BINOL at the same concentration.

The fluorescence response of dendrimer (S)-**5.8** in the presence of alaninol is studied. A mixed solvent of hexane and methylene chloride (9:1 v) is found to be necessary in order to dissolve these molecules and observe efficient fluorescence quenching. The fluorescence quenching of dendrimer (S)-**5.8** (1.0×10^{-7} M) by alaninol (0–6 mM) follows the Stern–Volmer equation. It is found that $K_{SV}^R = 243.5\,M^{-1}$ and $K_{SV}^S = 216.0\,M^{-1}$; thus $K_{SV}^R/K_{SV}^S = 1.13$. The fluorescence quenching is moderately enantioselective with the R enantiomer of the amino alcohol quenches more efficiently than the S enantiomer. The fluorescence queching of (S)-**5.6** and (S)-**5.7** by alaninol also follows the Stern–Volmer equation under the same conditions with $K_{SV}^R = 75.5\,\text{M}^{-1}$ and $K_{SV}^S = 63.5\,\text{M}^{-1}$ for (R)-**15** and $K_{SV}^R = 99.9\,\text{M}^{-1}$ and $K_{SV}^S = 88.4\,\text{M}^{-1}$ for (R)-**16**.

HO~~~NH$_2$ (alaninol)

alaninol

The much larger Stern–Volmer constants of the G2 dendrimer (S)-**5.8** than those of G0 and G1 shows that the higher generation dendrimer is a more sensitive fluorescent sensor than the lower generation ones. All the observed fluorescence quenchings probably occur while the amino alcohol binds with the hydroxyl groups of the sensor core. As the dendritic generation increases, more light that is harvested by the dendrons and funneled to the core gets quenched, leading to the increased quenching efficiency for the G2 dendrimer.

The fluorescence quenching of (S)-**5.8** by the amino alcohols leucinol and valinol is also studied under the same conditions as the use of alaninol. The Stren–Volmer constants are summarized in Table 5.6. As shown in the table, similar degree of enantioselectivity is observed for these amino alcohols.

5.3.2. Synthesis of a Phenylene-Binaphthyl Dendrimer with Periphery Hydroxyl Groups[22]

We have synthesized a binaphthyl core-based dendrimer that contains periphery hydroxyl groups as a unimolecular chiral micelle. Scheme 5.4 shows the synthesis of the phenylene dendron **5.9** that contains MOM-protected terminal hydroxyl groups. The Suzuki coupling of **5.9** with the tetrabromobinaphthyl compound (R)-**5.1** gives the second generation chiral dendrimer (R)-**5.10** in 67% yield (Scheme 5.5). The MOM-protecting groups of (R)-**5.10** are then completely removed by hydrolysis in the

Table 5.6. The Stern–Volmer constants of (S)-**5.8** in the presence of the chiral amino alcohols.

Quencher	(R)-alaninol	(S)-alaninol	(R)-leucinol	(S)-leucinol	(R)-valinol	(S)-valinol
K_{SV} (M^{-1})	243.5	216.0	191.5	162.5	145.5	133.5
$K_{SV}^R - K_{SV}^S$	27.5		29.0		12.0	
K_{SV}^R / K_{SV}^S	1.13		1.18		1.09	

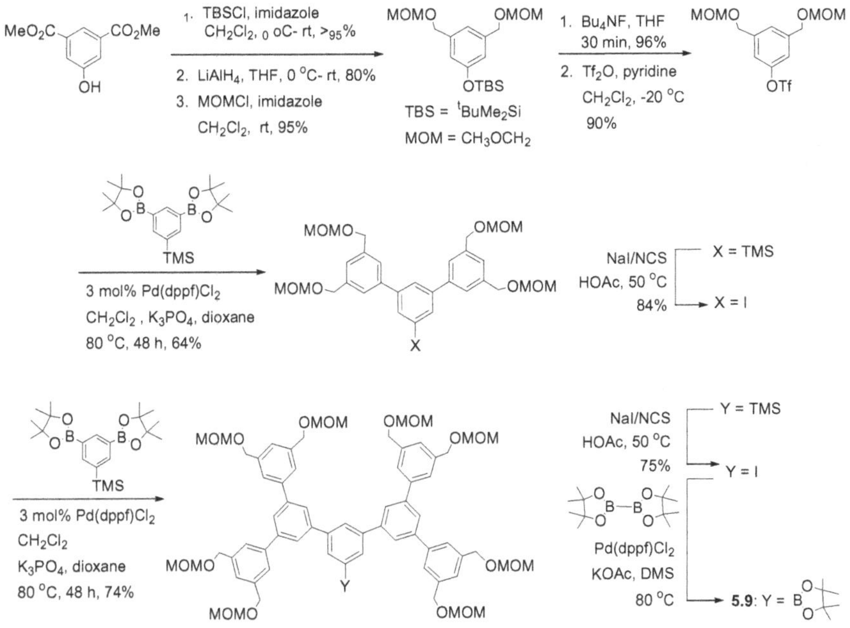

Scheme 5.4. Synthesis of phenylene dendrons with protected terminal hydroxyl groups.

presence of HCl to generate (R)-**5.11**. The dendrimer structure of (R)-**5.11** has been established by ^{1}H and ^{13}C NMR spectroscopic analysis and mass spectrometry. The 32 periphery hydroxyl groups of (R)-**5.11** render it soluble in a mixed H_2O and THF (3:1) solvent. Its solubility in this mixed solvent is higher than it in pure H_2O or THF. Dendrimer (R)-**5.11** is also soluble in DMSO, EtOH/THF and MeOH/THF but insoluble in EtOAc, CH_2Cl_2 and $CHCl_3$.

Dendrimer (R)-**5.11** displays UV absorptions at $\lambda_{max} = 260$ (sh), 288, 310 (sh) and 361 (sh) nm in THF solution. Changing the solvent from THF to H_2O/THF (3:1) significantly increases the half-height width of the UV signal from 52 to 89 nm, indicating a conformational difference for (R)-**5.11** in different solvents. There should be more intramolecular hydrogen bonds between the hydroxyl groups of (R)-**5.11** in THF, which may produce a tighter conformation for the dendrimer. In H_2O/THF solution, (R)-**5.11** should have significant intermolecular hydrogen bonds with H_2O and become better solvated. This dendrimer also exhibits strong fluorescence at 407 nm in both THF and H_2O/THF solutions. The chiral recognition property of this unimolecular micelle-like dendrimer has not been studied.

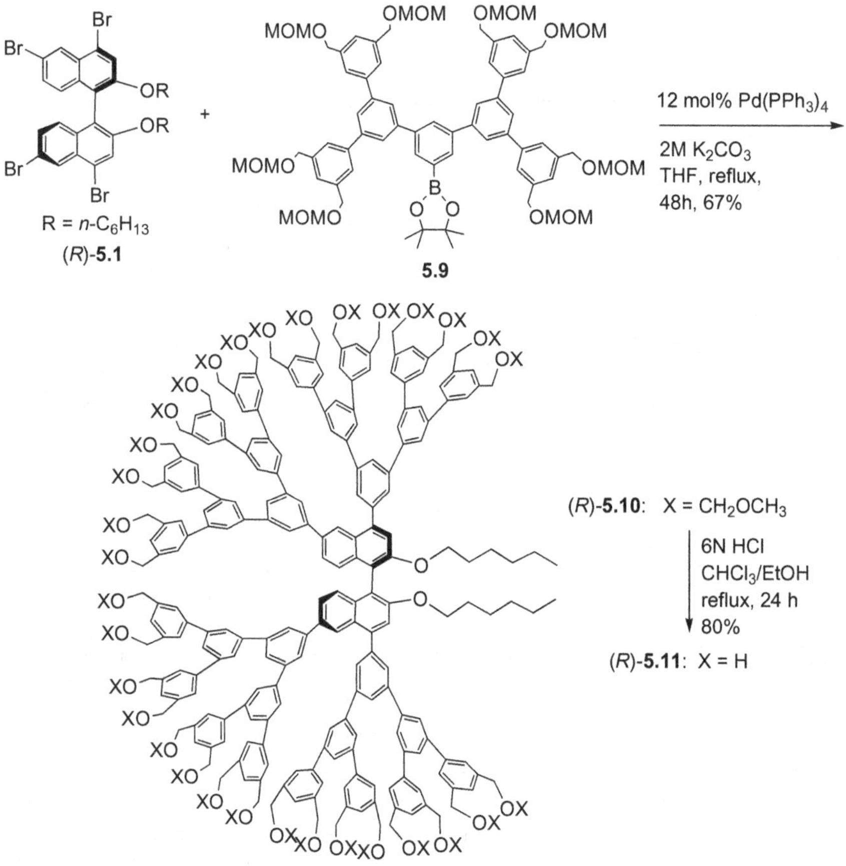

Scheme 5.5. Synthesis of unimolecular micelle-like dendrimer (R)-**5.11**.

5.4. Using Functionalized BINOLs for the Recognition of Amino Alcohols and Amines

In 1992, Iwanek and Mattay reported that the fluorescence of BINOL could be quenched by chiral amino alcohols and amines with a small degree of enantioselectivity for certain substrates.[19] Table 5.3 also shows that (S)-BINOL has very low enantioselective fluorescent response to the amino alcohol phenylalaninol. In order to improve the effectiveness of BINOL for the fluorescent recognition of chiral amino alcohols and amines, we have attempted to increase the intermolecular hydrogen binding capability of BINOL by introducing additional hydroxyl groups adjacent to the central diols. The tetrahydroxyl compound (R)-**5.12** and its derivatives (R)-**5.13** and (R)-**5.14** are obtained. The preparation of these compounds and their use in asymmetric catalysis are discussed in Section 4.2.5 in Chapter 4. In the ^{1}H NMR spectra of these compounds, the hydroxyl proton signals of (R)-**5.12** are at δ 6.53 and 4.64, those of (R)-**5.13** are at δ 8.49 and those of (R)-**5.14** are at δ 5.56. The most down field shift for the hydroxyl proton signal of (R)-**5.13** indicates strong intramolecular hydrogen bonds.

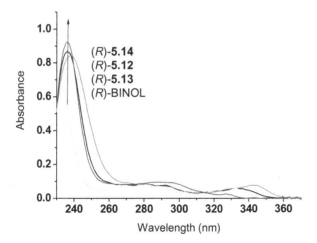

Figure 5.7 compares the UV absorptions of (R)-**5.12**, (R)-**5.13**, (R)-**5.14** and (R)-BINOL. Compound (R)-**5.13** gives the most red-shifted signal at the long wavelength absorption, which is attributed to the intramolecular hydrogen bonding between the naphthol hydroxyl protons

Figure 5.7. UV Spectra of (R)-**5.12**, (R)-**5.13**, (R)-**5.14** and (R)-BINOL (8.0×10^{-6} M) in CH_2Cl_2.

and the most basic methoxy oxygens. This is consistent with the NMR observation. Other compounds lack either the basic methoxy oxygens, such as compounds (R)-**5.12** and BINOL, or the acidic naphthol protons, such as (R)-**5.14**, and their UV signals are blue shifted at the long wavelength absorption. Compound (R)-**5.14** may have the weakest intramolecular hydrogen bonds because of its combination of the less acidic hydroxyl groups with the less basic aryl methyl ether oxygens.

The fluorescence spectra of compounds (R)-**5.12**, (R)-**5.13**, (R)-**5.14** and (R)-BINOL in methylene chloride are compared in Fig. 5.8. The emissions of (R)-**5.12** ($\lambda_{max} = 379$ nm) is very similar to that of (R)-BINOL ($\lambda_{max} = 372$ nm) with a small red shift. There is a small fluorescence enhancement from (R)-**5.12** and (R)-BINOL to (R)-**5.13**. Compound (R)-**5.14** gives a greatly enhanced fluorescence signal at $\lambda_{max} = 361$ nm. The much more acidic naphthol hydroxyl protons in (R)-**5.12**, (R)-**5.13** and (R)-BINOL than those in (R)-**5.14** might have quenched their fluorescence through hydrogen bonds and/or proton transfer at the excited state.

Figure 5.9 shows that (R)-**5.12** is more sensitive than (R)-BINOL when interacted with (R)-phenylalaninol. In a mixed solvent of methylene chloride/hexane (30:70), the fluorescence quenching of (R)-**5.12**

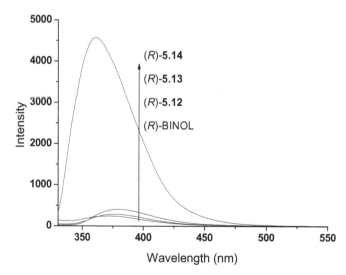

Figure 5.8. Fluorescence spectra of (R)-BINOL, (R)-**5.12**, (R)-**5.13** and (R)-**5.14** in CH_2Cl_2 (8.0×10^{-6} M, λ_{exc} at 294, 298, 298 and 302 nm, respectively, emission slits: 5.0 nm).

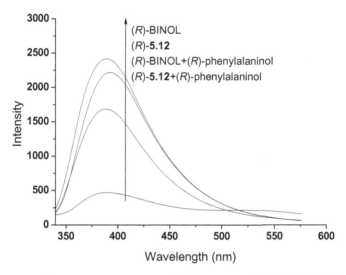

Figure 5.9. Fluorescence spectra of (R)-**5.12** and (R)-BINOL (both at 1.0×10^{-5} M) in CH_2Cl_2/hexane (30:70) with and without (R)-phenylalaninol $(8.0 \times 10^{-3}$ M) (λ_{exc} at 298 and 300 nm, respectively).

by (R)-phenylalaninol, I_0/I, is 4.0 and that of (R)-BINOL by (R)-phenylalaninol is only 1.4.

The fluorescence quenching of (R)-**5.12** (1.0×10^{-5} M in 30:70 CH_2Cl_2/hexane) by (R)- and (S)-phenylalaninol (2.0×10^{-3}–6.0×10^{-3} M) follows the Stern–Volmer equation with $K_{SV}^{R,R} = 389 \, M^{-1}$ ($K_{SV}^{R,R}$) and $K_{SV}^{R,S} = 309 \, M^{-1}$. That is, (R)-phenylalaninol quenches the fluorescence of (R)-**5.12** more efficiently than (S)-phenylalaninol with $K_{SV}^{R,R}/K_{SV}^{R,S} = 1.26$. The enantiomeric sensor (S)-**5.12** gives the mirror-image-related fluorescence responses in the presence of (R)- and (S)-phenylalaninol. The Stern–Volmer constants of (R)-BINOL in the presence of (R)- and (S)-phenylalaninol are found to be 106 and 91 M^{-1}, respectively, under the same conditions as the use of (R)-**5.12**. The ratio $K_{SV}^{R,R}/K_{SV}^{R,S}$ is 1.16. Thus, the binding of (R)-**5.12** with the amino alcohol is much stronger than (R)-BINOL with a slightly higher enantioselectivity. The fluorescence quenching of these BINOL-based compounds by the chiral amino alcohol can be attributed to both ground-state hydrogen bonding and excited-state deprotonation of individual BINOL molecules. When the concentration of the amino alcohol is further increased, a new emission signal starts to appear at above 500 nm, which might be due to the emission of the deprotonated (R)- or (S)-**5.12** at the excited state.

Table 5.7. Stern–Volmer constants and enantioselective factors of (S)-**5.12** in the presence of chiral amino alcohols.

Amino alcohol	K_{SV} (M^{-1})	Enantioselective factor
(S)-phenylalaninol	320	1.24
(R)-phenylalanino	258	
(S)-phenylglycinol	83	1.22
(R)-phenylglycinol	68	
(S)-leucinol	279	1.24
(R)-leucinol	226	
(S)-valinol	312	1.09
(R)-valinol	285	

The interactions of (S)-**5.12** with other amino alcohols including phenylglycinol, leucinol and valinol are also studied and their Stern–Volmer constants are summarized in Table 5.7. In all cases, the (S)-amino alcohols quenche the fluorescence of (S)-**5.12** more efficiently than the (R)-amino alcohols do.

phenylglycinol

No fluorescence quenching is observed for (R)-**5.14** in the presence of phenylalaninol. Thus, the interaction of the acidic naphthol hydroxyl group with the basic amine is important for the observed fluorescence quenching in (R)-**5.12** and (R)-BINOL.

To our surprise, fluorescence enhancement (nonmonotonic) rather than quenching is observed for (R)-**5.13** when it is treated with amino alcohol-like phenylalaninol and leucinol. Both ^{1}H NMR and UV spectroscopic analyses support the formation of the strongest intramolecular hydrogen bonds in (R)-**5.13** among these BINOL derivatives. This intramolecular hydrogen bond could rigidify the structure of this compound and give its slightly enhanced fluorescence intensity than those of (R)-**5.12** and (R)-BINOL as shown in Fig. 5.8. Complexation of (R)-**5.13** with the amino alcohols could further rigidify the structure and enhance the fluorescence. This process should be in competition with the amine-induced fluorescence quenching process.

When (R,R)- or (S,S)-1,2-diphenylethylenediamine is used to interact with (S)-**5.12** and (R)-**5.13**, fluorescence quenching is observed for both compounds. The two basic amine groups of the diamine might have overcome the effect of the intramolecular hydrogen bonds in (R)-**5.13** and give the expected fluorescence quenching. The Stern–Volmer constants of (S)-**5.12** (1.0×10^{-5} M in CH_2Cl_2) in the presence of the (R,R)- and (S,S)-diamine are 1175 and 795 M^{-1}, respectively, and those of (R)-**5.13** are 807 and 553 M^{-1}, respectively. Both (S)-**5.12** and (R)-**5.13** are very sensitive and moderately enantioselective in the fluorescent recognition of this chiral diamine. The four hydroxyl groups of (S)-**5.12** probably make it bind stronger with the chirality-matched (R,R)-diamine with higher fluorescent sensitivity than (R)-**5.13**.

(R,R)-1,2-diphenylethylenediamine

5.5. Using Acyclic Bisbinaphthyls for the Recognition of α-Hydroxycarboxylic Acids and Amino Acid Derivatives[24,25]

In order to conduct the enantioselective fluorescent recognition of α-hydroxycarboxylic acids, a class of biologically interesting and synthetically useful functional organic molecules, we have designed a bisbinaphthyl-based sensor (S)-**5.15**. Scheme 5.6 shows the interaction of this bisbinaphthyl compound with (S)-mandelic acid through three specific intermolecular hydrogen bonds to form an intermolecular complex. The molecular modeling structure is produced by using the PCSpartan-semiemperical PM3 program. The bisbinaphthyl molecule (S)-**5.15** contains a nitrogen atom whose lone pair electrons can quench the fluorescence of the naphthyl units in the compound through a photo-induced electron transfer process. In the complex with mandelic acid, this nitrogen atom forms a hydrogen bond with the carboxylic acid proton, which makes the lone pair electrons of the nitrogen unavailable for the photo-induced electron transfer process and enhances the fluorescence.

Scheme 5.7 shows the synthesis of (S)-**5.15** from (S)-BINOL. Protection of one of the hydroxyl groups of (S)-BINOL followed by reaction with

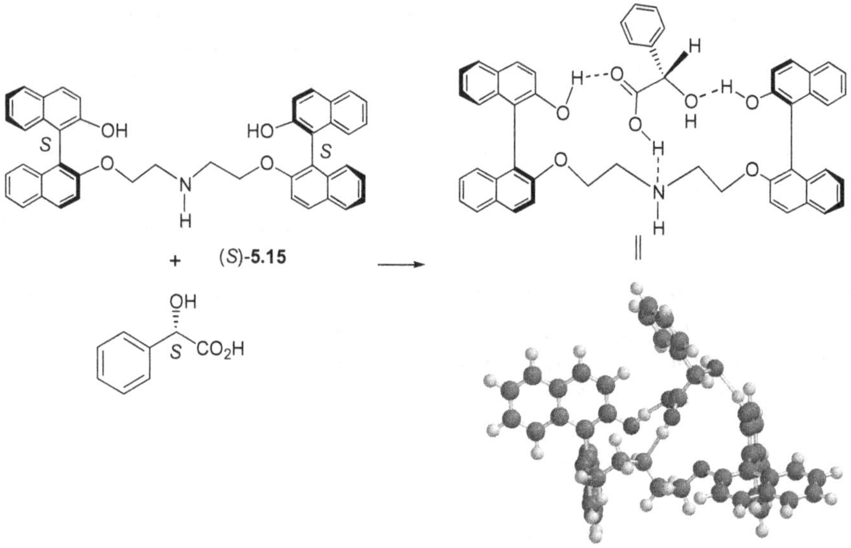

Scheme 5.6. Interaction of a bisbinaphthyl molecule with (*S*)-mandelic acid.

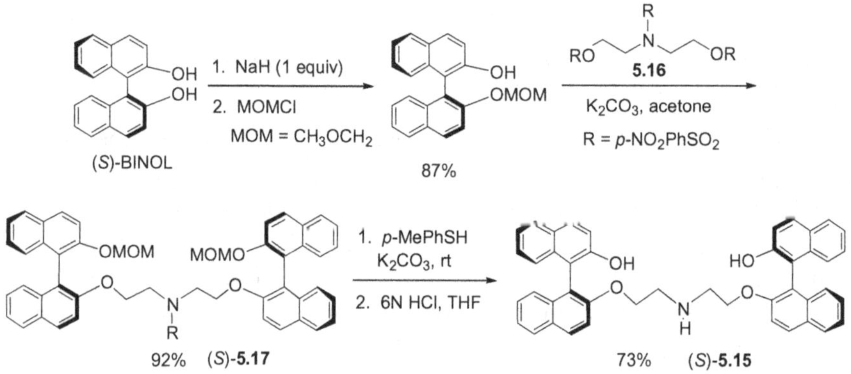

Scheme 5.7. Synthesis of chiral bisbinaphthyl compound (*S*)-**5.15**.

a tris(*p*-nitrophenylsulfonyl) compound **5.16** gives (*S*)-**5.17**. Removal of the *p*-nitrophenylsulfonyl group with *p*-methylbenzenethiol and the MOM groups with 6 N HCl converts (*S*)-**5.17** to (*S*)-**5.15**. Both the enantiomeric and diastereomeric purities of this compound are greater than 98% as determined by using HPLC-Chiracel AD column. In methylene chloride solution, this compound shows UV absorptions at $\lambda_{max} = 246, 280$ and 334 nm, and a fluorescence signal at 367 nm while excited at 310 nm.

Compound (S)-**5.15** (9.5×10^{-5} M) is interacted with (R)- and (S)-mandelic acid in benznene solution and 2% dimethoxyethylene (DME) is added to help dissolution of the acid. It is found that in the presence of 5.0×10^{-3} M of (S)-mandelic acid, the fluorescence intensity of (S)-**5.15** is increased to 2.87 times of the original value (Fig. 5.10). In the presence of (R)-mandelic acid at the same concentration, the fluorescence intensity is only increased to 1.75 times of the original. The enantiomeric fluorescence enhancement ratio, $ef \, [= (I_S - I_0)/(I_R - I_0)]$, is 2.49. Thus the fluorescence response of (S)-**5.15** toward the chiral α-hydroxycarboxylic acid is enantioselective.

The fluorescence enhancements of (S)-**5.15** in the presence of mandelic acid follow the Benesi–Hildebrand-type equation, which give an association constant of $348 \pm 3\%$ M^{-1} for the complex of (S)-**5.15** + (S)-mandelic acid and $163 \pm 6\%$ M^{-1} for the complex of (S)-**5.15**+(R)-mandelic acid. That is, the complex of (S)-**5.15**+(S)-mandelic acid is more stable than the complex of (S)-**5.15** + (R)-mandelic acid by $ca.$ 0.45 kcal/mol ($\Delta\Delta G$). In C$_6$D$_6$, the ^{1}H NMR signal of the methine proton of (S)-mandelic acid undergoes upfield shift ($\sim$0.1 ppm) in the presence of (S)-**5.15**. The Job plot obtained

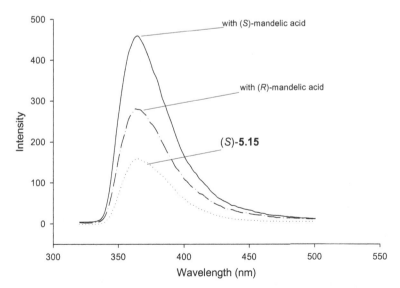

Figure 5.10. Fluorescence spectra of sensor (S)-**5.15** with and without mandelic acid ($\lambda_{exc} = 310$ nm).

by using ^{1}H NMR spectroscopy shows a 1:1 complexation between the sensor and mandelic acid.

Compound (R)-**5.15**, the enantiomer of (S)-**5.15**, is synthesized. The fluorescence responses of (R)-**5.15** in the presence of the two enantiomers of mandelic acid have a mirror image relationship with those of (S)-**5.15**. That is, (R)-mandelic acid increases the fluorescence intensity of (R)-**5.15** significantly greater than (S)-mandelic acid. This confirms the observed enantioselective fluorescent recognition. According to the Benesi–Hilderbrand-type equation, the association constants of (R)-**5.15** with (R)- and (S)-mandelic acids are 333 and 144 M^{-1}, respectively.

Mandelic acid of various enantiomeric compositions is interacted with (S)-**5.15**. The fluorescence response of (S)-**5.15** is found to have a linear relationship with the enantiomeric composition of mandelic acid. That is, the fluorescence intensity of the sensor can be used to determine the enantiomeric composition of the α-hydroxycarboxylic acid.

The interaction of (S)-**5.15** with O-acyl mandelic acid is studied. Although fluorescence enhancement is observed, there is no enantioselectivity. This indicates that the hydrogen bonding through the hydroxyl group of mandelic acid with the sensor should be important for the chiral recognition. When (S)-**5.15** is treated with methyl mandelate, no fluorescence enhancement is observed. This is consistent with the proposed fluorescence enhancement mechanism *via* the protonation of the nitrogen in (S)-**5.15** by the carboxylic acid.

O-acyl mandelic acid   methyl mandelate

As described in Sections 5.2 and 5.3, the attachment of dendritic arms to the BINOL core can greatly increase the fluorescent sensitivity of the sensors because of the light harvesting effect of the dendrimers. In order to improve the sensitivity of the sensor (S)-**5.15**, we have introduced dendritic units to this compound. Schemes 5.8 and 5.9 show the syntheses of compounds (R)-**5.18** (G0) and (R)-**5.21** (G1) that contain multiple phenylene units at the 6,6′-positions of the bisbinaphthyl core, respectively.

In Scheme 5.8, the mono-MOM-protected 6,6′-Br$_2$BINOL is used as the starting material. Reaction with compound **5.16** followed by Suzuki coupling with phenylboronic acid and then deprotection gives the G0 molecule (R)-**5.18**.

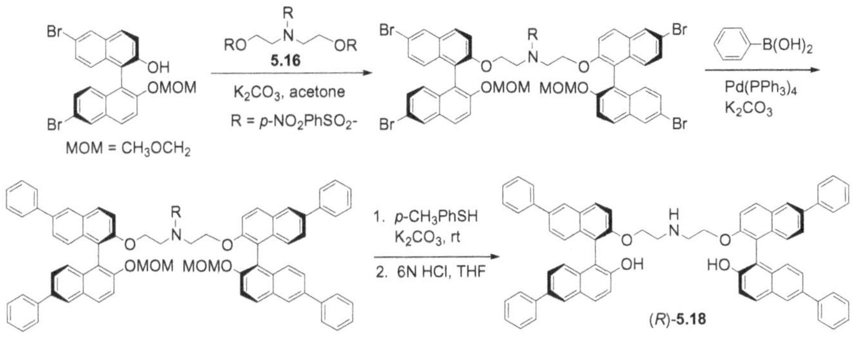

Scheme 5.8. Synthesis of G0 bisbinaphthyl compound (R)-**5.18**.

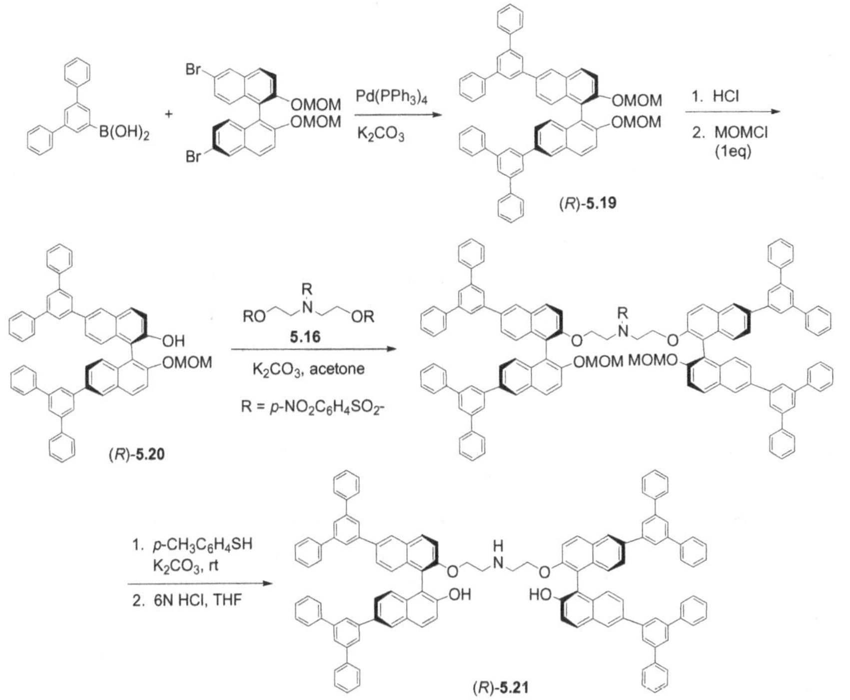

Scheme 5.9. Synthesis of G1 bisbinaphthyl compound (R)-**5.21**.

In Scheme 5.9, the MOM-protected 6,6'-Br$_2$BINOL is coupled with 3,5-diphenyl-phenylboronic acid to give compound (R)-**5.19**. Removal of the two MOM groups of (R)-**5.19** followed by reinstalling one MOM group gives (R)-**5.20**. Reaction of (R)-**5.20** with **5.16** followed by removal of the protecting groups produces the G1 molecule (R)-**5.21**.

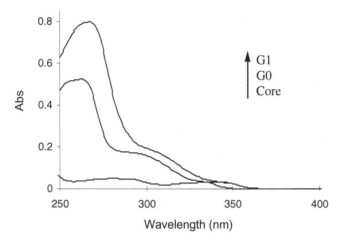

Figure 5.11. UV Spectra of the binaphthyl compounds in methylene chloride.

The UV spectra of compounds (R)-**5.15** (core), (R)-**5.18** (G0) and (R)-**5.21** (G1) are compared in Fig. 5.11. Introduction of the multiple phenylene units to the binaphthyl units of the core has greatly increased the absorption in the short-wavelength region.

The fluorescence responses of compounds (R)-**5.15** (core), (R)-**5.18** (G0) and (R)-**5.21** (G1) with (R)-mandelic acid are shown in Fig. 5.12. These compounds (3.1×10^{-6} M) are dissolved in benzene (0.1% DME) and the concentration of the acid is 1.0×10^{-3} M. The fluorescence signals of (R)-**5.18** (G0) and (R)-**5.21** (G1) are shifted to the red side of that of (R)-**5.15** due to the more extended conjugation of these compounds than the core. The increased light absorption of (R)-**5.18** (G0) and (R)-**5.21** (G1) leads to their increased fluorescence intensity. At a concentration of 3.1×10^{-6} M, the fluorescence enhancement of (R)-**5.15** in the presence of (R)-mandelic acid is low. This is different from that shown in Fig. 5.10 where the concentration of (S)-**5.15** is 30 times higher and its fluorescence enhancement in the presence of (S)-mandelic acid is much greater. When (R)-**5.18** (G0) and (R)-**5.21** (G1) are used to interact with (R)-mandelic acid, however, as shown in Fig. 5.12, there are much greater fluorescence enhancements. The fluorescence enhancement of (R)-**5.18** is 14 times that of the core and that of (R)-**5.21** is 22 times. Thus, attachment of the dendritic arms has significantly increased the fluorescence sensitivity.

Both (R)-**5.18** and (R)-**5.21** show enantioselective fluorescence response toward the enantiomers of mandelic acid. At an acid concentration

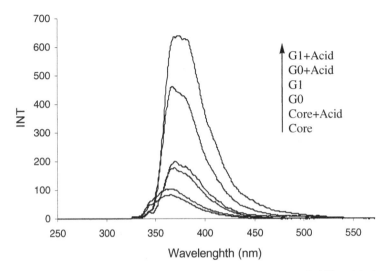

Figure 5.12. Comparison of the fluorescence spectra of (R)-**5.15** (core), (R)-**5.18** (G0) and (R)-**5.21** (G1) in the presence/absence of (R)-mandelic acid.

of 1.0×10^{-3} M (benzene containing 0.1% DME), the *ef* of (R)-**5.18** $(1.0 \times 10^{-5}$ M) is 2.05 and that of (R)-**5.21** is 1.49 $(3.1 \times 10^{-6}$ M). The effect of DME on the fluorescence response of (R)-**5.21** in the presence of mandelic acid is studied. It is found that when the concentration of DME increases greater than 0.1%, the enantioselectivity drops significantly. The polar DME probably disrupts the hydrogen bonding interaction between the sensor and the acid, leading to the reduced selectivity.

5.6. Using Diphenylethylenediamine-BINOL Macrocycles for the Recognition of α-Hydroxycarboxylic Acids[26-28]

5.6.1. *Study of the Macrocycles*

In order to further develop enantioselective fluorescent sensors for the recognition of α-hydroxycarboxylic acids, we have studied the use of a macrocyclic compound (S)-**5.22**. This compound was initially prepared by Brunner and Schiessling from the condensation of (S)-3,3′-diformylBINOL with (R,R)-1,2-diphenylethylenediamine followed by reduction with NaBH$_4$ (Scheme 5.10).[29] Using the chirality mismatched (S,S)-1,2-diphenylethylenediamine to react with (S)-3,3′-diformylBINOL led to the formation of a polymer rather than a macrocycle. The use of

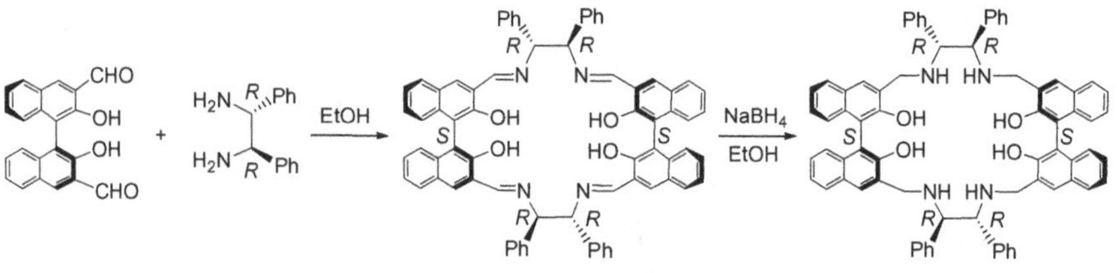

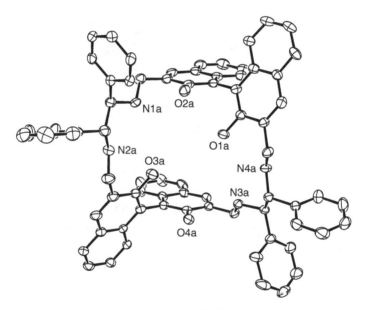

Scheme 5.10. Synthesis of bisbinaphthyl macrocycle (S)-**5.22**.

Figure 5.13. ORTEP drawing (30% ellipsoids) of (S)-**5.22**.

the unreduced macrocyclic salen in asymmetric catalysis is described in Section 4.2.6 in Chapter 4.

In collaboration with Sabat, a single crystal X-ray analysis of the chiral macrocycle (S)-**5.22** is conducted on a colorless crystal obtained from its deuterated acetone solution. The crystal is triclinic with the space group P1. Two independent molecules of slightly different conformation are found to exist in the unit cell of this molecule. Figure 5.13 shows the ORTEP drawing of one of those molecules. The central cavity of the macrocycle is flanked on two opposite sides by almost exactly parallel naphthol moieties separated by 7.65 Å. The long axes of the two groups are mutually perpendicular.

In addition to the molecules of (S)-**5.22**, there are six molecules of the deuterated acetone molecules. Each of the macrocycles contains one of the solvent molecules sandwiched between the parallel naphthol units. This indicates that small molecules can be included inside the cavity of this macrocycle. The distances between the C atoms of acetone and naphthol range between 3.80 and 4.05 Å, indicating possible C–H$\cdots\pi$ interactions. All of the O atoms of the naphthol units are on the same side of the macrocycle. The macrocyle rings are stabilized by intramolecular O–H$\cdots$N bonds, with the average O$\cdots$N donor acceptor distance of 2.68 Å in one molecule, which is slightly shorter than the corresponding distance of 2.75 Å in the other.

The UV spectrum of (S)-**5.22** in benzene shows absorptions at $\lambda_{max}(\varepsilon)$ = 270 (7760), 282 (17400), 292 (sh, 14200), 330 (sh, 10900) and 340 (13800) nm. When it is excited at 340 nm, it shows two emission signals at λ_{emi} = 365 and 424 nm (Fig. 5.14). It is found that the long wavelength emission of (S)-**5.22** decreases as its concentration decreases while the positions of the UV absorptions remain the same. This indicates that the emission of (S)-**5.22** at λ_{emi} = 424 nm is probably due to the formation of excimers. The fluorescence responses of (S)-**5.22** in the presence of (R)- and (S)-mandelic acid are studied. As shown in Fig. 5.14, when (S)-**5.22** is treated with (S)-mandelic acid, there is large fluorescence enhancement, especially at the excimer emission. The change in the fluorescence is much

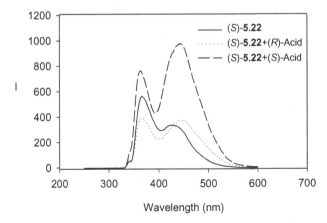

Figure 5.14. Fluorescence spectra of (S)-**5.22** (1.0×10^{-4} M in benzene containing 2% DME) with/without (R)- and (S)-mandelic acid (2.0×10^{-2} M) (λ_{exc} = 340 nm).

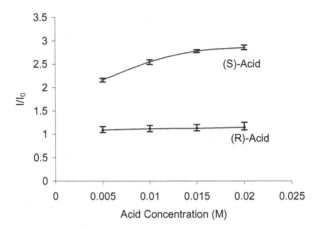

Figure 5.15. Fluorescence enhancement of (S)-**5.22** (1.0×10^{-4} M in benzene containing 2% DME) *versus* concentration of (R)- and (S)-mandelic acid.

smaller when (S)-**5.22** is treated with (R)-mandelic acid. That is, the fluorescence response of (S)-**5.22** in the presence of mandelic acid is highly enantioselective.

Figure 5.15 plots the fluorescence enhancement of (S)-**5.22** (1.0×10^{-4} M) at $\lambda_{emi} = 424$ nm when treated with (R)- and (S)-mandelic acid in the concentration range of 5.0×10^{-3} to 2.0×10^{-2} M. In this concentration range, (S)-mandelic acid increases the fluorescence intensity of (S)-**5.22** to 2–3 folds, but (R)-mandelic acid causes little change. The enantiomeric fluorescence difference ratio, ef [$ef = (I_S - I_0)/(I_R - I_0)$], is greater than 12 when the acid concentration is 2.0×10^{-2} M. This enantioselective fluorescent recognition is confirmed by using (R)-**5.22** to interact with the enantiomers of the acid. As expected, (R)-mandelic acid leads to large fluorescence enhancement for (R)-**5.22**, whereas (S)-mandelic acid does not change the fluorescence of (R)-**5.22** significantly.

Figure 5.16 plots the fluorescence enhancement (I/I_0) of (S)-**5.22** *versus* the enantiomeric composition of mandelic acid. As the composition of the S enantiomer increases, the fluorescence enhancement increases. That is, from the fluorescence enhancement of the sensor, the enantiomeric composition of mandelic acid can be determined.

The amount of DME in the benzene solution is found to influence the fluorescence spectrum of (S)-**5.22**. In the absence of a substrate, the ratio of the long wavelength emission of (S)-**5.22** (1.0×10^{-4} M in benzene) to its short wavelength emission increases from 0.61 to 1.7 as

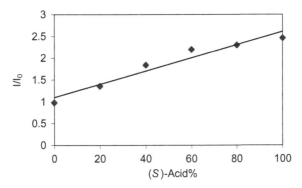

Figure 5.16. Fluorescence intensity change of (S)-**5.22** (1.0×10^{-4} M in benzene containing 1% DME) *versus* enantiomeric composition of mandelic acid (1.0×10^{-2} M).

the amount of DME increases from 0 to 20%. The increased polarity of the solvent with the addition of DME should encourage the association of the less polar aromatic rings of (S)-**5.22**, which should contribute to the increased excimer emission. Increasing the amount of DME leads to reduced enantioselective fluorescence response. This could be attributed to the reduced hydrogen bonding interaction between the sensor and the acid substrate.

The interaction of (R)-**5.22** with (R)-mandelic acid is studied by using ^{1}H NMR spectroscopy. In benzene-d_6, the methine proton signal of (R)-mandelic acid undergoes upfield shift in the presence of (R)-**5.22** with a maximum shift of 0.04 ppm when the ratio of (R)-**5.22** to (R)-mandelic acid is 1:4. This indicates that (R)-**5.22** may form a 1:4 complex with (R)-mandelic acid, that is, each of the four nitrogen atoms of (R)-**5.22** interacts with one (R)-mandelic acid molecule. This interaction should suppress the photo-induced electron transfer of the nitrogen lone pair electrons and enhances the fluorescence of the sensor.

The fluorescence response of (R)-**5.22** toward mandelic acid in methylene chloride solution is studied. As shown in Fig. 5.17, the fluorescence enhancement of (R)-**5.22** in the presence of mandelic acid and the enantioselective fluorescence responses occur at the short wavelength (the monomer emission) rather than at the long wavelength (the excimer emission) unlike that observed in benzene solution. The enantioselectivity is also smaller with $ef = 3.2$ when the concentration of (R)-**5.22** is 1.0×10^{-4} M and that of (R)-mandelic acid is 0.02 M.

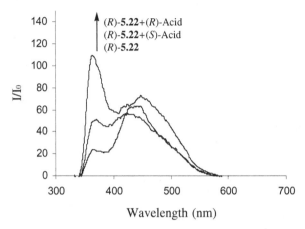

Figure 5.17. Fluorescence spectra of (R)-**5.22** (1.0×10^{-4} M in $CH_2Cl_2/2\%$ DME) in the presence of (R)- and (S)-mandelic acid (0.02 M) ($\lambda_{exc} = 327$ nm).

In benzene solution (2% DME), (S)-**5.22** is interacted with the enantiomers of hexahydromandelic acid. Similar to that observed with mandelic acid, (S)-**5.22** (1.0×10^{-4} M) shows up to 80% fluorescence enhancement at its excimer emission in the presence of (S)-hexahydromandelic acid (5.0×10^{-3} to 2.0×10^{-2} M), but almost no change in the presence of (R)-hexahydromandelic acid. The fluorescence enhancement of (S)-**5.22** is linear with the enantiomeric composition of hexahydromandelic acid.

The macrocycle (R)-**5.22** is also used to recognize the derivatives of amino acids. Figure 5.18 shows the fluorescence spectra of (R)-**5.22** in benzene (2% DME) after treated with D- and L-N-benzyloxycarbonylphenyl glycine (BPG). Highly enantioselective fluorescent enhancement at the excimer emission of (R)-**5.22** is observed, which is similar to that observed in the presence of mandelic acid. The enantiomer D-BPG increased the excimer emission of (R)-**5.22** to five-fold with $ef = 5.7$. The concentration effect of the N-protected amino acid on the fluorescence response is shown in Fig. 5.19. The high enantioselectivity is confirmed by studying the interaction of the enantiomeric macrocycle (S)-**5.22** with the amino acid derivative. A linear relationship is observed between the fluorescence enhancement of (R)-**5.22** and the enantiomeric composition of the acid.

BPG

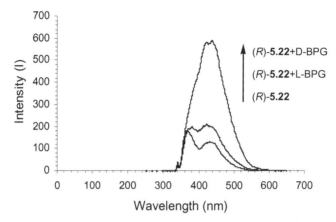

Figure 5.18. Fluorescence spectra of (R)-**5.22** (1.0×10^{-4} M in benzene containing 2% DME) with/without D- and L-BPG (4.0×10^{-3} M) ($\lambda_{exc} = 340$ nm).

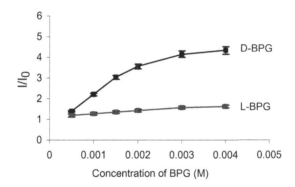

Figure 5.19. Fluorescence enhancement (I/I_0) at λ_{long} of (R)-**5.22** (1.0×10^{-4} M in benzene/2% DME) versus concentration of D- and L-BPG.

The interaction of (S)-**5.22** with L-BPG is investigated by using ^{1}H NMR spectroscopy. The methine proton signal of L-BPG ($\delta = 5.44$) undergoes downfield shift in the presence of (S)-**5.22** with the maximum shift ($\Delta\delta = 0.55$) observed at a 4:1 ratio of L-BPG versus (S)-**5.22**. This indicates the formation of a 4:1 complex between L-BPG and (S)-**5.22** probably through the protonation of the four nitrogen atoms of (S)-**5.22** by the acid. Other binding modes are also possible.

The fluorescence responses of (S)-**5.22** toward other amino acid derivatives are studied. As shown in Table 5.8, various degrees of enantioselectivity

Table 5.8. Results for the enantioselective fluorescent responses of (R)- or (S)-**5.22** (1.0×10^{-4} M) to various amino acid derivatives.

Entry	Amino acid	Receptor	Solvent	Concentration (M)	ef
1	BPG	(R)-**5.22**	C_6H_6/2%DME	4×10^{-3}	5.7[a]
2		(R)-**5.22**	C_6H_6/2%DME	2×10^{-3}	1.9[a]
3		(S)-**5.22**	C_6H_6/20%DME	2×10^{-2}	1.7[a]
4		(R)-**5.22**	C_6H_6/12%DME	4×10^{-3}	2.2[b]
5		(R)-**5.22**	C_6H_6/0.1%DME	2×10^{-3}	1.7[b]
6		(R)-**5.22**	C_6H_6/8%DME	2×10^{-2}	1.2[a]
7		(R)-**5.22**	C_6H_6/4%DME	5×10^{-3}	1.1[b]

[a] D/L.
[b] L/D.

are observed. The enantioselectivity is inverted by replacement of the benzyloxycarbonyl protecting group of the substrate in entry 2 with the acetyl group of the substrate in entry 3. The additional hydroxyl or carboxylic acid group in entries 6 and 7 greatly reduces the enantioselectivity.

5.6.2. Study of the 6,6'-Substituted Derivatives of the Macrocycles and Their Acyclic Analogs

We have introduced conjugated units to the binaphthyl macrocycle in order to improve its fluorescence property. Schemes 5.11 and 5.12 are the synthesis of the two macrocycles (S)-**5.24** and (S)-**5.26** that contain p-ethoxyphenyl and styryl substituents, respectively, at the 6,6'-positions of the binaphthyl units. In Scheme 5.11, the 6,6'-di(p-methoxyphenyl)binaphthyl dialdehyde (S)-**5.23** is prepared from the MOM-protected 6,6'-Br$_2$BINOL. Condensation of (S)-**5.23** with (R, R)-1,2-diphenylethylenediamine gives a macrocyclic Schiff base, which upon reduction with NaBH$_4$ produces the macrocycle (S)-**5.24** in 80% yield over the two steps. In Scheme 5.10, the 6,6'-distyrylbinaphthyl dialdehyde (S)-**5.25** is prepared from the MOM-protected 6,6'-Br$_2$BINOL. Condensation of (S)-**5.25** with (R, R)-1,2-diphenylethylenediamine followed by reduction with NaBH$_4$ produces (S)-**5.26** in 81% yield over the two steps. The syntheses of (S)-**5.24**

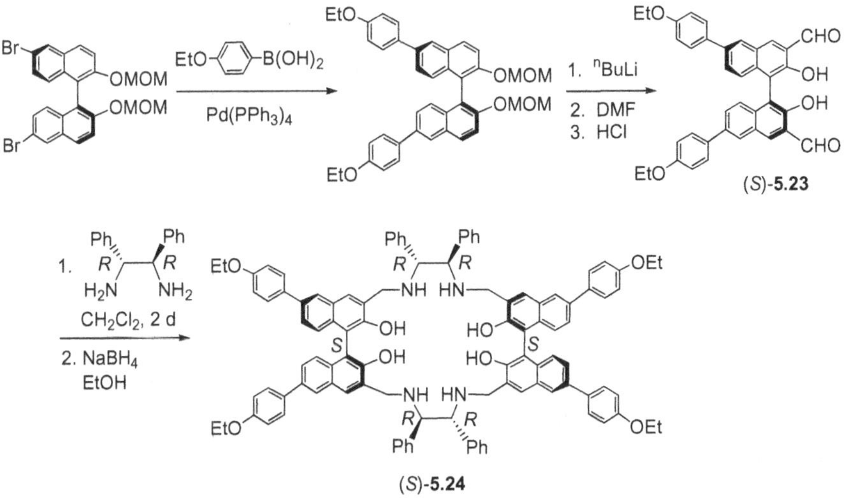

Scheme 5.11. Synthesis of 6,6'-p-ethoxyphenyl-substituted bisbinaphthyl macrocycle (S)-**5.24**.

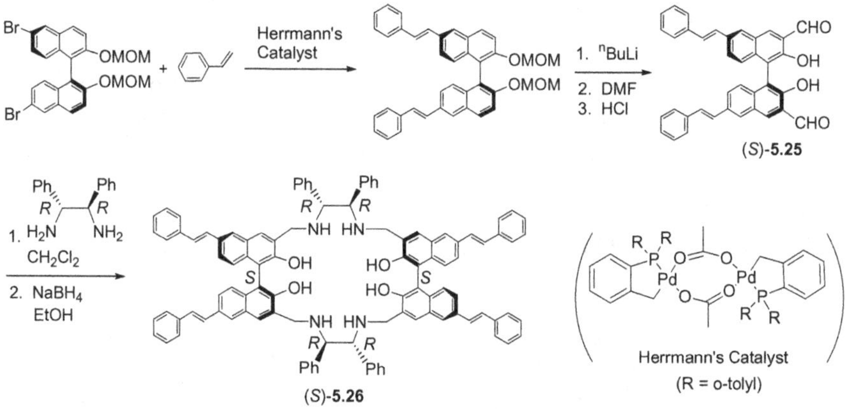

Scheme 5.12. Synthesis of 6,6'-styryl-substituted bisbinaphthyl macrocycles (*S*)-**5.26**.

and (*S*)-**5.26** demonstrate that the substituents at the 6,6'-positions of the binaphthyl-3,3'-dicarboaldehydes do not significantly influence the 4-component condensation to macrocycles.

As shown in Schemes 5.13 and 5.14, two acyclic analogs of the chiral macrocycles, compounds (*R*)-**5.28** and (*S*)-**5.30**, are prepared from the

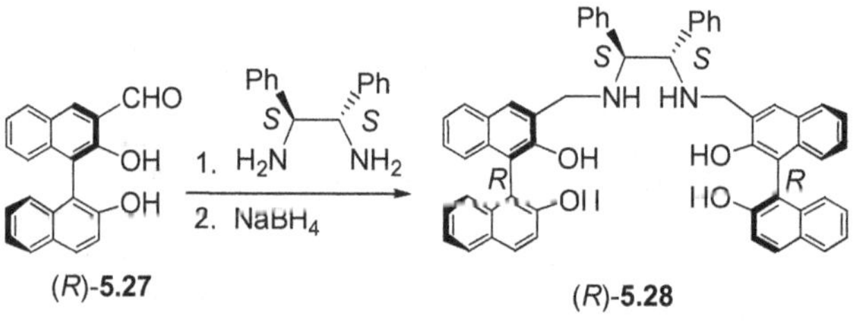

Scheme 5.13. Synthesis of acyclic bisbinaphthyl (*R*)-**5.28**.

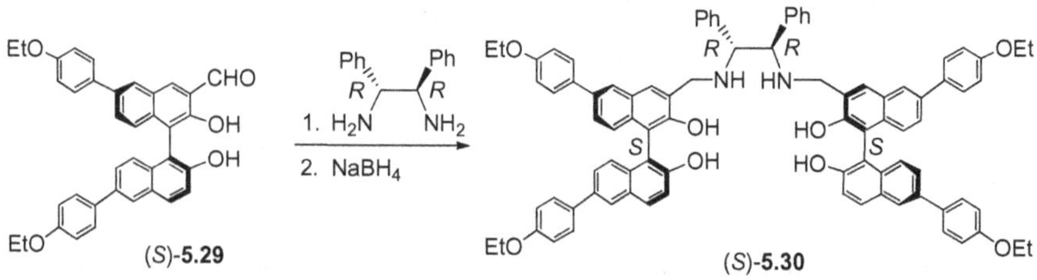

Scheme 5.14. Synthesis of acyclic bisbinaphthyl (*S*)-**5.30**.

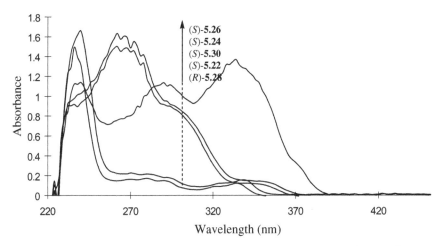

Figure 5.20. UV spectra of the binaphthyl macrocycles and their analogs in methylene chloride (1.0×10^{-5} M).

reaction of the binaphthyl monoaldehydes (R)-**5.27** and (S)-**5.29** with (R,R)- and (S,S)-1,2-diphenylethylenediamine, respectively.

The UV spectra of the macrocycles and acyclic compounds are compared. As shown in Fig. 5.20, the UV spectra of the macrocycle (S)-**5.22** and its acyclic analog (R)-**5.28** are very similar except a small red shift from the acyclic compound (R)-**5.28** to the macrocycle (S)-**5.22**. The more restricted rotation around the 1,1′-bond of the binaphthyl units in the macrocycle might have provided more partial conjugation between the naphthalene rings than the more flexible acyclic compound. The macrocycle (S)-**5.24** and its acyclic analog (S)-**5.30** contain p-ethoxyphenyl substituents on the binaphthyl units, which extend the conjugation and cause significant red shifts for the major absorptions. The styryl-substituted macrocycle (S)-**5.26** further extends the conjugation of the naphthalene rings and gives the most red-shifted UV signals.

The UV spectra of the macrocycle (S)-**5.22** and its acyclic analog (R)-**5.28** are compared with those of 2-naphthol and BINOL. As shown in Fig. 5.21, the increased intensity from 2-naphthol to BINOL, (S)-**5.22** and (R)-**5.28** is mostly due to the increased number of naphthol units. The slight red shifts in the UV signals from 2-naphthol to BINOL then to (S)-**5.22** and (R)-**5.28** may be contributed from a partial conjugation across the 1,1′-bonds in the binaphthyl compounds and the intramolecular N–H–O bonds in (S)-**5.22** and (R)-**5.28**.

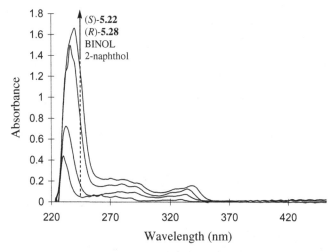

Figure 5.21. UV spectra of compounds (*S*)-**5.22**, (*R*)-**5.28**, 2-naphthol and BINOL in methylene chloride (1.0×10^{-5} M).

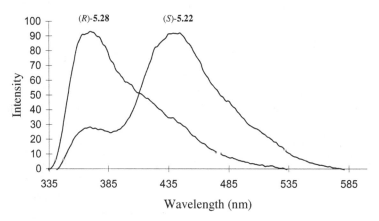

Figure 5.22. Fluorescence spectra of the macrocycle (*S*)-**5.22** and the acyclic analog (*R*)-**5.28** in methylene chloride (1.0×10^{-4} M, $\lambda_{exc} = 327$ nm).

Figure 5.22 compares the fluorescence spectra of the macrocycle (*S*)-**5.22** with the acyclic analog (*R*)-**5.28** in methylene chloride. In spite of their similar UV spectra, their fluorescence spectra show a major difference. The macrocycle (*S*)-**5.19** gives a very strong excimer emission but (*R*)-**5.28** does not. One explanation for this difference could be the different solvation ability of these two compounds. The more rigid cyclic structure of (*S*)-**5.22**

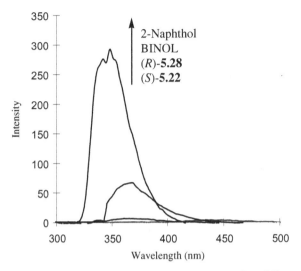

Figure 5.23. Fluorescence spectra of 2-naphthol, BINOL, (S)-**5.22** and (R)-**5.28** in methylene chloride (2.0×10^{-6} M, λ_{exc} = 274, 278, 305 and 305 nm, respectively).

probably does not allow it to be as well solvated as the flexible acyclic compound (R)-**5.28**. This would make it easier for (S)-**5.22** to form the excimer through intermolecular interaction than (R)-**5.28**.

Figure 5.23 compares the fluorescence spectra of (S)-**5.22** and (R)-**5.28** with those of 2-naphthol and BINOL. Although the absorptions of (S)-**5.22** and (R)-**5.28** are greater than those of 2-naphthol, the fluorescence intensities of (S)-**5.22** and (R)-**5.28** are about two orders of magnitude weaker than that of 2-naphthol (λ_{max} = 349 nm). This can be attributed to the photo-induced electron transfer of the nitrogen lone-pair electrons in (S)-**5.22** and (R)-**5.28**, which quenches the fluorescence of the naphthol units in these compounds. The lower fluorescence intensity of BINOL (λ_{max} = 368 nm) *versus* that of 2-naphthol is probably due to the rotation around the 1,1'-bond in BINOL, which could help relax the excited state energy.

In Fig. 5.24, the fluorescence spectra of (S)-**5.22** and (R)-**5.28** are compared with those of the 6,6'-substituted derivatives (S)-**5.24**, (S)-**5.26** and (S)-**5.30** in methylene chloride at 2.0×10^{-6} M. The *p*-ethoxyphenyl and styryl substituents in (S)-**5.24**, (S)-**5.26** and (S)-**5.30** have enhanced their fluorescence intensity by about two orders of magnitude over the

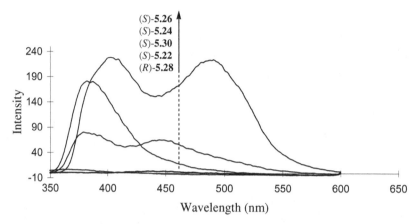

Figure 5.24. Comparison of the fluorescence spectra of the 6,6'-substituted compounds (S)-**5.24**, (S)-**5.26** and (S)-**5.30** with those of the unsubstituted compounds (S)-**5.22** and (R)-**5.28** in methylene chloride (2.0×10^{-6} M).

unsubstituted (S)-**5.22** and (R)-**5.28**. Similar to the macrocycle (S)-**5.22**, the 6,6'-substituted macrocycles (S)-**5.24** and (S)-**5.26** show intense excimer emissions with dual emission peaks at 379, 445 nm and 403, 491 nm, respectively. The excimer emissions decrease rapidly as the concentrations of (S)-**5.24** and (S)-**5.26** decrease. In contrast to the macrocycles, the more flexible 6,6'-substituted acyclic compound (S)-**5.30** does not show significant excimer emission just like the unsubstituted acyclic compound (R)-**5.28**. The excimer emissions of the macrocycles are found to decrease as the solvent is changed from methylene chloride to benzene in which the short wavelength emission of the macrocycles becomes the major peak. This supports the solvation hypothesis. That is, benzene is probably a better solvent for the aromatic rings of these compounds, which reduces the intermolecular interaction between the aromatic rings for the excimer formation.

Similar to the study of (S)-**5.22**, the fluorescence responses of the 6,6'-substituted bisbinaphthyl macrocycles (S)-**5.24** and (S)-**5.26** and the acyclic compound (S)-**5.30** in the presence of mandelic acids are investigated. As shown in Figs. 5.25 and 5.26, in methylene chloride (0.4% DME) solution, both (S)-**5.24** and (S)-**5.26** exhibit fluorescence enhancement at their monomer emission in the presence of mandelic acid with certain enantioselectivity ($ef \approx 2$). These observations are similar to those observed for the unsubstituted macrocycle (S)-**5.22** in

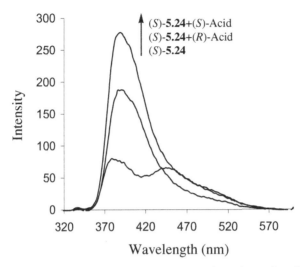

Figure 5.25. Fluorescence spectra of the 6,6′-p-ethoxyphenyl-substituted macrocycle (S)-**5.24** (2.0×10^{-6} M in CH_2Cl_2/0.4%DME) in the presence of (R)- and (S)-mandelic acid (4.0×10^{-3} M) ($\lambda_{exc} = 305$ nm).

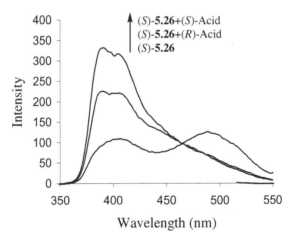

Figure 5.26. Fluorescence spectra of the 6,6′-styryl-substituted macrocycle (S)-**5.26** (2.0×10^{-6} M in CH_2Cl_2/0.4%DME) with (R)- and (S)-mandelic acid (4.0×10^{-3} M) ($\lambda_{exc} = 305$ nm).

methylene chloride except that the greatly increased fluorescent intensity of (S)-**5.24** and (S)-**5.26** over (S)-**5.22** allows them to be used at a concentration two orders of magnitude lower than that of (S)-**5.22**. In this sense, the two substituted macrocycles are more sensitive sensors.

The fluorescence responses of (S)-**5.24** and (S)-**5.26** toward mandelic acid in benzene are studied as well. However, fluorescence quenching is observed in both cases. The solution remains clear with no precipitate formation. The observed fluorescence quenching is very different from that observed for (S)-**5.22** in benzene solution, which shows highly enantioselective fluorescent enhancement at the excimer emission. No explanation could be provided yet for the unexpected behavior of (S)-**5.24** and (S)-**5.26** in benzene. The fluorescence of the acyclic compound (S)-**5.30** in methylene chloride solution is also enhanced by mandelic acid, but no enantioselectivity is observed. This demonstrates that the macrocyclic structure is important for the enantioselectivity.

The fluorescence responses of the 6,6′-substituted bisbinaphthyl compounds (S)-**5.24** and (S)-**5.26** to the N-protected amino acid BPG are studied. It is found that the fluorescence study can be conducted with the concentrations of (S)-**5.24** and (S)-**5.26** at two orders of magnitude lower than the unsubstituted (S)-**5.22** (10^{-6} M $versus$ 10^{-4} M). Figures 5.27 and 5.28 are the fluorescence spectra of (S)-**5.24** and (S)-**5.26** in methylene chloride (0.4% DME) solution in the presence of L- and D-BPG. Both (S)-**5.24** and (S)-**5.26** show fluorescence enhancement at the short wavelength

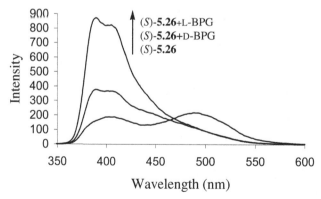

Figure 5.27. Fluorescence spectra of (S)-**5.26** (2.0×10^{-6} M in CH_2Cl_2/0.4% DME) in the presence of D- and L-BPG (4.0×10^{-3} M) ($\lambda_{exc} = 336$ nm).

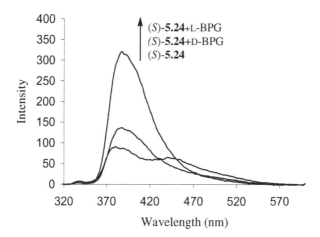

Figure 5.28. Fluorescence spectra of (S)-**5.24** (2.0×10^{-6} M in CH_2Cl_2/0.4% DME) in the presence of D- and L-BPG (4.0×10^{-3} M) ($\lambda_{exc} = 305$ nm).

in the presence of the N-protected amino acid. The enantioselectivity is high with ef up to 10. In both cases, L-BPG leads to much greater fluorescence enhancement than D-BPG.

5.6.3. Study of the Macrocyclic Analogs Containing Less Functional Groups

We have synthesized a macrocyclic analog of (S)-**5.22** that contains only half of the amine and hydroxyl functional groups. As shown in Scheme 5.15, the reaction of the mono MOM-protected BINOL with 1,4-dibromobutane gives (S)-**5.31**. Treatment of (S)-**5.31** with ^{n}BuLi then DMF gives the dialdehyde (S)-**5.32**. Condensation of (S)-**5.32** with (R,R)-1,2-diphenylethylendiamine followed by reduction gives the macrocycle (S)-**5.33**. This macrocycle contains only half of the hydroxyl and amine groups of the macrocycle (S)-**5.22**, and the size the ring is also smaller and more flexible.

The UV spectrum of (S)-**5.33** is compared with those of macrocycle (S)-**5.22** and acyclic compound (S)-**5.28** in Fig. 5.29. The two macrocycles are very similar but that of the acyclic compound (S)-**5.28** is shifted slightly to blue probably due to less conjugation across the 1,1'-bond. The fluorescence spectrum of (S)-**5.33** is also very similar to that of (S)-**5.22** with significant excimer emission (Fig. 5.30).

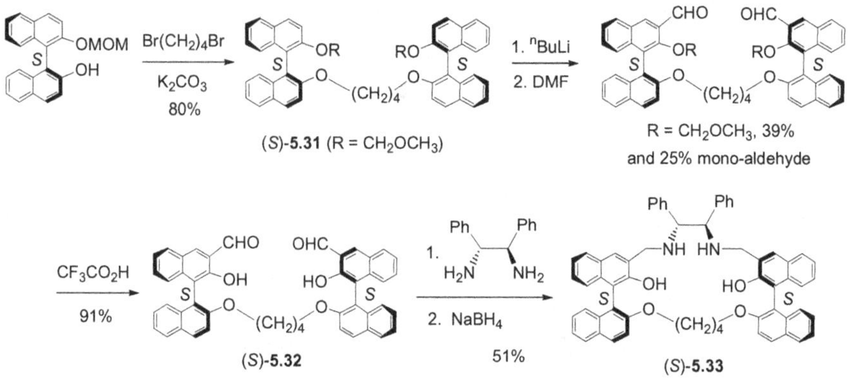

Scheme 5.15. Synthesis of bisbinaphthyl macrocycle (S)-**5.33**.

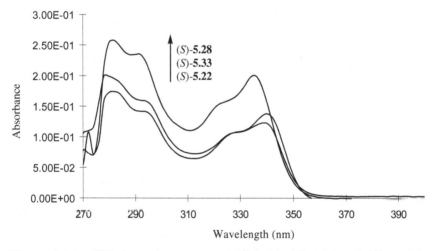

Figure 5.29. UV absorption spectra of (S)-**5.22**, (S)-**5.28** and (S)-**5.33** in benzene at 1.0×10^{-5} M.

Figure 5.31 shows the fluorescence responses of (S)-**5.33** in the presence of (R)- and (S)-mandelic acid. In benzene solution, an enantioselective fluorescence enhancement is observed at the excimer emission. This is similar to that observed for (S)-**5.22** in benzene solution but the enantioselectivity of (S)-**5.33** is much smaller. The highest *ef* observed for (S)-**5.33** in the presence of the enantiomers of mandelic acid is 4. In the presence of (S)-mandelic acid, the excimer emission of (S)-**5.33** undergoes

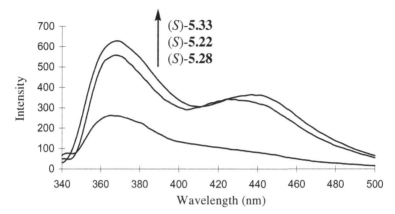

Figure 5.30. Fluorescence spectra of (S)-**5.22**, (S)-**5.28** and (S)-**5.33** in benzene at 1.0×10^{-5} M ($\lambda_{\text{exc}} = 332$ nm. ex/em slits = 3.5/6.5 nm).

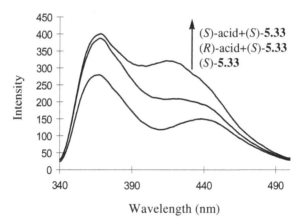

Figure 5.31. Fluorescence spectra of (S)-**5.33** (1.0×10^{-4} M in benzene/ 0.04% DME) in the presence of (R)- and (S)-mandelic acid (1.6×10^{-4} M) ($\lambda_{\text{exc}} = 322$ nm. ex/em slits = 3.5/4.5 nm).

blue shift. This is opposite to that observed for (S)-**5.22** whose excimer emission shifts to red in the presence of (S)-mandelic acid.

The fluorescence quantum yields of compounds (S)-**5.22**, (S)-**5.28** and (S)-**5.33** in benzene (1.0×10^{-5} M, 0.05% DME) are found to be 0.79, 0.19 and 0.89%, respectively, by using a quinine sulfate solution in 1 N H_2SO_4 ($\phi = 55\%$) as the standard. The more rigid macrocyclic structures of (S)-**5.22** and (S)-**5.33** lead to their much higher fluorescence quantum yields than the acyclic compound (S)-**5.28**. The higher quantum yield of (S)-**5.33** than (S)-**5.22** can be attributed to its less number of nitrogen atoms. At a higher concentration (1.0×10^{-4} M in benzene/0.04% DME), the fluorescence quantum yield of (S)-**5.33** is decreased by seven-fold to 0.12%. In the presence of (R)- and (S)-mandelic acid (1.6×10^{-4} M), the quantum yield of (S)-**5.33** (1.0×10^{-4} M in benzene/0.04% DME) is increased to 0.16 and 0.20%, respectively. ^{1}H NMR spectroscopy shows that in the presence of (S)-**5.33**, the methine proton signal of (S)-mandelic acid in benzene-d_6 (2% DME) only shifts ~0.01 ppm.

5.7. Using Cyclohexanediamine-BINOL Macrocycles and Their Derivatives for the Recognition of α-Hydroxycarboxylic Acids and Amino Acid Derivatives[30]

5.7.1. *Synthesis and Characterization of the Binaphthyl Compounds*

In Section 5.6, the study of the macrocyclic and acyclic bisbinaphthyl compounds incorporating chiral 1,2-diphenylethylenediamine is discussed. We have also synthesized a series of macrocyclic and acyclic bisbinaphthyl compounds by incorporating chiral cyclohexane-1,2-diamine units. The macrocycle (S)-**5.34** is prepared in the same way as (S)-**5.22** (Scheme 5.16). Condensation of (S)-3,3'-diformylBINOL with (R,R)-cyclohexane-1,2-diamine followed by reduction with NaBH$_4$ gives (S)-**5.34** in 55% yield over the two steps. The enantiomer (R)-**5.34** is obtained by starting with (R)-BINOL and (S,S)-cyclohexane-1,2-diamine.

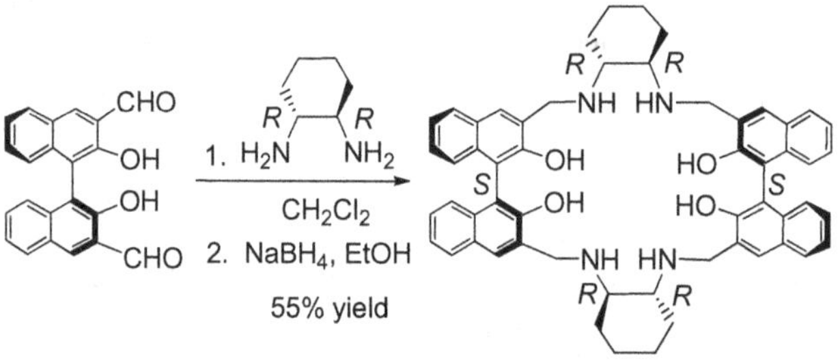

Scheme 5.16. Synthesis of cyclohexanediamine-BINOL macrocycle (S)-**5.34**.

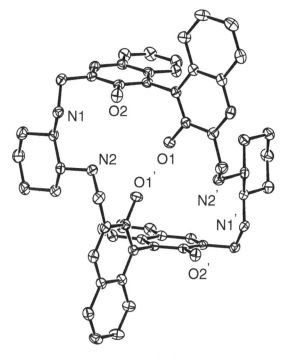

Figure 5.32. ORTEP drawing (30% ellipsoids) of (R)-**5.34**.

A single crystal of (R)-**5.34** is obtained from its acetone/hexane solution and its X-ray analysis has established the molecular structure. As shown in Fig. 5.32, the central cavity of the molecule is defined by two almost parallel naphthol units separated by 8.13 Å. The cavity is large enough to host an acetone solvent molecule, which is sandwiched between the two naphthol rings. The distances between the C atoms of acetone and naphthol range between 3.77 and 4.20 Å. All of the O atoms of the naphthol units are on the same side of the macrocycle. The macrocycle rings are stabilized by intramolecular O–H···N hydrogen bonds with the average O···N donor–acceptor distance of 2.74 Å.

The acyclic compound (S)-**5.35** is obtained from the reaction of the binaphthyl monoaldehyde and (R, R)-cyclohexane-1,2-diamine followed by reduction (Scheme 5.17).

The macrocyclic analog (S)-**5.36** that contains only half of the hydroxyl and amine functional groups of (S)-**5.31** is synthesized according to Scheme 5.18. This synthesis is similar to the preparation of (S)-**5.33** shown in Scheme 5.15.

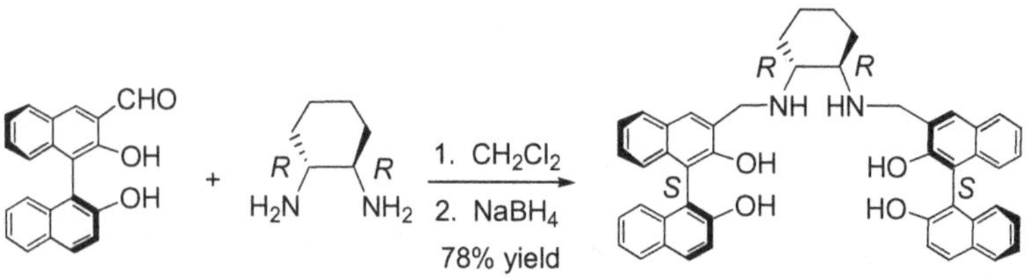

Scheme 5.17. Synthesis of acyclic bisbinaphthyl compound (S)-**5.35**.

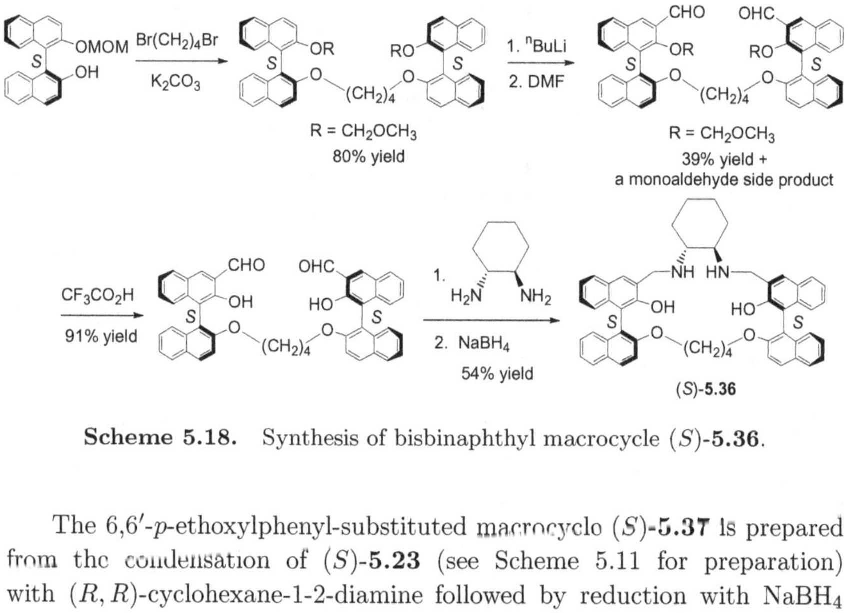

Scheme 5.18. Synthesis of bisbinaphthyl macrocycle (S)-**5.36**.

The 6,6'-p-ethoxylphenyl-substituted macrocyclo (S)-**5.37** is prepared from the condensation of (S)-**5.23** (see Scheme 5.11 for preparation) with (R,R)-cyclohexane-1-2-diamine followed by reduction with NaBH$_4$ (Scheme 5.19).

The UV spectra of compounds (S)-**5.34**, (S)-**5.35**, (S)-**5.36** and (S)-**5.37** in benzene (1.0 × 10^{-5} M) are compared in Fig. 5.33. The UV absorptions of the acyclic compound (S)-**5.35** is shifted to blue in comparison with those of the two macrocycles (S)-**5.34** and (S)-**5.36**, which is similar to that observed for the 1,2-diphenylethylenediamine-based acyclic compound (S)-**5.28**. This indicates that there is probably less conjugation across the 1,1'-binaphthyl bond or weaker intramolecular O–H···N hydrogen bond interaction in the acyclic compound (S)-**5.35** than those in the macrocycles (S)-**5.34** and (S)-**5.36**. The 6,6'-substituted compound (S)-**5.37** has more extended conjugation and it shows red

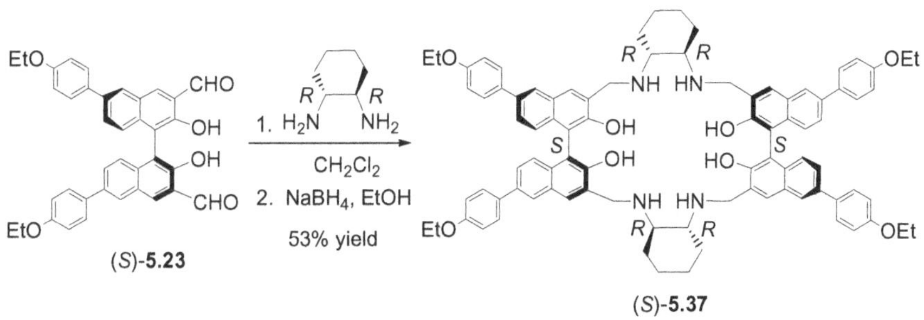

Enantioselective Fluorescent Sensors

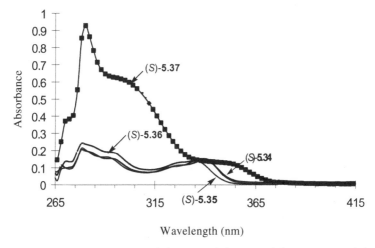

Scheme 5.19. Synthesis of 6,6'-substituted bisbinaphthyl macrocycle (S)-**5.37**.

Figure 5.33. UV spectra of (S)-**5.34**, (S)-**5.35**, (S)-**5.36** and (S)-**5.37** (1.0×10^{-5} M in benzene).

shift and greatly increased absorption compared to the unsubstituted compounds.

The fluorescence spectra of these compounds are compared in Fig. 5.34. All the three macrocycles (S)-**5.34**, (S)-**5.36** and (S)-**5.37** show major excimer emission at the long wavelength, but the acyclic compound (S)-**5.35** shows mainly the monomer emission. This is similar to that observed in the 1,2-diphenylethylenediamine-based analogs.

The CD spectra of (S)-**5.34**, (S)-**5.35**, (R)-**5.36** [the enantiomer of (S)-**5.36**] and (S)-**5.37** in methylene chloride are given in Fig. 5.35. Although both (S)-**5.34** and (S)-**5.35** have the same configuration at the binaphthyl

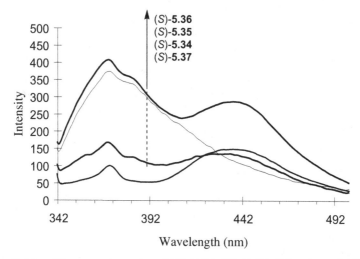

Figure 5.34. Fluorescent spectra of (S)-**5.34**, (S)-**5.35**, (S)-**5.36** and (S)-**5.37** in benzene at 1.0×10^{-5} M. ($\lambda_{exc} = 332$ nm, slit = 3.5; 6.5 nm).

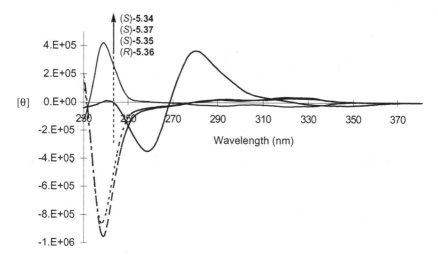

Figure 5.35. CD spectra of compounds (S)-**5.34**, (S)-**5.35**, (R)-**5.36** and (S)-**5.37** (1.0×10^{-5} M in CH_2Cl_2).

units and the carbon centers, (S)-**5.34** shows an intense *positive* Cotton effect at 239 nm opposite to that of (S)-**5.35**, which shows an intense *negative* Cotton effect at 237 nm. This indicates very different conformations at the binaphthyl units for these two compounds. The binaphthyl units in the macrocycle (S)-**5.36** should be conformationally similar to those of (S)-**5.34** as indicated by the opposite CD signals of the enantiomer (R)-**5.36**. The more extended conjugation in (S)-**5.37** significantly shifts its CD signals to red.

5.7.2. Enantioselective Fluorescent Recognition

The fluorescence response of (S)-**5.34** in the presence of mandelic acid is studied. Figure 5.36 shows that when (S)-**5.34** (1.0×10^{-5} M in benzene/0.05% DME) is treated with (S)-mandelic acid (5.0×10^{-4} M), there is a dramatic fluorescence enhancement at the monomer emission. However, very little fluorescence enhancement is observed when (S)-**5.34** is treated with (R)-mandelic acid. This extremely high enantioselectivity is confirmed by using the enantiomer of (S)-**5.34** to interact with the enantiomers of mandelic acid, which gives a mirror image relationship in the observed fluorescence enhancements.

The fluorescence enhancements of (S)-**5.34** in the presence of various concentrations of the enantiomers of mandelic acid are shown in Fig. 5.37. (S)-Mandelic acid greatly enhances the fluorescence of (S)-**5.34** at various concentrations but (R)-mandelic acid does not. The fluorescence enhancement can be over 20-folds and the enantioselectivity *ef* is up to 46.

In Fig. 5.38, Curve A shows the relationship between the fluorescence enhancement of (R)-**5.34** (1.0×10^{-5} M in benzene/0.05%DME) and the enantiomeric composition of mandelic acid (5.0×10^{-4} M). It shows that as the R enantiomer in the R/S mixture of mandelic acid increases, the fluorescence intensity of (R)-**5.34** increases nonlinearly. Curve B in Fig. 5.38 shows the relationship between the fluorescence enhancement

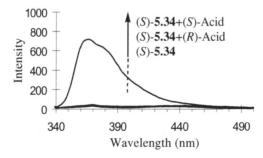

Figure 5.36. Fluorescence spectra of (S)-**5.34** with/without (R)- and (S)-mandelic acid ($\lambda_{exc} = 332$ nm, slit $= 3.5/3.5$ nm).

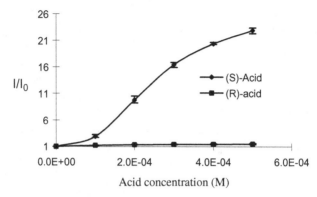

Figure 5.37. Fluorescence enhancement of (S)-**5.34** (1.0×10^{-5} M in benzene/0.05% DME) *versus* concentration of (R)- and (S)-mandelic acid ($\lambda_{\text{exc}} = 332$ nm).

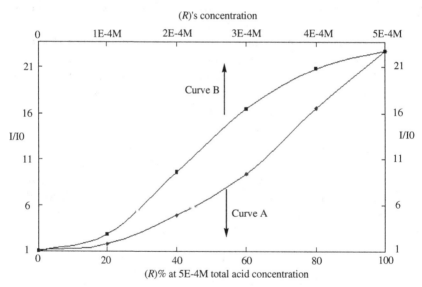

Figure 5.38. Fluorescence enhancement of (R)-**5.34** in the presence of (A) mandelic acid at various enantiomeric compositions and (B) the optically pure (R)-mandelic acid.

of (R)-**5.34** (1.0×10^{-5} M in benzene/0.05%DME) and pure (R)-mandelic acid at concentrations that are the same as the R composition in the enantiomeric mixture of mandelic acid. Comparison of Curve A with Curve B indicates that the presence of (S)-mandelic acid in the R/S mixture can reduce the fluorescence enhancement even though pure (S)-mandelic acid cannot appreciably change the fluorescence intensity of (R)-**5.34**. One of the explanations for this could be the intermolecular interaction of (R)- and (S)-mandelic acid, which could reduce the effective concentration of (R)-mandelic acid in the R/S mixture, leading to the lower fluorescence enhancement.

In Section 5.6, it is discussed that the 1,2-diphenylethylenediamine-based macrocycle (S)-**5.22** at 10^{-4} M in benzene exhibits two- to three-fold fluorescence enhancement at its excimer emission in the presence of (S)-mandelic acid. The fluorescence enhancement of (S)-**5.22** at its monomer emission is much lower. The fluorescence responses of (R)-**5.34** at 1.0×10^{-4} M in benzene (0.1% DME), an order of magnitude higher concentration than that in Fig. 5.37, in the presence of (R)- and (S)-mandelic acid are examined. It is found that at low concentration of (R)-mandelic acid, (R)-**5.34** shows fluorescence enhancement at both the monomer and excimer emissions. At higher concentration of (R)-mandelic acid, the fluorescence enhancement at the monomer emission of (R)-**5.34** predominates. Figure 5.39 shows that the fluorescence enhancement at the monomer emission of (R)-**5.34** can be over 41-folds when treated with (R)-mandelic acid while very little fluorescence enhancement in the presence of (S)-mandelic acid. This demonstrates that this cyclohexane-1,2-diamine-based compound is a much more sensitive and enantioselective fluorescent sensor than (S)-**5.22** derived from 1,2-diphenylethylenediamine.

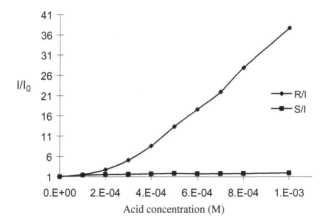

Figure 5.39. Fluorescence enhancement of (R)-**5.34** (1.0×10^{-4} M in benzene/0.1% DME) in the presence of (R)- and (S)-mandelic acid ($\lambda_{\text{exc}} = 332$ nm).

The fluorescence quantum yield of (R)-**5.34** (1.0×10^{-5} M in benzene) is 0.38% by using a quinine sulfate solution in 1 N H_2SO_4 ($\phi_F = 0.55$) as the standard. It is increased to 5.3% when treated with (R)-mandelic acid (5.0×10^{-4} M in benzene/0.05% DME), and only to 0.6% with (S)-mandelic acid under the same conditions. The quantum yields of the acyclic compound (S)-**5.35** and the macrocycle (S)-**5.36** are 0.6 and 0.8%, respectively, which are greater than that of (R)-**5.34** because they contain less nitrogen atoms to quench the fluorescence.

^{1}H NMR spectroscopic analysis shows that there is a large upfield shift of ~1.1 ppm for the methine proton signal of (S)-mandelic acid when it interacts with (S)-**5.34** in a 1:1 mixture in benzene-d_6/4% acetone-d_6. Under the same conditions, there is only 0.25 ppm upfield shift when the chirality mismatched (R)-mandelic acid interacts with (S)-**5.34**. This indicates that (S)-mandelic acid should bind (S)-**5.34** stronger and be included deeper in the chiral pocket of the macrocycle than (R)-mandelic acid. When (S)-mandelic acid binds with (S)-**5.34**, it also causes a significant shift for the methylene proton signals at the 3,3′-positions of the binaphthyl units in (S)-**5.34**. This further supports the inclusion of (S)-mandelic acid inside the chiral cavity of (S)-**5.34**. The NMR titration experiment suggests the formation of a 1:1 complex between (S)-mandelic acid and (S)-**5.34** with an estimated association constant of over 2000.

Figure 5.40 gives the fluorescence responses of (R)-**5.34** (1.0×10^{-5} M in benzene/0.4% DME) in the presence of (R)- and (S)-hexahydromandelic

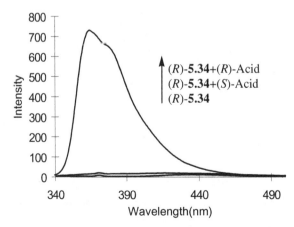

Figure 5.40. Fluorescence spectra of (R)-**5.34** (1.0×10^{-5} M in benzene/0.4% DME) with/without (R)- and (S)-hexahydromandelic acid (4.0×10^{-3} M) ($\lambda_{exc} = 332$ nm, slit = 2.5/2.5 nm).

acid (4.0×10^{-3} M), which shows even higher sensitivity and enantioselectivity than in the recognition of mandelic acid. The fluorescence enhancement at the monomer emission is over 80-fold with an *ef* of 64. The fluorescence quantum yield of (R)-**5.34** (1.0×10^{-5} M in benzene/0.4% DME) in the presence of (R)-hexahydromandelic acid is increased to 12.9% and in the presence of (S)-hexahydromandelic acid only to 0.9%.

Hexahydromandelic acid

The interactions of (R)-**5.34** with other chiral acids are further studied and the results are summarized in Table 5.9. The highest sensitivity and enantioselectivity are observed for mandelic acid and hexahydromandelic acid. In spite of the lower enantioselectivity and sensitivity for other chiral acids, the observed fluorescence enhancement and chiral discrimination are still useful for the enantioselective recognition of these compounds.

Table 5.9. Fluorescence responses of (R)-**5.34** toward chiral acids.

Acid	Acid concentration	(R)-**5.34** concentration	I/I_0	$ef \left[= \frac{(I_R - I_0)}{(I_S - I_0)} \right]$
(OH, CO₂H phenyl)	5.0×10^{-4} M	1.0×10^{-5} M (benzene/0.05% DME)	20	46
(OH, CO₂H cyclohexyl)	4.0×10^{-3} M	1.0×10^{-5} M (benzene/0.4% DME)	80	64
(HO, OH with benzyl)	8.0×10^{-3} M	1.0×10^{-5} M (benzene/1% DME)	6	3
(Cbz-NH-CH(Ph)-CO₂H)	8.0×10^{-3} M	1.0×10^{-5} M (benzene/1% DME)	5	7
(Cbz-NH-CH(CH₂Ph)-CO₂H)	8.0×10^{-3} M	1.0×10^{-5} M (benzene/1% DME)	5	2

When (S)-**5.34** is treated with either (R)- or (S)-O-acetyl mandelic acid, there is almost no fluorescence enhancement. Even when acetic acid is used, there is only very small fluorescence enhancement in (S)-**5.34**. This demonstrates that the large fluorescence enhancement observed for (S)-**5.34** in the presence of (S)-mandelic acid or (S)-hexahydromandelic acid is not simply due to the protonation of the nitrogen atoms of (S)-**5.34** by the acid to suppress the photo-induced electron transfer.

In order to understand the much higher fluorescence sensitivity and enantioselectivity of the bisbinaphthyl macrocycle (R)-**5.34** *versus* (S)-**5.22**, the structure of (R)-**5.34** is compared with that of (S)-**5.22**. Figure 5.41 are the space-filling models of the single crystal X-ray structures of (R)-**5.34** than (S)-**5.22**. Both of the compounds contain a pair of parallel naphthalene rings across the macrocycle with a face to face distance of 7–8 Å. However, there is a major difference between these two structures. The distance between the two oxygen atoms on the pair of the *unparallel* naphthalene rings in (R)-**5.34** is 2.80 Å, but that in (S)-**5.22** is 4.63 Å. This difference makes the chiral cavity of (R)-**5.34** like a bucket with a sealed bottom and that of (S)-**5.22** like a tube with an open bottom. A small molecule could pass through the ring of (S)-**5.22** but not that of (R)-**5.34**.

When (R)-**5.34** is treated with the chirality matched (R)-mandelic acid, the hydroxyl and carboxylic acid groups of (R)-mandelic acid could bind with the amine and hydroxyl groups of (R)-**5.34** inside the macrocycle with the phenyl ring of the acid sandwiched in between the two parallel

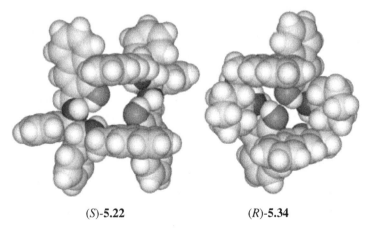

(S)-**5.22** (R)-**5.34**

Figure 5.41. The space-filling models for the X-ray structures of (R)-**5.34** and (S)-**5.22**.

naphthalene rings of (R)-**5.34**. This is supported by the ^{1}H NMR study of (R)-mandelic acid with (R)-**5.34**, which shows a large upfield shift for the methine proton signal of the acid and the formation of a 1:1 complex. It is proposed that this 1:1 complex is structurally much more rigid than the uncomplexed (R)-**5.34** and therefore should be an inherently much stronger fluorophore. However, in the presence of one equivalent of (R)-mandelic acid, there is very little fluorescence enhancement. This can be attributed to the fluorescence quenching by the multiple nitrogen atoms of the macrocycle. Thus, additional binding of the (R)-**5.34** + (R)-mandelic acid 1:1 complex with the acid is necessary in order to suppress the photo-induced electron transfer fluorescence quenching by the uncomplexed nitrogen atoms. Thus, this two-stage interaction of (R)-**5.34** with (R)-mandelic acid leads to the observed large fluorescence enhancement. The mismatched chirality of (R)-**5.34** with (S)-mandelic acid cannot produce a structurally rigid complex, leading to the very low fluorescence enhancement.

When the acyclic compound (S)-**5.35** is treated with either (R)- or (S)-mandelic acid, there is only very small fluorescence enhacement with little enantioselectivty. This is attributed to its flexible and undefined binding sites.

When the macrocycle (R)- or (S)-**5.36** that contains only half of the functional groups of (S)-**5.34** is used to interact with mandelic acid, fluorescence enhancement is observed mainly at the excimer emission of this compound. There is about two- to three-fold fluorescence enhancement with an *ef* of 1.7. Thus, this macrocycle shows much lower fluorescence sensitivity and enantioselectivity than (S)-**5.34**.

^{1}H NMR spectroscopic analysis shows that when (S)-mandelic acid is treated with (S)-**5.36** in benzene-d_6 (2% DME), the methine proton signal of the acid gives a maximum of 0.025 ppm downfield shift. This shift is not only much smaller than the use of (S)-**5.34** ($\Delta\delta \approx 1.1$) but also to the opposite direction. Thus the interaction of (S)-mandelic acid with (S)-**5.36** is very weak and also very different from it with (S)-**5.34**. A molecular modeling structure of (S)-**5.36** is shown in Fig. 5.42 (PC-Spartan-Pro Semi-Emperical-AM1 program). In this structure, the C–O bonds of the four naphthol units in (S)-**5.36** are pointing away from each other. There are no parallel naphthalene rings in (S)-**5.36** that could produce a sandwich complex with the benzene ring of mandelic acid. This leads to the much lower enantioselectivity and fluorescence enhancement of (S)-**5.36** than those of (S)-**5.34**.

Figure 5.42. Molecular modeling structure of (S)-**5.36** (the hydrogen atoms are omitted for clarity).

The 6,6'-p-ethoxyphenyl substituted macrocycle (S)-**5.37** shows over nine-fold fluorescence enhancement at its monomer emission with *ef* up to 2.8 when it is treated with the enantiomers of mandelic acid. Such sensitivity and enantioselectivity of (S)-**5.37**, though useful, are much lower than those of (S)-**5.36**. That is, the introduction of the 6,6'-substituents has significantly changed the sensor–substrate interaction. It is not clear from the molecular model of (S)-**5.37** as to why there is such a decrease in fluorescence enhancement and enantioselectivity.

5.8. Application of the Cyclohexanediamine-BINOL Macrocycle Sensor in Catalyst Screening[31]

The highly enantioselective and sensitive recognition of mandelic acid and hexahydromandelic acid by (R)-**5.34** makes this compound potentially useful for high throughput chiral catalyst screening. In order to use this macrocycle sensor to conduct chiral catalyst screening, we have synthesized a soluble supported mandelic acid. It is found that various alkoxy groups can be incorporated into the para position of mandelic acid to control the solubility of this compound in different solvents. From the reaction of p-hydroxymandelic acid with alkylbromide in the presence of base, the alkylated compounds C6, C18 and C22 are obtained (Scheme 5.20). The solubility of these compounds in various solvents is listed in Table 5.10. Compound C22 has very poor solubility in methanol, diethyl ether, methylene chloride and ethylacetate and is soluble in THF. These properties make compound C22 useful for catalyst screening. Its poor solubility in

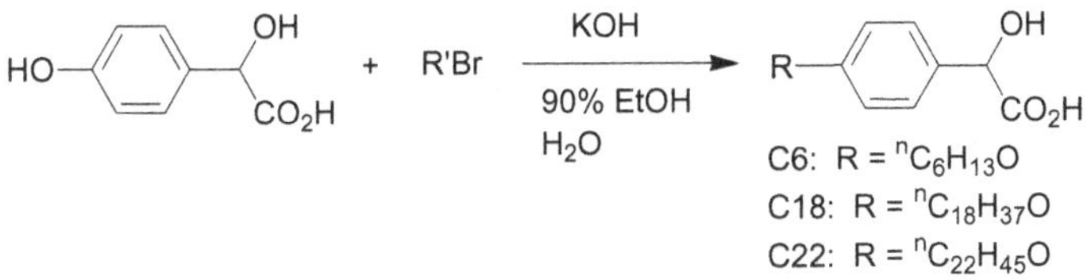

Scheme 5.20. Synthesis of alkylated 4-hydroxy mandelic acids.

C6: R = $^nC_6H_{13}O$
C18: R = $^nC_{18}H_{37}O$
C22: R = $^nC_{22}H_{45}O$

Table 5.10. Solubility of compounds C6, C18 and C12 in various solvents.

Compound		Methanol	Diethyl ether	Methylene chloride	Ethyl acetate	THF
$C_6H_{13}O$–	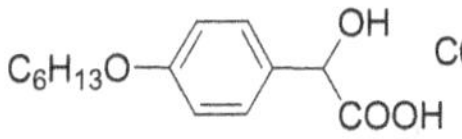 C6	Good	Good	Good	Good	Good
$C_{18}H_{37}O$–	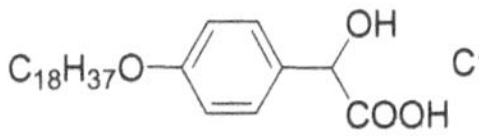 C18	Moderate	Moderate	Good	Good	Good
$C_{22}H_{45}O$–	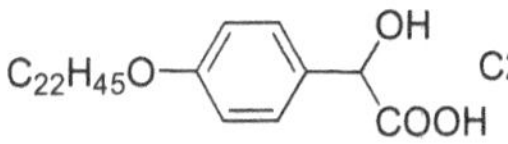 C22	Poor ($< \frac{1\,mg}{9\,mL}$)	Poor	Poor ($< \frac{1\,mg}{10\,mL}$)	Poor	Good

various solvents makes it easy to purify after it is produced from the catalyst screening reactions. Its good solubility in THF will allow it to interact with the fluorescent sensor in solution.

The asymmetric cyanohydrin formation of p-$^nC_{22}H_{25}OC_6H_4CHO$ with TMSCN in the presence of Ti(O^{i}Pr)$_4$ and (+)-/(−)-diisopropyltartrate (DIPT) (70 mol%) is conducted at room temperature (Scheme 5.21). Treatment of the cyanohydrin product with HCl (g) in the presence of methanol generates an α-hydroxy methyl ester, which is soluble in many organic solvents. HPLC analysis (Chiracel OD column) of the α-hydroxy methyl ester shows an ee of 79.5% (90% S enantiomer) from (+)-DIPT, and an ee of 72% (86% R enantiomer) from (−)-DIPT. Treatment of the α-hydroxy methyl ester with KOH in dioxane at 60°C followed by acidification give compound C22 as a precipitate. No racemization is found in this hydrolysis step as determined by converting C22 back to the α-hydroxy methyl ester followed by HPLC analysis.

The cyclohexanediamine-based macrocyclic sensors (S)- and (R)-**5.34** are used to interact with the enantiomerically enriched (S)- and (R)-C22.

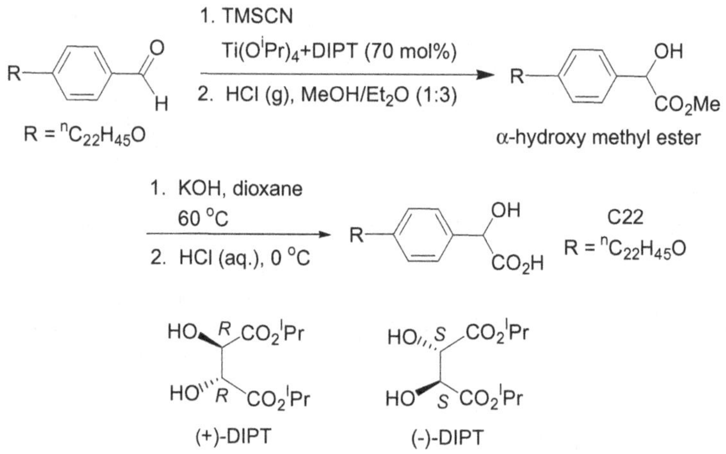

1,1′-Binaphthyl-Based Chiral Materials

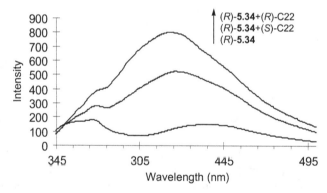

Scheme 5.21. Asymmetric reaction of $p\text{-}^nC_{22}H_{25}OC_6H_4CHO$ with TMSCN in the presence of $(+)\text{-}/(-)\text{-DIPT}$ and $Ti(O^iPr)_4$.

Figure 5.43. Fluorescence spectra of (R)-**5.34** (2.0×10^{-5} M) with (R)- and (S)-C22 (2.0×10^{-4} M) ($\lambda_{\text{exc}} = 332$ nm, slit = 3.5; 6.5 nm).

It is found that a ternary solvent of THF:hexane:benzene (1:5.5:18.5) is necessary for the enantioselective fluorescent recognition. Figure 5.43 shows that in this solvent system, the fluorescence of (R)-**5.34** (2.0×10^{-5} M) responses differently to the two enantiomers of C22. The enantioselective fluorescence enhancement of (R)-**5.34** is observed at its excimer emission rather than at the monomer emission as observed in the presence of (R)-mandelic acid. Thus, the fluorescence response of (R)-**5.34** toward C22 is very different from it toward mandelic acid. This could be attributed to two factors: (a) the long chain alkoxy group of C22 might significantly disturb its interaction with the chiral cavity of the sensor and (b) the use of a significant amount of THF to dissolve C22 might also reduce the specific binding of the sensor (R)-**5.34** with the R enantiomer of the α-hydroxycarboxylic acid, giving reduced enantioselective response. Nevertheless, the observed enantioselectivity of (R)-**5.34** is still useful to determine the ee of C22.

Figure 5.44 shows the relationship of the fluorescence enhancement of (R)-**5.34** in the presence of C22 (2.0×10^{-4} M) at various enantiomeric compositions. A mirror image relationship is observed when (S)-**5.34** is used to interact with C22.

We have used both (R)- and (S)-**5.34** to interact with C22 and established the relationship of ΔI with the ee of C22 (Fig. 5.45).

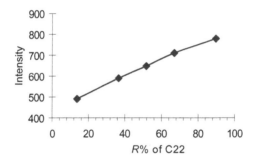

Figure 5.44. Fluorescence responses of (R)-**5.34** (2.0×10^{-5} M) with C22 (2.0×10^{-4} M) at various R compositions ($\lambda_{\text{exc}} = 332$ nm, slit = 3.5; 6.5 nm).

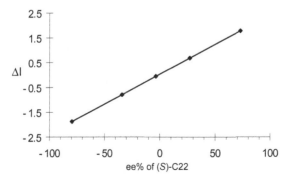

Figure 5.45. The relationship between ΔI and the enantiomeric composition of mandelic acid.

$\Delta I = (I_S/I_{S0}) - (I_R/I_{R0})$. I_S: Fluorescence intensity of (S)-**5.34** in the presence of the acid. I_R: Fluorescence intensity of (R)-**5.34** in the presence of the acid. I_{S0} and I_{R0}: Fluorescence intensity of (R)-/(S)-**5.34** without the acid. Thus, using Fig. 5.45, one should be able to determine the enantiomeric composition of a given sample of C22 from ΔI.

We have studied the conversion of $p\text{-}^nC_{22}H_{25}OC_6H_4CHO$ to C22 by TMSCN addition in the presence of the bisbinaphthyl salen ligands (S)-/(R)-BBS and (+)-/(−)-DIPT in combination with $Ti(O^iPr)_4$ according to Scheme 5.21 under various conditions. In Section 4.5.2 in Chapter 4, we have discussed the use of the bisbinaphthyl salen ligands to catalyze the asymmetric TMSCN addition to aldehydes. As the results summarized in Table 5.11 show, under various conditions the product C22 is obtained in the range of 60–70% yield. The product isolation involves simply centrifugation and filtration because of the insolubility of C22 in the absence of THF. The ten samples of C22 obtained from the above catalyst screening are treated with the fluorescent sensors (R)- and (S)-**5.34** under the same conditions as those in Fig. 5.44. The ee's of the samples are determined according to Fig. 5.45 by measuring ΔI.

Table 5.11. Asymmetric TMSCN addition to $p\text{-}^nC_{22}H_{25}OC_6H_4CHO$ and fluorescent ee determination of the product C22.

Entry	Solvent	Ligand (mol%)	Ti(O^{i}Pr)$_4$ (mol%)	$\Delta I = \left(\frac{I_S}{I_{S0}} - \frac{I_R}{I_{R0}}\right)$	ee (%)
1	CH$_2$Cl$_2$	(+)-DIPT (40)	40	1.92	79
2	CH$_2$Cl$_2$	(−)-DIPT (40)	40	−1.81	−77
3	CH$_2$Cl$_2$	(S)-BBS (10)	8	2.06	84
4	CH$_2$Cl$_2$	(S)-BBS (10)	10	1.33	54
5	CH$_2$Cl$_2$	(S)-BBS (10)	12	0.38	14
6	THF	(S)-BBS (10)	10	0.99	39
7	Ether	(S)-BBS (10)	10	0.11	3
8	Toluene	(S)-BBS (10)	10	0.95	38
9	CH$_2$Cl$_2$	(R)-BBS (10)	10	−1.26	−54
10	CH$_2$Cl$_2$	(R)-BBS (10)	2	−1.31	−56

Table 5.12. Comparison of the ee's from HPLC analysis with those from the fluorescence measurement.

Sample	1	2	3	4	5	6	7	8	9	10
FL	79	−77	84	54	14	39	3	38	−54	−56
HPLC	73	−69	72	49	23	50	8	44	−47	−54

We have also used HPLC-chiral OD column to analyze the ee's of the α-hydroxy methyl esters obtained from the screening experiments in Table 5.11. Table 5.12 compares the HPLC data with those from the fluorescence measurement and a good correlation is observed. Thus, the fluorescence measurement can be used for chiral catalyst screening.

5.9. Using Monobinaphthyl Compounds for the Recognition of α-Hydroxycarboxylic Acids and Amino Acid Derivatives[32]

Besides the development of the nitrogen containing bisbinaphthyl sensors, monobinaphthyl molecules functionalized with amine quenchers are also investigated. As shown in Scheme 5.22, reaction of (S)-3,3′-diformylBINOL with benzylamine followed by reduction with NaBH$_4$ gives (S)-3,3′-bisbenzylaminmethyl BINOL (S)-**5.38**. The specific optical rotation of (S)-**5.38** is $[\alpha]_D = -70.4$ (C$_6$H$_6$, c = 0.215). The UV absorption signals of (S)-**5.38** (2.0 × 10^{-5} M in benzene) are observed at $\lambda_{max}(\varepsilon)$ = 282 (10,300), 294 (sh, 8600), 330 (sh, 7000), 342 (9000) nm. While excited at 320 nm, (S)-**5.38** (1.0 × 10^{-4} M in benzene containing 2% DME) shows emission at 360 nm. It also shows broad excimer emission around 430 nm. The enantiomer (R)-**5.38** is obtained starting from (R)-BINOL.

In the presence of (R)-mandelic acid (2.0 × 10^{-2} M), over 30-fold fluorescence enhancement at the monomer emission is observed for (S)-**5.38**

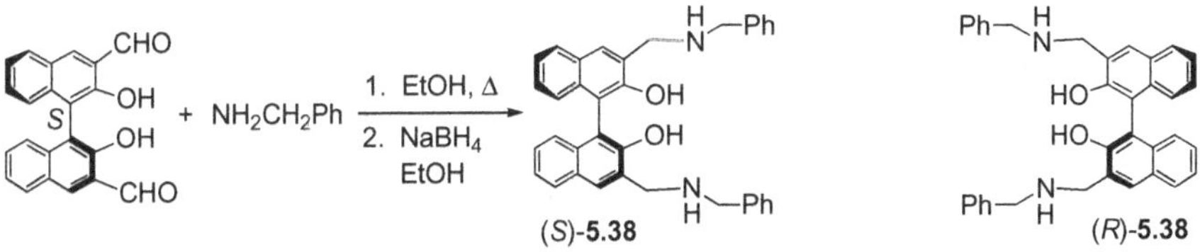

Scheme 5.22. Synthesis of 3,3′-benzylaminomethyl substituted BINOL (S)- and (R)-**5.38**.

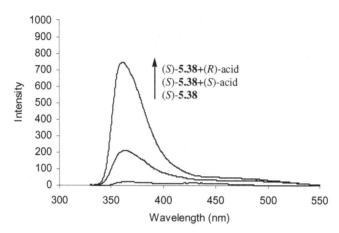

Figure 5.46. Fluorescence spectra of (S)-**5.35** (1.0×10^{-4} M in benzene containing 2% DME) with/without (R)- and (S)-mandelic acid (2.0×10^{-2} M) ($\lambda_{exc} = 320$ nm).

(1.0×10^{-4} M in benzene containing 2% DME) (Fig. 5.46). Under the same conditions in the presence of (S)-mandelic acid, the fluorescence enhancement for (S)-**5.38** is much smaller. Thus, the fluorescent recognition has significant enantioselectivity. The enantiomer (R)-**5.38** gives the mirror image related responses when treated with (R)- and (S)-mandelic acid. Figure 5.47 gives the fluorescence enhancement of (R)-**5.38** in the presence

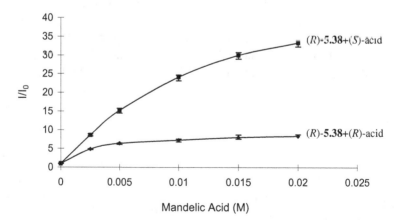

Figure 5.47. Fluorescence enhancement of (R)-**5.38** (1.0×10^{-4} M in benzene/2% DME) versus concentration of (R)- and (S)-mandelic acid (2.5×10^{-3}–2.0×10^{-2} M) ($\lambda_{exc} = 320$ nm).

of (R)- and (S)-mandelic acid at various concentrations. At the acid concentration of 2.0×10^{-2} M, the *ef* is 4.2.

Compound (R)-**5.38** shows enantioselective fluorescence response to the N-protected amino acid D/L-BPG (N-benzyloxycarbonylphenylglycine). At 4.0×10^{-3} M of D-BPG, the fluorescence intensity of (R)-**5.38** (1.0×10^{-4} M in benzene/2% DME) increases over 11.5-fold in the presence of D-BPG and only 6.1-fold by L-BPG. Thus, the *ef* (D/L) is 2.0.

The ^{1}H NMR spectrum of (S)-mandelic acid in the presence of (R)-**5.38** in benzene-d_6 (2% DME) shows upfield shift ($\Delta\delta$ up to 0.5) for its methine proton signal. Although this change in the chemical shift of (S)-mandelic acid is smaller than that in the presence of the cyclohexanediamine-based macrocycle (S)-**5.34**, it is much bigger than those caused by other cyclic and acyclic amines. It indicates that the acid is located quite deep in the chiral cavity generated by one or more of the binaphthyl molecules. The Job plot obtained from the ^{1}H NMR study indicates a 2:3 acid/(R)-**5.38** complex as well as other binding modes. In the presence of (R)-**5.38**, the ^{1}H NMR spectrum of D-BPG (R) also shows an upfield shift ($\Delta\delta$ up to 0.025) for its methine signal. The Job plot shows the formation of a 4:1 acid/(R)-**5.38** complex and other binding modes. Thus, the interaction of D-BPG with (R)-**5.38** is very different from that of (S)-mandelic acid with (R)-**5.38**.

(R)-α-Methylbenzylamine is used to condense with (R)- and (S)-3,3'-diformylBINOL which upon reduction with NaBH$_4$ gives compounds (R)-**5.39** and (S)-**5.40**, respectively. The specific optical rotation of (R)-**5.39** is $[\alpha]_D = +19.3$ (C$_6$H$_6$, c = 0.19) and that of (S)-**5.40** is $[\alpha]_D = -18.5$ (C$_6$H$_6$, c = 0.205). The chirality of the BINOL unit in these compounds controls the sign of the optical rotation. The additional chiral carbon centers in these two compounds have significantly reduced the magnitude of the optical rotation in comparison with that of (R)- and (S)-**5.38**. The fluorescence spectra of both (R)-**5.39** and (S)-**5.40** in benzene (2% DME) show emission at 365 nm while excited at 320 nm.

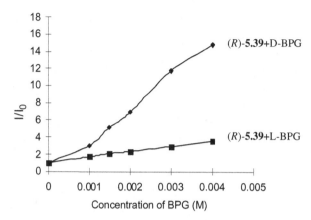

Figure 5.48. Fluorescence enhancement of (R)-**5.39** (1.0×10^{-4} M in benzene/2% DME) *versus* concentration of D/L-BPG ($\lambda_{exc} = 320$ nm).

When (R)-**5.39** (1.0×10^{-4} M in benzene/2% DME) is treated with (R)-mandelic acid (2.0×10^{-2} M), there is a 4.8-fold fluorescence enhancement. Using (S)-mandelic acid leads to a 3.0-fold fluorescence enhancement, that is, *ef* is 1.9. Both the fluorescence sensitivity and enantioselectivity of (R)-**5.39** are much lower than those of (R)-**5.38** in the recognition of mandelic acid. Their enantioselectivities are also opposite. When (R)-**5.39** is used to interact with D-/L-BPG, its enantioselectivity is significantly higher than that of (R)-**5.38**. As shown in Fig. 5.48, the fluorescence of (R)-**5.39** increases to 14.8-fold in the presence of D-BPG at 4.0×10^{-3} M while only to 3.6-fold in the presence of L-BPG at the same concentration. This corresponds to an *ef* of 5.1.

Compound (S)-**5.40** shows smaller fluorescence enhancement and enantioselectivity in the recognition of both mandelic acid and BPG. When (S)-**5.40** (1.0×10^{-4} M in benzene/2% DME) is treated with (S)-mandelic acid, it shows up to three-fold fluorescence enhancement with *ef* (S/R) = 1.4. D-BPG increases the fluorescence of (S)-**5.40** up to 4.8-fold with *ef* (D/L) = 1.9. In the recognition of the amino acid BPG, both (R)-**5.39** and (S)-**5.40** show greater fluorescence enhancement with D-BPG than with L-BPG. This demonstrates that the chiral amine centers in these compounds are more important than the chirality of the BINOL unit. However, (R)-**5.39** and (S)-**5.40** have the opposite enantioselectivity when they are used to recognize mandelic acid. (R)-Mandelic acid is the better enhancer for (R)-**5.39**, whereas (S)-mandelic acid is the better enhancer

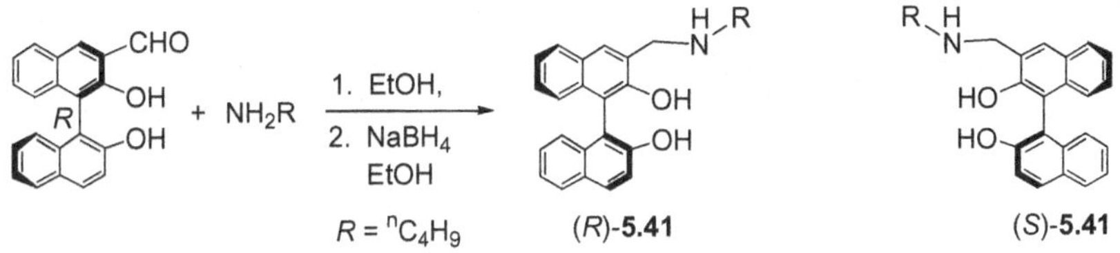

[Scheme 5.23 illustration]

Scheme 5.23. Synthesis of 3-alkylaminomethyl substituted BINOL (*R*)- and (*S*)-**5.41**.

for (*S*)-**5.40**. Thus, in the case of mandelic acid recognition, the chirality of the BINOL unit is more important than that of the chiral carbon centers.

The monoaminomethyl-substituted compounds (*R*)- and (*S*)-**5.41** are prepared from the reaction of 3-formylBINOL with butylamine followed by reduction (Scheme 5.23). In the presence of mandelic acid (2.5×10^{-3}–2.0×10^{-2} M), there is over ten-fold fluorescence enhancement for (*R*)- or (*S*)-**5.41** (1.0×10^{-4} M in benzene/1.25% DME) at $\lambda_{emi} = 363$ nm ($\lambda_{exc} = 336$ nm). However, the enantioselectivity is low with an *ef* of less than 1.4.

5.10. Enantioselective Precipitation and Solid-State Fluorescence Enhancement in the Recognition of α-Hydroxycarboxylic Acids by Using Monobinaphthyl Compounds[33]

In collaboration with Hou, we have used the reaction of (*S*)-3,3'-diformylBINOL with (1*R*,2*S*)-2-amino-1,2-diphenylethanol followed by reduction with NaBH$_4$ to prepare the BINOL-amino alcohol (*S*)-**5.42** (Scheme 5.24). The specific optical rotation $[\alpha]_d$ of this compound is −24.5

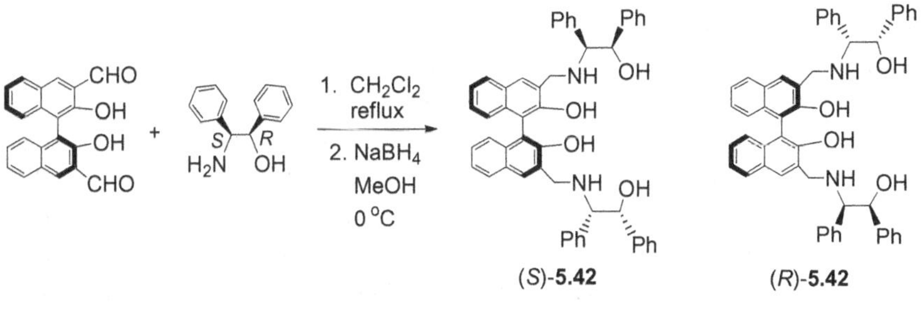

[Scheme 5.24 illustration]

Scheme 5.24. Preparation of the BINOL-amino alcohol compounds (*S*)- and (*R*)-**5.42**.

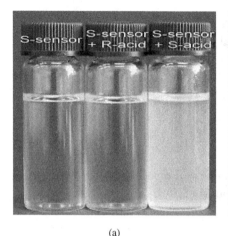 

(a) (b)

Figure 5.49. The images of (S)-**5.42** (a) and (R)-**5.42** (b) at 5.0×10^{-4} M with (R)- and (S)-mandelic acid (4.0×10^{-3} M) in benzene/0.4% v DME.

($c = 1.15$, CH$_2$Cl$_2$). The enantiomer (R)-**5.42** is prepared from (R)-3,3′-diformylBINOL and (1S,2R)-2-amino-1,2-diphenylethanol.

The BINOL-amino alcohol (S)-**5.42** is used to interact with the enantiomers of mandelic acid. In benzene (0.4% v DME) solution, (S)-**5.42** (5.0×10^{-4} M) forms a white precipitate with (S)-mandelic acid ($\geq 3.0 \times 10^{-3}$ M) (Fig. 5.49(a)). However, when (S)-**5.42** is treated with (R)-mandelic acid (3.0×10^{-3}–8.0×10^{-3} M), the solution remains to be clear. This enantioselective precipitation is confirmed with the use of (R)-**5.42** to interact with (R)- and (S)-mandelic acid. In this case, (R) **5.42** forms precipitate with (R)-mandelic acid but not with (S)-mandelic acid (Fig. 5.49(b)). Although precipitate formation is one of the most convenient visual methods in the classical chemical analysis, enantioselective precipitation was never developed as an analytical tool for chiral recognition. The observed enantioselective precipitation provides a new direction for the development of visual methods for chiral analysis.

Filtration of the (S)-**5.42**+(S)-mandelic acid suspension leads to the isolation of the precipitate. This precipitate can be dissolved in acetone-d_6 and the ^{1}H NMR spectrum of this solution is obtained. It shows that the ratio of (S)-**5.42** *versus* (S)-mandelic acid is 1:4. This ratio is found to be independent of the original ratio of (S)-**5.42**/(S)-mandelic acid. Powder X-ray diffraction analysis of the (S)-**5.42**+(S)-mandelic acid precipitate indicates that it has certain crystalline order. The X-ray diffraction pattern

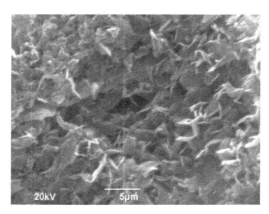

Figure 5.50. The SEM image of the precipitate generated from (S)-**5.42** (5.0×10^{-4} M) and (S)-mandelic acid (4.0×10^{-3} M).

of the (S)-**5.42**+(S)-mandelic acid precipitate is different from that of the crystalline (S)-mandelic acid. Compound (S)-**5.42** alone is found to be amorphous. The (S)-**5.42**+(S)-mandelic acid precipitate has a porous structure as shown by its scanning electron microscopy (SEM) image (Fig. 5.50). The X-ray analysis and SEM study demonstrates that the self-assembly of (S)-**5.42** with (S)-mandelic acid forms an ordered structure very different from the closely packed crystals of (S)-mandelic acid.

The fluorescence spectrum of (S)-**5.42** in benzene solution displays emissions at $\lambda = \sim 370$ (sh) and 439 nm. Similar to the other BINOL molecules we have studied, the short wavelength signal is attributed to its monomer emission and the long wavelength signal to the excimer emission. Addition of methanol to the benzene solution of (S)-**5.42** leads to increase in the ratio of the excimer emission *versus* the monomer emission while the UV absorption spectrum does not change. This indicates that the excimer of (S)-**5.42** is probably formed between the less polar aromatic rings of the molecules. Increasing the polarity of the solvent should increase the interaction of the polar OH and NH groups of (S)-**5.42** with the solvent and push the less polar aromatic rings together. A similar increase in excimer emission is also observed when DME is added to the benzene solution of the macrocycle (S)-**5.22** as described in Section 5.6.1. In the concentration range of 10^{-3}–10^{-7} M in benzene, the excimer emission of (S)-**5.42** dominates.

When (S)-**5.42** is treated with (S)-mandelic acid, the resulting precipitate shows a large fluorescence enhancement at the monomer emission

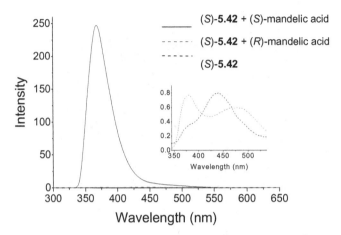

Figure 5.51. The fluorescence spectra of (S)-**5.42** (5.0×10^{-4} M) with (R)- and (S)-mandelic acid (4.0×10^{-3} M) in benzene (0.4% v DME) ($\lambda_{exc} = 341$ nm, slit $= 5.0/5.0$ nm).

(Fig. 5.51). As shown in Fig. 5.51, while the fluorescence intensities of (S)-**5.42** and (S)-**5.42** + (R)-mandelic acid are very close to the baseline, the fluorescence intensity of (S)-**5.42** + (S)-mandelic acid at $\lambda_{emi} = 367$ nm is increased by over 950-fold! When (S)-**5.42** is used, the same fluorescence enhancement is observed in the presence of (R)-mandelic acid while there is only a small change with (S)-manedlic acid. This demonstrates that the enantioselective precipitation is associated with highly enantioselective fluorescence enhancement.

The solution shows diminished fluorescence after the (S)-**5.42** + (S)-mandelic acid precipitate is separated. The fluorescence microscopy image of the (S)-**5.42** + (S)-mandelic acid precipitate is obtained (Fig. 5.52). It shows that the solid particles of the (S)-**5.42** + (S)-mandelic acid precipitate are strongly fluorescent. The particles of (S)-mandelic acid are non-fluorescent and those of (S)-**5.42** have weaker fluorescence.

In the study of light-emitting materials, it is generally observed that the solution of a compound or polymer exhibits much stronger fluorescence than when they are in the solid state.[34] This is attributed to the much stronger intermolecular interaction in the solid state, which leads to self-quenching and excimer/exciplex formation. Recently, fluorescence enhancement has been observed in inclusion complexes and gels.[35] However, no enantioselective solid-state fluorescence enhancement was observed before. According to the NMR study of the (S)-**5.42**+(S)-mandelic acid precipitate, one

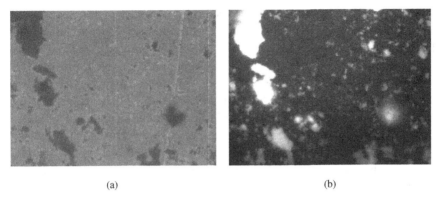

Figure 5.52. The topography image (a) and the fluorescence image (b) of the precipitate generated from (S)-**5.42** (5.0×10^{-4} M) and (S)-mandelic acid (4.0×10^{-3} M).

molecule of (S)-**5.42** is associated with four molecules of (S)-mandelic acid in the solid state. The solid particles show predominantly the monomer emission of (S)-**5.42**. On the basis of these results and the observations by X-ray diffraction, SEM and fluorescence microscopy, we propose that there is a highly organized self-assembly in the (S)-**5.42**+(S)-mandelic acid precipitate, and each molecule of (S)-**5.42** is separated by an excess amount of (S)-mandelic acid molecules. Thus, there should be no intermolecular interaction between the molecules of (S)-**5.42** to form excimer upon excitation. In the precipitate, the nitrogen atoms of (S)-**5.42** are expected to form hydrogen bonds with the acidic proton of (S)-mandelic acid. This should suppress the photo-induced electron transfer fluorescence quenching by the nitrogen lone-pair electrons, leading to the fluorescence enhancement of (S)-**5.42**. In addition, the complexation of (S)-**5.42** with (S)-mandelic acid in the solid state should also restrict the motion of this molecule and make it a stronger fluorophore than it in solution.

When (S)-**5.42** is treated with the enantiomers of hexahydromandelic acid under the same conditions as the use of mandelic acid, it also forms precipitate with (S)-hexahydromandelic acid but not with (R)-hexahydromandelic acid. Strong fluorescence enhancement is also observed for the (S)-**5.42**+(S)-hexahydromandelic acid precipitate but not for the (S)-**5.42**+(R)-hexahydromandelic acid solution. Thus, (S)-**5.42** is useful as an enantioselective sensor for both aromatic and aliphatic α-hydroxycarboxylic acids.

Compound (*S*)-**5.43** is prepared as a diastereomer of (*S*)-**5.42** from the condensation of (*S*)-3,3′-diformylBINOL with (1*S*,2*R*)-2-amino-1,2-diphenylethanol followed by reduction. When (*S*)-**5.43** is interacted with the enatiomers of mandelic acid under the same conditions as the use of (*S*)-**5.42**, no precipitate formation is observed. This indicates that matching the chiral configuration of the BINOL and amino alcohol units is important for the chiral discrimination. Compound (*S*)-**5.44** is a methylated derivative of (*S*)-**5.42**. This compound does not form precipitate with the enantiomers of mandelic acid. The hydrogen bonding of the hydroxyl groups in the amino alcohol units of (*S*)-**5.42** should be involved in the enantioselective precipitation.

(*S*)-**5.43**   (*S*)-**5.44**

References

1. Pu L, *Chem Rev* **104**:1687–1716, 2004.
2. James TD, Sandanayake KRAS, Shinkai S, *Nature* **374**:345–347, 1995.
3. Grady T, Harris SJ, Smyth MR, Diamond D, Hailey P, *Anal Chem* **68**:3775, 1996.
4. Pagliari S, Corradini R, Galaverna G, Sforza S, Dossena A, Marchelli R, *Tetrahedron Lett* **41**:3691, 2000.
5. Abe Y, Fukui S, Koshiji Y, Kobayashi M, Shoji T, Sugata S, Nishizawa H, Suzuki H, Iwata K, *Biochim Biophys Acta* **1433**:188, 1999.
6. Hembury GA, Borovkov VV, Inoue Y, *Chem Rev* **108**:1–73, 2008.
7. Mei XF, Wolf C, *J Am Chem Soc* **128**:13326–13327, 2006.
8. Heo J, Mirkin CA, *Angew Chem Int Ed* **45**:941–944, 2006.
9. (a) Zhu L, Zhong ZL, Anslyn EV, *J Am Chem Soc* **127**:4260–4269, 2005. (b) Dai ZH, Xu XD, Canary JW, *Chirality* **17**:S227–S233, 2005.
10. Reetz MT, Sostmann S, *Tetrahedron* **57**:2515, 2001.
11. Hu Q-S, Pugh V, Sabat M, Pu L, *J Org Chem* **64**:7528–7536, 1999.
12. Pugh V, Hu Q-S, Pu L, *Angew Chem Int Ed* **39**:3638–3641, 2000.
13. Pugh V, Hu Q-S, Zuo X-B, Lewis FD, Pu L, *J Org Chem* **66**:6136–6140, 2001.

14. (a) Newkome GR, Moorefield CN, Vögtle F, *Dendritic Macromolecules: Concepts, Syntheses, Perspectives*, VCH, Weinheim, Germany, 1996. (b) Fischer M, Vögtle F, *Angew Chem Int Ed* **38**:884, 1999.
15. Tomalia DA, Durst HD, *Top Curr Chem* **165**:193, 1993.
16. Fréchet JMJ, *Science* **263**:1710, 1994.
17. (a) Zeng F, Zimmerman SC, *Chem Rev* **97**:1681, 1997. (b) Smith DK, Diederich F, *Chem Eur J* **4**:1353, 1998. (c) Gorman C, *Adv Mater* **10**:295, 1998.
18. Devadoss C, Bharathi P, Moore JS, *J Am Chem Soc* **118**:9635, 1996.
19. Iwanek W, Mattay J, *J Photochem Photobiol A Chem* **67**:209–226, 1992.
20. Gong L-Z, Hu Q-S, Pu L, *J Org Chem* **66**:2358–2367, 2001.
21. Miller TM, Neenan TX, Zayas R, Bair HE, *J Am Chem Soc* **114**:1018–1025, 1992.
22. Gong L-Z, Pu L, *Tetrahedron Lett* **42**:7337–7340, 2001.
23. Wang Q, Chen X, Tao L, Wang L, Xiao D, Yu X-Q, Pu L, *J Org Chem* **72**:97–101, 2007.
24. Lin J, Hu Q-S, Xu MH, Pu L, *J Am Chem Soc* **124**:2088–2089, 2002.
25. Xu M-H, Lin J, Hu Q-S, Pu L, *J Am Chem Soc* **124**:14239–14246, 2002.
26. Lin J, Zhang H-C, Pu L, *Org Lett* **4**:3297–3300, 2002.
27. Li Z-B, Lin J, Zhang H-C, Sabat M, Hyacinth M, Pu L, *J Org Chem* **69**:6284–6293, 2004.
28. Li Z-B, Pu L, *J Mater Chem* **15**:2860–2864, 2005.
29. (a) Brunner H, Schiessling H, *Angew Chem Int Ed Engl* **33**:125, 1994. (b) Brunner H, Schiessling H, *Bull Soc Chim Belg* **103**:119, 1994.
30. (a) Li Z-B, Lin J, Pu L, *Angew Chem Int Ed* **44**:1690–1693, 2005; *Angew Chem* **117**:1718–1721, 2005. (b) Li Z-B, Lin J, Sabat M, Hyacinth M, Pu L, *J Org Chem* **72**:4905–4916, 2007.
31. Li Z-B, Lin J, Qin Y-C, Pu L, *Org Lett* **7**:3441–3444, 2005.
32. Lin J, Li Z-B, Zhang H-C, Pu L, *Tetrahedron Lett* **45**:103–106, 2004.
33. Liu H-L, Hou X-L, Pu L, *Angew Chem* **48**:382–385, 2009.
34. Fan X, Sun J-L, Wang F-Z, Chu Z-Z, Wang P, Dong Y-Q, Hu RR, Tang B-Z, Zou D-C, *Chem Commun* 2989–2991, 2008, and references therein.
35. (a) Yoshida K, Uwada K, Kumaoka H, Bu L, Watanabe S, *Chem Lett* **30**:808, **2001**. (b) Ooyama Y, Yoshida K, *New J Chem* **29**:1204–1212, 2005. (c) Bua L, Sawadaa T, Kuwaharaa Y, Shosenjib H, Yoshida K, *Dyes Pigments* **59**:43–52, 2003. (d) Leong WL, Tam AY-Y, Batabyal SK, Koh LW, Kasapis S, Yam VW-W, Vittal JJ, *Chem Commun* 3628–3630, 2008. (e) Leong WL, Batabyal SK, Kasapis S, Vittal JJ, *Chem Eur J* **14**:8822–8829, 2008.

Chapter 6

Miscellaneous Studies on Materials Related to 1,1′-Binaphthyls

6.1. Chiral Molecular Wires[1]

Oligo-phenylene-ethynylene-dithiol (OPE) molecules can undergo self-assembly on the surface of gold. This allows the study of these conjugated molecules as molecular wires for electron transport. We have incorporated the chiral 1,1′-binaphthyl units into the OPE molecules to construct the unprecedented chiral molecular wires.

As shown in Scheme 6.1, the bromine atoms of (S)-6,6′-dibromo-2,2′-diethoxy-1,1′-binaphthyl are replaced with two less reactive chlorine atoms by treatment with ^{n}BuLi and C_2Cl_6. The resulting 6,6′-dichloro compound (S)-**6.1** then reacts with bromine to give the 4,4′-dibrominated compound (S)-**6.2**. Since the 4,4′-bromine atoms in (S)-**6.2** are more reactive than the 6,6′-chloroine atoms, the coupling of (S)-**6.2** with p-ethynyl-benzenethiol acetate in the presence of $Pd(PPh_3)_4$ and CuI gives the chiral OPE analog (S)-**6.3**. The specific optical rotation $[\alpha]_D$ of (S)-**6.3** is −60.1 (c = 1.0, CH_2Cl_2). The enantiomer (R)-**6.3** is prepared by using the (R)-binaphthyl starting material. The distance between the two ending sulfur atoms in these compounds is calculated to be 2.4 nm.

On the basis of our studies on the binaphthyl-based materials, we expect that the conjugation in (S)-**6.3** should be determined by half of the molecule and there should be little conjugation across the 1,1′-bond of the binaphthyl unit. When the UV spectrum of (S)-**6.3** is compared with that of the diacetyl OPE molecule **6.4**, compound (S)-**6.3** exhibits absorptions at longer wavelengths (Fig. 6.1). This indicates that a substituted naphthalene

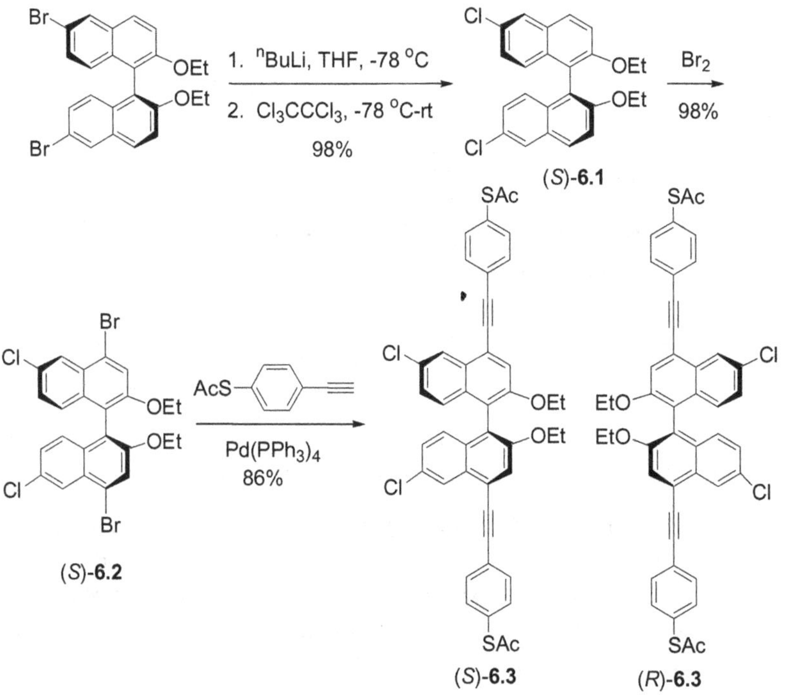

Scheme 6.1. Synthesis of chiral OPE analog (S)- and (R)-**6.3**.

unit in (S)-**6.3** provides more effective conjugation than a diphenylethylene unit in **6.4**.

Figure 6.2 gives the fluorescence spectra of (S)-**6.3** and **6.4**. The binaphthyl compound (S)-**6.3** emits at a longer wavelength than the OPE molecule **6.4**. In methylene chloride solution, while excited at 330 nm, (S)-**6.3** gives emission at 439 nm. The emission maximums of **6.4** are at 376, 400 and 424 (sh) nm while excited at 334 nm. The fluorescence spectrum of compound **6.4** shows more vibrational structure.

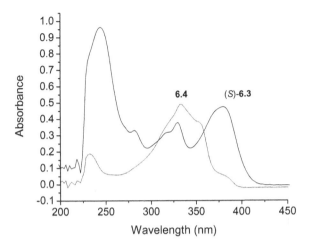

Figure 6.1. UV spectra of (S)-**6.3** and **6.4** in methylene chloride (10^{-6} M).

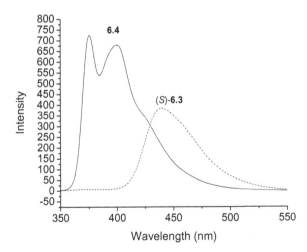

Figure 6.2. Fluorescence spectra of (S)-**6.3** and **6.4** in methylene chloride (10^{-5} M).

Figure 6.3 gives the CD spectra of the two enantiomers (S)- and (R)-**6.3** in solution, which are mirror images of each other. The bisigned signal centered at ∼250 nm is attributed to the exciton coupling in the binaphthyl unit and the long wavelength signal is due to the extended conjugation with the incorporation of the phenyleneethynylene units.

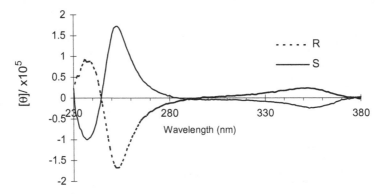

Figure 6.3. Circular dichroism spectra of (*S*)- and (*R*)-**6.3** in methylene chloride (10^{-5} M).

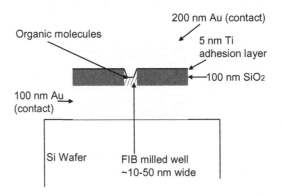

Figure 0.4. A nanowell device.

The chiral conjugated molecules (*S*)- and (*R*)-**6.3** are incorporated into a nanowell device as shown in Fig. 6.4. This device contains a silicon wafer patterned with gold and covered with silicon dioxide. A well is milled down through the silicone dioxide layer by using a focused ion beam. Compound (*S*)- or (*R*)-**6.3** is dissolved in ethanol in the presence of sulfuric acid to remove the acetyl-protecting groups. Then, the nanowell chip is placed in the solution of the chiral conjugated molecule for a minimum of 48 h. A monolayer of the compound is expected to self-assemble on the gold surface in the nanowell through the bonding between the ending sulfur atom and gold (Fig. 6.5). Then, titanium and gold are deposited on top of the nanowell by evaporation. A voltage is applied between the top and bottom gold layers

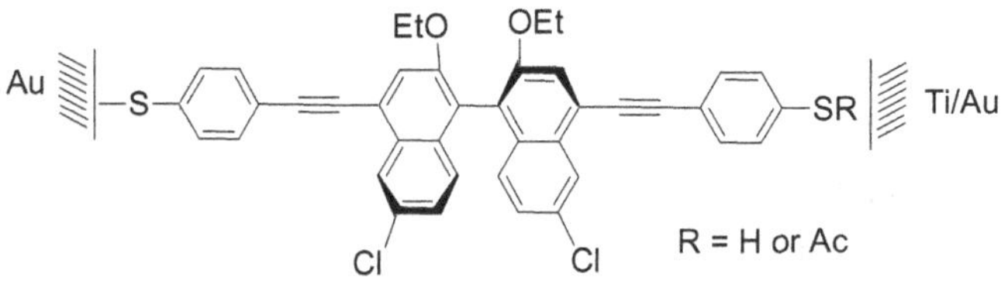

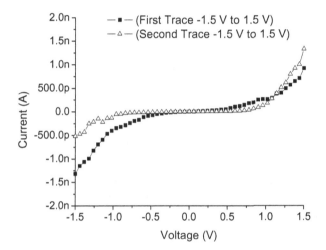

Figure 6.5. An optically active molecular electronic wire prepared from (S)-**6.3** in a nanowell.

Figure 6.6. I–V plot of (R)-**6.3** in nanowell.

of the device to probe the electrical transporting properties of the chiral molecular wire.

Figure 6.6 gives an example of the current–voltage (I–V) curves of the nanowell devices made of the self-assembled (R)-**6.3**. When the voltage is 2.5 V and over, there is permanent electrical breakdown of the device. At 1.5 V, the median current value is 1.3 nA. When (S)-**6.3** is used, similar I–V curves are obtained. The median current for the devices containing either (R)- or (S)-**6.3** is about one order of magnitude smaller than the nanowell devices using the self-assembled simple OPE molecules. This indicates that the non-planar binaphthyl unit in these chiral molecular wires has reduced their conductivity.

Samples with various enantiomeric compositions of (S)- and (R)-**6.3** are also used to prepare the nanowell devices and their I–V curves are

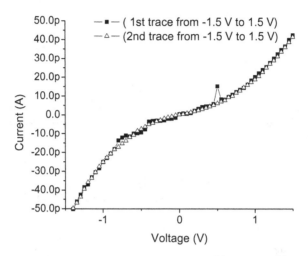

Figure 6.7. I–V plot of the devices containing 80% R and 20% S in nanowell.

obtained (Fig. 6.7). The conductivities of these devices are found to be significantly lower than those made of the enantiomerically pure (R)- or (S)-**6.3**. The packing of the homochiral (R)- or (S)-**6.3** in the self-assembled monolayer should be quite different from that of the materials containing both enantiomers. This could contribute to their difference in conductivity.

6.2. A Biphenol Polymer[2]

2,2' Biphenol is an analog of BINOL but with very low rotation barrier around the biaryl bond. The chiral conformations of 2,2'-biphenol interconvert rapidly. Polymerization of a substituted biphenol has been conducted. As shown in Scheme 6.2, bromination of 2,2'-biphenol followed by reaction with acetic anhydride gives compound **6.5**. In **6.5**, the two bromine atoms at the 5,5'-positions are more reactive than the two at the 3,3'-positions because of less steric hindrance. Treatment of **6.5** with 1-octyne in the presence of Pd(PPh$_3$)$_2$Cl$_2$ and CuI gives **6.6**. Introduction of the alkynyl groups at the 5,5'-positions of the monomer not only protects these two positions but can also render the subsequent polymer soluble in organic solvents. Polymerization of **6.6** with Ni(COD)$_2$ followed by hydrolysis gives the poly(o-phenol) **6.7** as a brown solid. This polymer is soluble in common organic solvents such as chloroform, methylene chloride and THF.

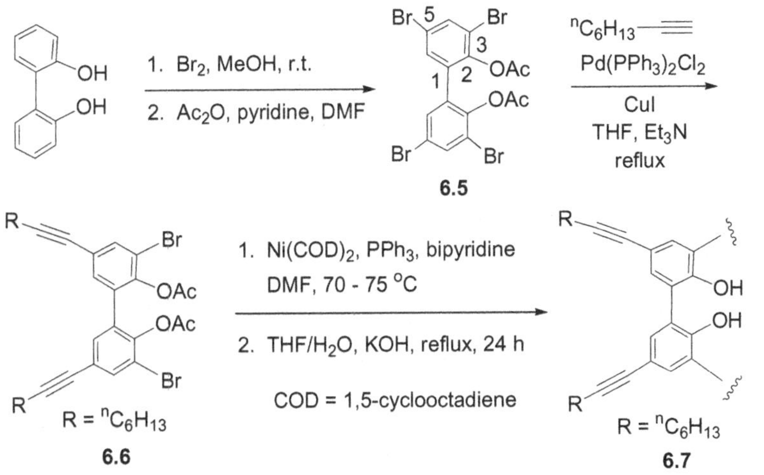

Scheme 6.2. Polymerization of a 2,2'-biphenol monomer to generate a poly(o-phenol).

GPC shows the molecular weight of the polymer as $M_w = 10{,}400$ and $M_n = 4500$ (PDI = 2.3).

In methylene chloride solution, polymer **6.7** shows UV absorptions at $\lambda_{max} = 246$ ($\varepsilon = 30{,}100$) and 302 (sh, $\varepsilon = 8100$) nm. The longest wavelength absorption of the polymer matches very well with that calculated for *p*-hydroxyphenylacetylene. This indicates that there is almost no extended conjugation beyond the *p*-hydoxy-alkynylphenylene unit. Although the IR spectrum of **6.7** shows the disappearance of the original ester carbonyl stretch at $1762\,\text{cm}^{-1}$ in the monomer, a broad absorption at $\sim 1700\,\text{cm}^{-1}$ is observed. This is attributed to a possible partial oxidation of the phenol units to quinone in the polymer chain. The ^{1}H and ^{13}C NMR spectra of the polymer give well-defined alkyl signals, but the aromatic and alkynyl carbon signals are broad and weak.

Polymer **6.7** is used to catalyze the reaction of phenylacetylene with benzaldehyde in the presence of Ti(O^{i}Pr)$_4$ and ZnEt$_2$. The reaction is conducted by treating the polymer with Ti(O^{i}Pr)$_4$ followed by combination with phenylacetylene, ZnEt$_2$ and benzaldehyde at room temperature. It is found that when 1/4 equivalent of Ti(O^{i}Pr)$_4$ (relative the phenol unit in the polymer) is used, the catalytic activity is much higher than when one equivalent of Ti(O^{i}Pr)$_4$ is used. In the presence of 20 mol% of the polymer (the phenol unit *versus* benzaldehyde) and 5 mol% of Ti(O^{i}Pr)$_4$, the product 1,3-diphenyl-2-propyn-1-ol is obtained in 71% yield after 80 h at room temperature. Under the same conditions when 100 mol% of Ti(O^{i}Pr)$_4$ is used, only 16% yield of the product is obtained. The product is also observed in low yields when the polymer and/or Ti(O^{i}Pr)$_4$ are not used.

It is proposed that when the polymer is treated with 1/4 equivalent of Ti(O^{i}Pr)$_4$, it could generate a Ti(IV) complex such as **6.8** where each Ti center can be coordinated with a tetradentated phenoxide ligand. Such a type of BINOL-based ligands in combination with Ti(IV) has been found to be an efficient Lewis acid catalyst in other reactions. When one equivalent of Ti(O^{i}Pr)$_4$ is used, complex **6.9** that contains Ti centers with only monodentated phenoxide ligands could be generated. This complex should have a catalytic property very different from that of **6.8**. Polymer **6.7** can also have two types of conformations as shown by **6.10** and **6.11** in which each o-biphenol unit adopts either a cisoid or a transoid conformation. The interconversion of these conformations and their optical and induced chiral optical properties could be useful for sensor development.

6.3. Supramolecular Chemistry of Self-Assembly of Racemic and Optically Active Propargylic Alcohols[3–5]

As described in Section 4.2 of Chapter 4, BINOL-based catalysts have been developed for the asymmetric synthesis of a variety of propargylic alcohols. In collaboration with Sabat, we have found that both the racemic and enantiomerically enriched propargylic alcohols have exhibited interesting supramolecular structures. Particularly, we have focused on the study of the supramolecular chemistry of the 3-aryl-1-pentafluorophenyl substituted propargylic alcohols. The supramolecular assembly of the propargylic alcohols is potentially useful in the construction of novel materials.

6.3.1. Phenyl- and Pentafluorophenyl-Substituted Propargylic Alcohols

As shown in Scheme 6.3, racemic 1-pentafluorophenyl-3-phenyl-2-propyn-1-ol, *rac*-**6.12**, is prepared from the reaction of phenylacetylene and pentafluorobenzaldehyde in the presence of ZnEt$_2$ and HMPA at room temperature. A single crystal of *rac*-**6.12** is obtained and its X-ray analysis shows a supramolecular hexameric structure (Fig. 6.8). The hexamer contains alternating *R* and *S* enantiomers linked mainly by the O–H···O hydrogen bonds. The intermolecular phenyl and pentafluorophenyl rings are almost exactly parallel to each other separated by 3.52 Å between the two planes. There are also intermolecular C–H···F hydrogen bonds between a *ortho*-H of the phenyl and a *ortho*-F of the pentafluorophenyl.

The hexamers of *rac*-**6.12** stack on top of each other with alternating phenyl–pentafluorophenyl π–π interaction along the unit cell axis to form

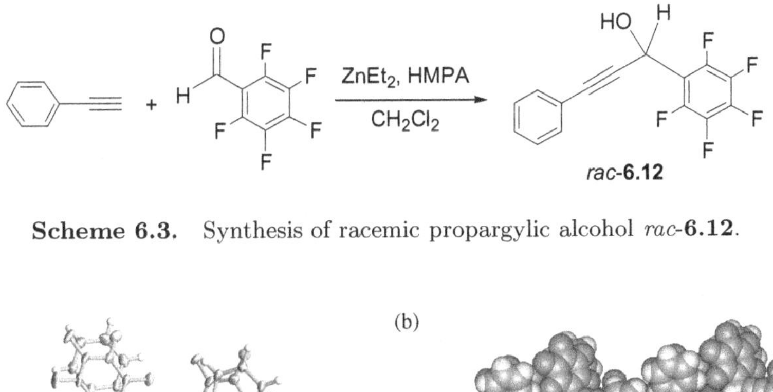

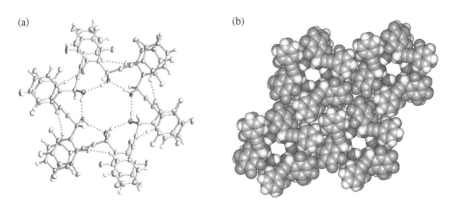

Scheme 6.3. Synthesis of racemic propargylic alcohol *rac*-**6.12**.

(a)   (b)

Figure 6.8. (a) Network of O–H···O hydrogen bonds, face-to-face π–π stacking interactions, and C–H···F hydrogen bonds, cooperatively forming the hexameric structure of *rac*-**6.12**. (b) Channels formed along the crystallographic axis *a*.

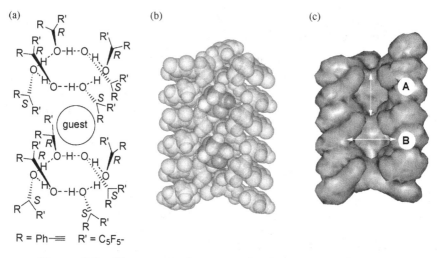

Figure 6.9. The cages in the supramolecular structure of *rac*-**6.12**.

infinite channels. Figure 6.9(a) depicts the stacking of two hexamers and a solvent molecule, such as dioxane and methylene chloride, is found to be trapped inside the cage between the two hexamers. Figure 6.9(b) shows a cross section of one channel in the crystal structure of *rac*-**6.12** with two cage-trapped molecules of dioxane. The dimensions of the cages are given in Fig. 6.9(c) with diameter A of 7.85 Å and diameter B of 8.35 Å for each cage.

The optically active (S)-**6.12** is prepared from the reaction of phenylacetylene with pentafluorobenzaldehyde in the presence of (S)-BINOL, Et_2Zn, $Ti(O^iPr)_4$ and HMPA at room temperature, which gives (S)-**6.12** with 88% *ee* (Scheme 6.4). However, this compound has not produced single crystals that are suitable for X-ray analysis yet.

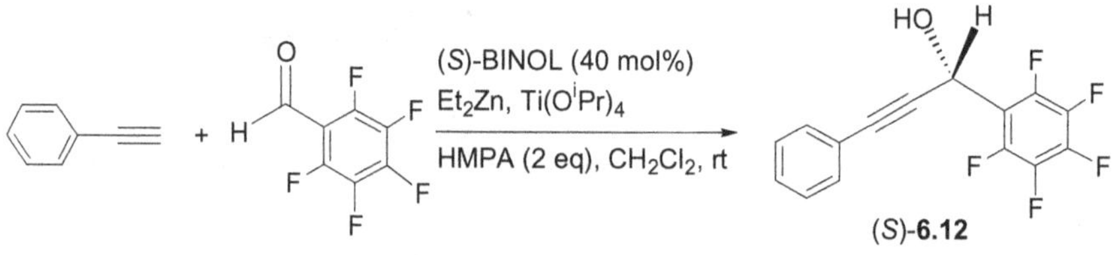

Scheme 6.4. Asymmetric synthesis of enantiomerically enriched (S)-**6.12**.

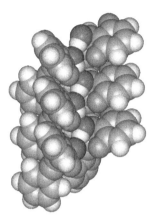

Figure 6.10. Space-filling representation for the supramolecular structure of rac-**6.13**.

Compound rac-**6.13** that contains only two fluorine atoms is prepared. The X-ray analysis of the crystal of this compound shows that it has a linear assembly with the opposite enantiomers held together by O–H···O hydrogen bonds (Fig. 6.10). No hexameric structure is observed. In the structure of rac-**6.13**, there are weak C–H···F interactions but no π–π stacking.

rac-**6.13**

6.3.2. Naphthalen-2-yl- and Pentafluorophenyl-Substituted Propargylic Alcohols

Racemic naphthalen-2-yl- and pentafluorophenyl-substituted propargylic alcohol rac-**6.14** is prepared from the reaction of 2-ethynylnaphthalene with pentafluorobenzaldehyde in the presence of $ZnEt_2$ and HMPA at room temperature. An X-ray analysis of the crystals of rac-**6.14** shows that this compound has a hexameric channel-like structure similar to that of rac-**6.12** (Fig. 6.11). The hexamers are stabilized by the cooperation of the O–H···O hydrogen bonds, naphthyl–pentafluorophenyl π–π stacking and

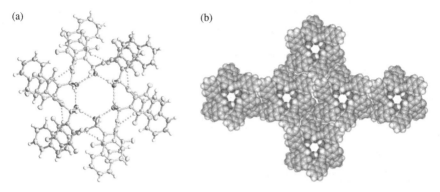

Figure 6.11. (a) Network of O–H···O hydrogen bonds, face-to-face π–π stacking interactions, and C–H···F hydrogen bonds, cooperatively forming the hexameric structure of *rac*-**6.14**. (b) Channels formed along the crystallographic axis *a*.

C–H···F hydrogen bonds. The cage in the structure of *rac*-**6.14** shows the diameter A of 7.76 Å and the diameter B of 8.46 Å, close to those of *rac*-**6.12** shown in Fig. 6.9.

rac-**6.14**

Similar to that shown in Scheme 6.4, the optically active enantiomer is synthesized from the reaction of 2-ethynylnaphthalene with pentafluorobenzaldehyde in the presence of (*S*)-BINOL, Et$_2$Zn, Ti(O^{i}Pr)$_4$ and HMPA, which gives (*S*)-**6.14** with 82% *ee*. The single crystals of (*S*)-**6.14** are obtained from slow evaporation of its benzene solution and their X-ray analysis has established the supramolecular structure of (*S*)-**6.14**.

(*S*)-**6.14**

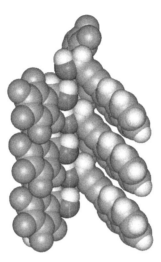

Figure 6.12. Space-filling representation for the supramolecular structure of (S)-**6.14**.

Figure 6.12 gives the space-filling representation of the X-ray structure of (S)-**6.14**, which is very different from that of rac-**6.14**. It shows that the optically active (S)-**6.14** has a chain-like structure linked by weak O–H···O hydrogen bonds. Almost no face-to-face naphthyl-pentafluorophenyl π–π interaction is observed.

Compound rac-**6.15** is prepared in which a methoxy group is introduced to rac-**6.14** in order to study how a substituent influences the supramolecular assembly. The X-ray structure of this compound is obtained. As shown in Fig. 6.13, the structure of rac-**6.15** is like (S)-**6.14** but very different from rac-**6.14**. It has a chain-like supramolecular assembly with weak O–H···O hydrogen bonds. In rac-**6.15**, the R and S enantiomers are alternatively linked head-to-head along the chain.

rac-**6.15**

Compound (S)-**6.16**, an analog of (S)-**6.14** containing no fluorine atoms, is prepared with 76% ee in the same way as the preparation of

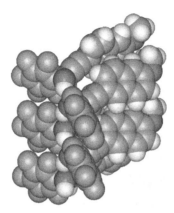

Figure 6.13. Space-filling representation for the supramolecular structure of rac-**6.15**.

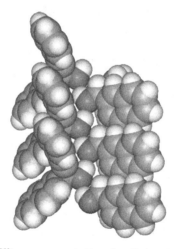

Figure 6.14. Space-filling representation for the supramolecular structure of (S)-**6.16**.

(S)-**6.12** and (S)-**6.14** by using (R)-BINOL in combination with ZnEt$_2$, Ti(O^{i}Pr)$_4$ and HMPA. The crystal structure of (R)-**6.16** shows a chain-like structure similar to (S)-**6.15** (Fig. 6.14). The configuration of (S)-**6.16** is in fact the opposite of that of (S)-**6.12** and (S)-**6.14** because the priority rank of the phenyl group in (S)-**6.16** is lower than that of the pentafluorophenyl groups in (S)-**6.12** and (S)-**6.14**. Therefore, the synthesis of (S)-**6.16** needs to use (R)-BINOL rather than (S)-BINOL used in the preparation of (S)-**6.12** and (S)-**6.14**.

(S)-**6.16**

6.3.3. Naphthalen-1-yl- and Pentafluorophenyl-Substituted Propargylic Alcohols

Compound rac-**6.17** that contains a napthalen-1-yl substituent is prepared. In comparison with compounds rac-**6.12** and rac-**6.14**, this compound has one less ortho-H atom on the non-fluorinated aromatic ring and is expected to have reduced intermolecular C–H···F hydrogen bonding. The crystal structure of rac-**6.17** shows that this compound does not generate a hexameric assembly as rac-**6.12** and rac-**6.14**. Instead, rac-**6.17** forms a tetrameric supramolecular structure stabilized by the O–H···O hydrogen bonds and the naphthyl–pentafluorophenyl π–π stacking (Fig. 6.15). No significant C–H···F interaction is observed. The enantiomers in the tetramer are arranged as *RRSS*.

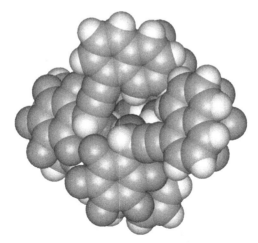

Figure 6.15. Space-filling representation for the supramolecular structure of rac-**6.17**.

rac-**6.17**

In the presence of (S)-BINOL, Et$_2$Zn, Ti(O^{i}Pr)$_4$ and HMPA, the optically active (S)-**6.17** is obtained from the reaction of 1-ethynylnaphthalene with pentafluorobenzaldehyde with 75% *ee*. The crystal structure of (S)-**6.17** shows that it contains a spiral chain of strong O–H···O hydrogen bonds (Fig. 6.16). No naphthyl–pentafluorophenyl stacking is observed, but columns of naphthyl–naphthyl and pentafluorophenyl–pentafluorophenyl are formed.

(S)-**6.17**

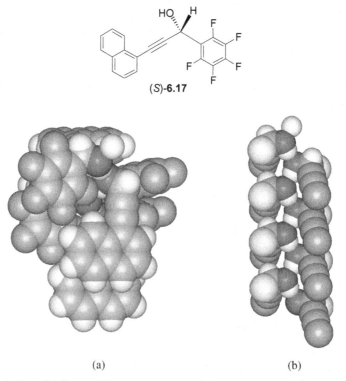

(a) (b)

Figure 6.16. (a) Space-filling presentation of the structure of (S)-**6.17**. (b) The hydrogen bond spine along the *b* axis of the unit cell (only the chiral carbon atom and the triple bond carbon atoms as well the connecting carbon atom of the phenyl ring are shown).

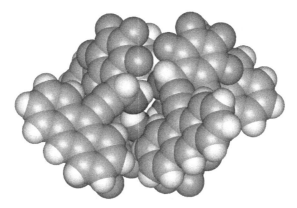

Figure 6.17. Space-filling presentation for the supramolecular structure of rac-**6.18**.

6.3.4. Anthracen-9-yl- and Pentafluorophenyl-Substituted Propargylic Alcohols

The anthracen-9-yl-substituted compound rac-**6.18** that contains no ortho H atom on the non-fluorinated aromatic ring is prepared. The crystal structure of rac-**6.18** shows that this compound has a tetrameric structure similar to that of rac-**6.17** (Fig. 6.17). In this structure, there is a centrosymmetric $RRSS$ arrangement stabilized by strong O–H···O hydrogen bonds.

rac-**6.18**

In the presence of (S)-BINOL, Et$_2$Zn, Ti(O^{i}Pr)$_4$ and HMPA, the optically active (S)-**6.18** is prepared with 88% ee from 9-ethynylanthracene and pentafluorobenzaldehyde. The X-ray analysis of this compound shows a chain-like structure linked by strong O–H···O hydrogen bonds (Fig. 6.18).

(S)-**6.18**

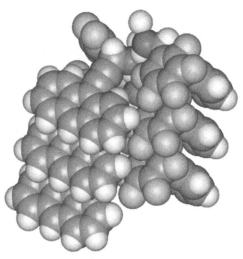

Figure 6.18. Space-filling presentation for the supramolecular structure of (S)-**6.18**.

6.3.5. A Pyridin-2-yl- and Pentafluorophenyl-Substituted Propargylic Alcohol

An additional hydrogen bond acceptor, a pyridine group, is incorporated into the propargylic alcohol for supramolecular assembly study. From the reaction of 2-ethynylpyridine with pentafluorobenzaldehyde in the presence of ZnEt$_2$ and HMPA, compound rac-**6.19** is obtained. Figure 6.19 shows the X-ray structure of the supramolecular assembly of rac-**6.19**. In this structure, the molecules are linked by the O–H···N hydrogen bonds between an (R)-enantiomer and an (S)-enantiomer. Additional interactions including the pentafluorophenyl–pyridyl π–π stacking, C–H···O interactions and C–H···F interactions are also observed in the structure.

rac-**6.19**

6.3.6. Gel Formation of the Propargylic Alcohols[5]

The gel formation of rac-**6.12** is discovered in collaboration with Caran. Cooling a hot alkane solution of rac-**6.12** with a concentration of 3–4%

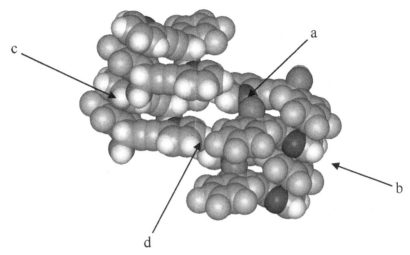

Figure 6.19. Space-filling representation of the structure of *rac*-**6.19** showing two neighboring stacks of molecules. Several interactions are specified: (a) the O–H···N hydrogen bond; (b) the pentafluorophenyl–pyridyl ring stacking; (c) the C–H···O hydrogen bond; (d) the C–H···F hydrogen bond between the neighboring stacks.

Figure 6.20. The SEM image of the desolvated gel of *rac*-**6.12** (scale bar = 20 μm).

by weight to room temperature is found to produce a unmovable gel. Figure 6.20 shows a SEM (scanning-electron-macroscopy) image of the gel of *rac*-**6.12** formed from cyclohexane after the removal of the solvent. It reveals the fibrous aggregate of this material.

As described in Section 6.3.1, the crystals of *rac*-**6.12** obtained from its solution form hexameric supramolecular structure. Sublimation of the

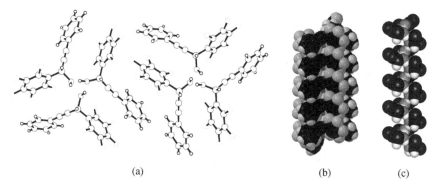

Figure 6.21. The structure of the sublimed crystals of *rac*-**6.12**: (a) two adjacent homochiral trimers with the opposite configuration; (b) the stacked trimers forming a column *via* extended hydrogen bonds; (c) the hydrogen bonds in the core of the column forming a helical chain.

powder of *rac*-**6.12** also generates single crystals. The X-ray analysis of these crystals reveals a very different structure. In the sublimed crystal of *rac*-**6.12**, molecules are arranged into pseudoinfinite columns. Each column contains homochiral trimers (all R or all S) stacked through phenyl–pentafluorophenyl rings (Fig. 6.21). The molecules in the trimer are linked by strong O–H···O hydrogen bonds. The adjacent columns have the opposite stereochemistry.

The powder X-ray diffraction pattern of the gel of *rac*-**6.12** is found to match that of the sublimed *rac*-**6.12** and very different from that of the solution recrystalized. That is, in the gel of *rac*-**6.12**, the molecules should assemble in a similar way as that found in the structure of the sublimed crystals of *rac*-**6.12** as shown in Fig. 6.21.

Compound *rac*-**6.20** containing a butoxy group also forms gel when its alkane solutions are cooled lower than the room temperature. However, compound *rac*-**6.21** with a longer chain alkoxy substituent stays in solution and cannot form a gel. Compound *rac*-**6.22** without the oxygen atom of the alkoxy group of *rac*-**6.20** also remains to be soluble in alkane solutions with no gel formation. The optically active enantiomer (R)-**6.12** crystallizes out of solution rather than forming a gel.

References

1. Zhu YL, Gergel N, Majumdar N, Harriott LR, Bean JC, Pu L, *Org Lett* **8**: 355–358, 2006.
2. Xu MH, Lin ZM, Pu L, *Tetrahedron Lett* **42**: 6235–6238, 2001.
3. Hyacinth M, Chruszcz M, Lee KS, Sabat M, Gao G, Pu L, *Angew Chem Int Ed* **45**: 5358–5360, 2006.
4. Hyacinth M, Sabat M, Caran KL, Pu L, unpublished results.
5. Borges AR, Hyacinth M, Lum M, Dingle CM, Hamilton PL, Chruszcz M, Pu L, Sabat M, Caran KL, *Langmuir* **24**: 7421–7431, 2008.

Index

α-hydroxycarboxylic acids, 257, 263
γ-hydroxy-α,β-acetylenic esters, 162
π–π attraction, 246
1,1'-binaphthyl-2,2'-dicarboxylic acid, 2
1,1'-binaphthyl-8,8'-dicarboxylic acid, 2
1,3-dipolar cycloaddition, 136
3,3'-bismorpholinomethyl BINOL, 204
3,3'-bismorpholinomethyl H$_8$BINOL, 194

absolute configuration, 2
alaninol, 249
alkylzinc addition to aldehydes, 216
alkyne addition to ketones, 196
alkyne additions to aldehydes, 151
amino acid derivatives, 269
amino alcohols, 244
ammonium salts, 5
anionic polymerization, 8
anti interaction, 1
arylzinc addition to aldehydes, 198
asymmetric epoxidation of α,β-unsaturated ketones, 125
asymmetric reduction of prochiral ketones, 122

Benesi–Hildebrand-type equation, 259
Bergman cyclization, 61
BINAP, 7, 141
binaphthyl dendrimers, 47
binaphthyl salens, 186

Binaphthyl-2,2'-diamine (BINAM), 37
BINOL, 8
BINOL–BINAP copolymer, 146
BINOL-salen, 186, 230
biphenyl, 1
bromination at the 6,6'-positions of (R)-BINOL, 17

catalyst screening, 294
chiral amplification, 15
chiral axis, 1
chiral-conjugated polymers, 13
chiral molecular wires, 311
chiral stationary phase, 15
circularly polarized light, 68
cisoid, 3
cisoid conformation, 10, 104
conductivity, 19
conformation, 4
conjugated polymers, 13
crown ether, 4–6, 30
crystal structure of *rac*-1,1'-binaphthyl, 4
crystals of (R)-BINOL, 10
crystals of racemic BINOL, 10
cyanohydrins, 225

dendrimers, 240
dendritic BINOL ligand, 223
dialkylzinc addition to aldehydes, 111
diamagnetic square planar Ni(II)-salophen, 72
Diels–Alder reaction, 133

diethylzinc addition to aldehydes, 101
dihedral angle, 2, 4, 10, 205, 219
diphenylzinc addition, 120

electroluminescence, 36, 37, 50
enantioselective permeable membranes, 15
enantioselective precipitation, 303
ene reaction, 8, 99
enediyne, 61
excimer, 36, 265
excimer emission, 36, 50, 265, 274, 276, 285
external quantum efficiency, 37, 50, 51

fluorescence enhancements, 287
fluorescence quantum yield, 19, 33, 47, 244, 281
Fréchet-type dendrons, 43
functional arylzinc addition, 210

gel, 329

H_8BINOL, 194
HDA reaction, 99
Heck coupling, 38
helical chain conformation, 14
helical conformation, 15
helical ladder polybinaphthyls, 76
helical polymer chain, 51
helical polysalophen, 71
hetero-Diels–Alder (HDA) reaction, 99
hetero-Diels–Alder reaction, 233
hexahydromandelic acid, 268, 291
hydration of the chiral
 γ-hydroxy-α,β-acetylenic esters, 166
hydrogenation, 8
hydrogenation of dehydroamino esters or acids, 142
hydrogenation of enanmides, 7
hydrogenation of ketones, 145

Job plot, 259

K_{SV}, 245

Langmuir–Blodgett (LB) films, 67
leucinol, 244, 256
light harvesting antenna, 244
light harvesting properties, 240
luminance, 37
luminescence, 50
luminous, 51

main chain chiral-conjugated polymers, 17
main chain chiral polymers, 88
major groove, 89
major-groove poly(BINOL)s, 89, 90, 95, 99, 101, 104, 105, 125, 127, 132, 138
mandelic acid, 257, 259, 262, 265, 280, 287, 299, 304
Michael addition, 139
minor groove, 89
minor-groove poly(BINOL)s, 90, 102–105, 129, 131
Mukaiyama aldol reaction, 95–98

NLO material, 68
non-planar Ni(II) coordination, 74, 76

organozinc addition to aldehydes, 105

paramagnetic tetrahedral Ni(II), 72
partially substituted PA, 14
PAs, 13
phenylalaninol, 244, 254
phenylglycinol, 256
photo-induced electron transfer, 257, 267, 275, 293
poly(BINAP), 142
polyacetylene (PA), 13
polydendrimers, 43
polyisocyanates, 15
polypeptides, 88

polythiophenes, 15
propargylic alcohols, 151, 152, 318

racemization, 2
racemization half-life, 1
racemization half-life of BINOL, 9
racemization pathways, 1
resolve racemic BINOL, 8
rhodium complex of BINAP, 7
ring-opening-metathesis-
  polymerization,
  14
rotation barrier, 1
ruthenium complex of BINAP, 7

second-harmonic CD effect, 69
second-harmonic efficiency, 68
second-harmonic response, 67
second-order non-linear optical
  properties, 67
solid-state fluorescence enhancement,
  303
soluble chiral PAs, 13

Sonogashira coupling, 23, 240
specific optical rotation $[\alpha]_D$ of
  (R)-BINOL, 8
spontaneous resolution, 2
Stern–Volmer constants, 245
Stern–Volmer equation, 245
structures of $(+)$-(R)- and racemic
  BINOL, 10
supramolecular structures, 318
Suzuki coupling, 17, 20, 48, 58, 76,
  94, 111, 117, 134, 179, 218, 246,
  250, 260
Suzuki coupling polymerization, 142
Suzuki polymerization, 102
syn interaction, 1

thermal racemization of BINOL, 9
TMSCN addition to aldehydes, 225
transoid conformation, 4, 104, 205

unimolecular chiral micelle, 250

valinol, 244, 256